Stem Cell Biology and Regenerative Medicine

Series Editor
Kursad Turksen, Ph.D.
kursadturksen@gmail.com

More information about this series at http://www.springer.com/series/7896

Martin K. Childers

Editor

Regenerative Medicine for Degenerative Muscle Diseases

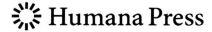 Humana Press

Editor
Martin K. Childers
Department of Rehabilitation Medicine
University of Washington
Seattle, WA, USA

Institute for Stem Cell and Regenerative
  Medicine, School of Medicine
University of Washington
Seattle, WA, USA

ISSN 2196-8985   ISSN 2196-8993   (electronic)
Stem Cell Biology and Regenerative Medicine
ISBN 978-1-4939-3227-6   ISBN 978-1-4939-3228-3   (eBook)
DOI 10.1007/978-1-4939-3228-3

Library of Congress Control Number: 2015956772

Springer New York Heidelberg Dordrecht London

Humana Press is a brand of Springer
Springer Science+Business Media LLC New York is part of Springer Science+Business Media
(www.springer.com)

Preface

Regenerative medicine—an emerging multidisciplinary field composed of clinicians, bioengineers, pharmacologists, surgeons, and others—seeks to develop new ways of repairing and replacing cells, tissues, and organs. A great example of regenerative medicine technology was pioneered by Anthony Atala, MD, who pioneered the world's first urinary bladder made from the patient's own cells and grown in a laboratory before being implanted into the patient. This process of regenerative medicine differs radically from conventional medicine by not only treating symptoms but by recreating damaged tissues and organs in the human body. Unthinkable only a few years ago, clinicians and scientists are rapidly exploiting nature's secrets to replace damaged tissue or to stimulate the body's own natural repair mechanisms to heal previously irreparable tissues or organs. For degenerative muscle diseases, this is a tough order, since skeletal muscle comprises the largest tissue type in the human body. Not only is skeletal muscle the largest "organ" in the body, but most degenerative diseases, such as the muscular dystrophies, are caused by rare inherited genetic mutations. So in this case, regenerative technologies must overcome two very difficult problems. The first problem is to create enough "medicine" to treat the entire volume of skeletal muscle in a human body, and the second is to replace defective genes. Both problems may seem insurmountable at first glance, but the history of medical progress, particularly in the field of genetics and gene therapy, tells us otherwise.

In this book, the first three chapters are dedicated to an overarching view of how genetics and regenerative technologies can synergistically work together to target skeletal muscle genetic diseases. The next five chapters explore how an understanding of cellular biology (particularly stem cell biology and cellular reprogramming) proffers alternative methods to gene replacement. Chapters 8 and 9 examine new ways to look at our conventional approaches (rehabilitation medicine and nutrition) when combined with regenerative technologies. And the final chapters introduce the concepts of "micro-RNAs" and how fish and dog models can be used to enhance our understanding of these devastating and often fatal diseases of skeletal muscle.

Altogether, these new concepts, technologies, and approaches should give patients, families, and doctors cutting-edge information about incredible regenerative advances already under way in laboratories and clinics worldwide.

Seattle, WA, USA   .   Martin K. Childers

Contents

Contributors

Berkcan Akpinar Department of Orthopaedic Surgery, University of Pittsburgh School of Medicine, Pittsburgh, PA, USA

Fabrisia Ambrosio Department of Physical Medicine and Rehabilitation, University of Pittsburgh, Pittsburgh, PA, USA

McGowan Institute for Regenerative Medicine, University of Pittsburgh, Pittsburgh, PA, USA

Alan H. Beggs Division of Genetics and Genomics, The Manton Center for Orphan Disease Research, Boston Children's Hospital, Harvard Medical School, Boston, MA, USA

Nicholas Boulis Department of Neurosurgery, Emory University, Atlanta, GA, USA

Martin K. Childers Department of Rehabilitation Medicine, University of Washington, Seattle, WA, USA

Institute for Stem Cell and Regenerative Medicine, School of Medicine, University of Washington, Seattle, WA, USA

Zoe E. Davidson Nutrition and Dietetics, Monash University, Notting Hill, VIC, Australia

Amy Lynnette Van Deusen Regenerative Medicine Strategy Group, LLC, Los Angeles, CA, USA

Robert W. Grange Human Nutrition, Foods, and Exercise, Virginia Tech, Blacksburg, VA, USA

Xuan Guan Institute for Stem Cell and Regenerative Medicine, School of Medicine University of Washington, Seattle, WA, USA

Vandana A. Gupta Division of Genetics and Genomics, The Manton Center for Orphan Disease Research, Boston Children's Hospital, Harvard Medical School, Boston, MA, USA

Angela J. Hasemann School of Clinical Sciences, University of Virginia Children's Hospital, Charlottesville, VA, USA

Johnny Huard Department of Orthopaedic Surgery, University of Pittsburgh School of Medicine, Pittsburgh, PA, USA

Joe N. Kornegay Department of Veterinary Integrative Biosciences (Mail Stop 4458), College of Veterinary Medicine, Texas A&M University, College Station, TX, USA

Jason Lamanna Department of Neurosurgery, Emory University, Atlanta, GA, USA

Department of Biomedical Engineering, Georgia Institute of Technology, Emory University, Atlanta, GA, USA

Aiping Lu Department of Orthopaedic Surgery, University of Pittsburgh School of Medicine, Pittsburgh, PA, USA

David Mack Department of Rehabilitation Medicine, University of Washington, Seattle, WA, USA

Institute for Stem Cell and Regenerative Medicine, School of Medicine, University of Washington, Seattle, WA, USA

Lisa Maves Center for Developmental Biology and Regenerative Medicine, Seattle Children's Research Institute, Seattle, WA, USA

Department of Pediatrics, University of Washington, Seattle, WA, USA

Davi A.G. Mázala Department of Kinesiology, School of Public Health, University of Maryland College Park, College Park, MD, USA

Michael Earl McGary Regenerative Medicine Strategy Group, LLC, Los Angeles, CA, USA

Khalid Medani Department of Neurosurgery, Emory University, Atlanta, GA, USA

Cynthia Moore Nutrition Counselling Center, University of Virginia, Charlottesville, VA, USA

Diem-Hang Nguyen-Tran Departments of Biochemistry, Biology, Bioengineering, Genome Sciences, Institute for Stem Cell and Regenerative Medicine, University of Washington, School of Medicine, Seattle, WA, USA

Department of Neurosurgery, McKnight Brain Institute, University of Florida, Gainesville, FL, USA

Carol Papillon Human Nutrition, Foods and Exercise, Virginia Polytechnic Institute and State University, Blacksburg, VA, USA

Elizabeth U. Parker Center for Developmental Biology and Regenerative Medicine, Seattle Children's Research Institute, Seattle, WA, USA

Jonathan Riley Department of Neurosurgery, Emory University, Atlanta, GA, USA

Greg Rodden Human Nutrition, Foods, and Exercise, Virginia Tech, Blacksburg, VA, USA

Hannele Ruohola-Baker Departments of Biochemistry, Biology, Bioengineering, Genome Sciences, Institute for Stem Cell and Regenerative Medicine, University of Washington, School of Medicine, Seattle, WA, USA

Laura L. Smith Division of Genetics and Genomics, The Manton Center for Orphan Disease Research, Boston Children's Hospital, Harvard Medical School, Boston, MA, USA

Richard O. Snyder Department of Molecular Genetics and Microbiology, University of Florida, College of Medicine, Gainesville, FL, USA

Atlantic Gene Therapies, INSERM UMR 1089, Université de Nantes, CHU de Nantes, Nantes, France

Center of Excellence for Regenerative Health Biotechnology, University of Florida, Alachua, FL, USA

Elizabeth C. Stahl Department of Orthopaedic Surgery, University of Pittsburgh School of Medicine, Pittsburgh, PA, USA

Kristen Stearns-Reider Department of Integrative Biology and Physiology, University of California Los Angeles, Los Angeles, CA, USA

Helen Truby School of Clinical Sciences, Monash University, Melbourne, VIC, Australia

Zejing Wang Medicine/Clinical Research, University of Washington/Fred Hutchinson Cancer Research Center, Seattle, WA, USA

About the Editor

Martin K. (Casey) Childers, D.O., Ph.D. is a professor in the Department of Rehabilitation Medicine and Investigator at the Institute for Stem Cell & Regenerative Medicine, University of Washington. He is a graduate of Seattle Pacific University (B.A., Music Performance), Western University (D.O., Medicine, Osteopathic), and The University of Missouri (Ph.D., Physiology & Pharmacology; residency, Rehabilitation Medicine). The Childers' laboratory works in two areas of investigation. In a series of preclinical studies, they address the hurdles required for systemic gene replacement delivery for patients with X-Linked Myotubular Myopathy (XLMTM). In other studies, they use a "disease-in-a-dish" approach with induced pluripotent stem (iPS) cells to study heart disease in patients with Duchenne muscular dystrophy (DMD). Dr. Childers' clinical medicine practice at the University of Washington Medical Center is dedicated to serve patients with neuromuscular diseases.

Chapter 1
Regenerative Medicine Approaches to Degenerative Muscle Diseases

Martin K. Childers and Zejing Wang

1.1 Background

1.1.1 What Is Regenerative Medicine?

An innovative interdisciplinary scientific field, termed *regenerative medicine*, focuses on new approaches to repairing and replacing cells, tissues, and organs and may involve the use of gene therapy. Derived from the fields of biomedical engineering, developmental biology, nanotechnology, physiology, molecular and cellular biology, and surgery, regenerative medicine holds new promise for previously incurable conditions. This emerging multidisciplinary field is rapidly bringing advances in cell therapy, bioengineering, and surgery to transform health care for those currently underserved by transplantation medicine.

Potentially, any disease or condition (acquired or genetic) that results in damaged, failing, or malfunctioning tissue may be amenable to regenerative medicine technologies. These technologies fall into two general categories: (1) a focus on growing tissues and organs in vitro (i.e., tissue engineering) and subsequently implanting them into the patient and (2) a focus to fully leverage the host's regenerative in vivo capacity using cellular and/or gene therapies.

M.K. Childers (✉)
Department of Rehabilitation Medicine, University of Washington, Seattle, WA, USA

Institute for Stem Cell and Regenerative Medicine, School of Medicine,
University of Washington, Seattle, WA, USA
e-mail: mkc8@uw.edu

Z. Wang
Medicine/Clinical Research, University of Washington/Fred Hutchinson Cancer Research
Center, 1100 Fairview Ave N, Seattle, WA, USA
e-mail: zwang@fredhutch.org

© Springer Science+Business Media New York 2016
M.K. Childers (ed.), *Regenerative Medicine for Degenerative Muscle Diseases*,
Stem Cell Biology and Regenerative Medicine,
DOI 10.1007/978-1-4939-3228-3_1

Regenerative medicine can be defined as the "process of creating living, functional tissues to repair or replace tissue or organ function lost due to age, disease, damage, or congenital defects" [1]. This new field holds the promise of regenerating damaged tissues and organs in the body by either replacing damaged tissue or by stimulating the body's own repair mechanisms to heal previously irreparable tissues or organs. As regenerative medicine becomes integrated into medical practice, new approaches will address the root cause of disease and offer prospects of tissue repair previously not possible.

The roots of regenerative medicine draw from the field of transplantation medicine that together with implantable medical devices have profoundly altered the trajectory of patients suffering from chronic and end-stage organ failure [2]. Indeed, the transplantation field has become so successful that the demand for organs now vastly exceeds the supply of donors. The US Department of Health and Human Services Organ Procurement and Transplantation Network cites the current US waiting list for all organs at 122,439 patients as of May 1, 2014 [3]. Although organ transplantation has become standard medical practice, in the United States more than 28,000 patients received transplanted organs in 2009 [4]. But, the gap between available organ donation and recipients is rapidly widening. Recipients, although fortunate to receive donor organs, are placed at risk for tissue rejection and must receive lifelong immunosuppressive drug therapy. Transplant complications plague certain tissues. For example, compared with recipients of other solid organ transplants [5–8], lung transplant recipients experienced the highest rates of rehospitalization for transplant complications: 43.7 per 100 patients in the first year [9]. Together, these data indicate that transplantation needs far exceed the supply, and even when transplanted organs are received, patients face lifelong challenges of tissue rejection.

The emerging field of regenerative medicine was created, in part, to address this unmet medical need for organ and tissue replacement. The multidisciplinary field of regenerative medicine involves close collaboration between clinicians and bench scientists to generate new tissues, not from human donors, but from the patient's own body. Regenerative technologies are aimed at growing organs and tissues in the laboratory to safely implant them into patients. This can potentially solve the problem of the shortage of organs available for donation and also address the problem of organ transplant rejection if the organ's cells are derived from the patient's own tissue or cells. Indeed, the world's first successful laboratory-grown organ (a urinary bladder) was generated from the patient's own cells, shaped by a laboratory bioreactor, and subsequently surgically implanted back into the patient [10–15]. In 2006, Atala and colleagues created engineered bladder tissues by seeding autologous urothelial and muscle cells onto collagen–polyglycolic acid scaffolds. After 7 weeks of culture in laboratory bioreactors, the autologous engineered bladders were used in a surgical reconstruction for seven myelomeningocele patients. The laboratory-grown bladders were implanted either with or without an omental wrap. Long-term follow-up in patients treated by this regenerative technique demonstrated improvement in bladder leak point pressure, volume, and compliance. This proof-of-concept technology began a new era in an emerging field that is rapidly gaining traction around the world.

A few examples of regenerative medicine technologies already tested in patients are presented in Table 1.1. A wide array of engineered organs (urinary bladder [10–15], trachea [17, 18], cornea [20, 22, 23]) or tissues (skin [16, 24–26], cartilage [19, 27–30], muscle [21, 31, 32]) are developed to the point of clinical applications. Many other regenerative technologies are under development in preclinical stages prior to human trials. For example, a breakthrough reported in the journal *Nature* from the laboratory of Charles Murry, MD, PhD described the use of human embryonic stem cells (hESCs) to regenerate damaged heart tissue in nonhuman primates [33]. Investigators injected one billion heart muscle cells derived from hESCs into the infarcted muscle of pigtail macaques. This was ten times more of these types of cells than researchers have ever been able to previously generate. This report demonstrated that hESCs can be grown, differentiated into cardiomyocytes, and cryopreserved at a scale sufficient to treat a large-animal model of myocardial infarction. The group found that over subsequent weeks, the stem cell-derived heart muscle cells infiltrated into the damaged heart tissue, then matured, assembled into muscle fibers, and began to beat in synchrony with the macaque heart cells. After 3 months, the cells appeared to have fully integrated into the macaque heart. Future efforts by this group will work to reduce the risk of arrhythmias (seen in the monkeys after stem cell transplantation), perhaps by using more electrically mature stem cells. They also will investigate if the stem cells strengthen the contractile strength of the recipient hearts.

1.1.2 Degenerative Muscle Diseases Are Rare, but Rare Diseases Are Common

The majority of degenerative muscle diseases are inherited genetic diseases passed from one generation to the next. These genetic degenerative diseases of muscle are rare, but altogether, rare diseases are common. In fact, one in ten Americans has a rare disease [34]. If all of the individuals with rare diseases were to move to an unpopulated continent—like Greenland—it would instantly become the third most populated country in the world. Therefore, "rare" diseases make up a large group of collected disorders that are not rare, but rather they are common among our population.

Table 1.1 Regenerative medicine technologies reported in patients

Engineered organ or tissue	Types of cells used	Scaffold	References
Skin	Epithelial or none	Complex; multiple in use	[16]
Urinary bladder	Urothelial, muscle	Collagen; collagen/polyglycolic acid	[10]
Trachea	Epithelial, chondrocytes	Cadaver trachea	[17, 18]
Cartilage	Mesenchymal stem cells, others	None	[19]
Cornea	Limbal stem cells	Amniotic membrane	[20]
Muscle	None	Porcine intestine submucosa	[21]

1.2   What Disorders Lead to Chronic Muscle Degeneration?

1.2.1   Myopathies

The term, myopathy, simply means muscle disease and is derived from two Greek terms, "myo" (muscle) and "pathos" (suffering). A myopathy implies that the primary defect is within the muscle, as opposed to a peripheral nerve or within the central nervous system. Chronic myopathies are generally inherited (familial), progressive, and clinically distinguished by the age of onset (congenital versus adult). In contrast, acquired myopathies result from different disease processes including endocrine, inflammatory, paraneoplastic, infectious, drug- and toxin-induced, critical illness, and metabolic disorders.

1.2.2   Acquired Myopathies

A common acquired myopathy results from administration of anti-inflammatory glucocorticoids (anti-inflammatory steroids) [35]. Referred to as "glucocorticoid-induced myopathy," administration of high doses of glucocorticoids in animals results in both a decrease in muscle mass and also muscle weakness. In patients, limb and respiratory muscle weakness in pulmonary disease can be attributed to glucocorticoid use. Limb muscle weakness has also been observed in patients with Cushing's syndrome who exhibit high levels of endogenous glucocorticoids [36]. Chronic glucocorticoid administration causes atrophy by a decrease in the rate of protein synthesis and increase in the rate of protein breakdown. Interestingly, atrophy is selective for type II (fast-twitch) muscle fibers with little effect on type I (slow-twitch) fibers [37]. The mechanism of such fiber specificity might be related to higher glucocorticoid receptor expression in type II fibers [35]. In patients, short-term glucocorticoid excess blunts the insulin-induced protein anabolism, suggesting that muscle loss occurs most likely through inhibition of the anabolic response to insulin [38].

1.2.3   Congenital Myopathies

Defined by childhood onset, congenital myopathies make up a wide spectrum of inherited muscle disorders. As an example, a form of congenital myopathy, termed centronuclear myopathy (CNM), has variable features that are most often present in childhood [39] and follow a progressive clinical course. Infants may present as "floppy" and hypotonic and many succumb to respiratory failure. The typical features of a CNM include proximal muscle weakness, atrophy, hypotonia, and elongated facies with a high-arched palate in young children [39].

Because of the overlap between clinicopathological features associated with individual gene mutations in CMs, classification of these disorders has been based on the appearance of abnormal structures on muscle biopsies examined under the light microscope. Further classification of CMs was based on the appearance of centrally located nuclei (central nuclear myopathies), rod-like structures, and other distinguishing features of organelles. Thus, "structured" CMs include central core disease (CCD), multi-minicore disease (MMD), centronuclear myopathies, nemaline myopathies, actin aggregate myopathy, desminopathy, and hyaline body myopathy. In contrast, "unstructured" CMs include congenital fiber-type disproportion (CFTD), a nonprogressive childhood neuromuscular disorder without prominent accumulation of abnormal structures visible under the light microscope. While these characteristic features provide some distinguishing characteristics that help differentiate congenital myopathies from metabolic myopathies and muscular dystrophies, such structural features are not informative about the underlying pathophysiology or genetic mutations associated with each disease.

Genetic abnormalities in the CNMs give rise to characteristic pathological changes within the sarcomere, the contractile apparatus of the myofiber. Examples of such changes include aberrant calcium handling and alterations of the normal excitation–contraction coupling machinery leading to inefficient muscle contraction. In contrast to muscular dystrophies where recurring cycles of myofiber degeneration and regeneration occur, in congenital myopathies, the myofibers do not undergo recurring degeneration/regeneration. Replacement of contractile tissues with noncontractile connective tissue or fat in muscular dystrophies or accumulation of glycogen depots in metabolic myopathies is usually not a typical feature of congenital myopathies.

1.2.4 Muscular Dystrophies

The muscular dystrophies are yet another collection of myopathies generally distinguished by ongoing muscle fiber degeneration with incomplete regeneration. The most common example is Duchenne muscular dystrophy (DMD). This progressive, lethal, X-linked disease of skeletal and cardiac muscle affects nearly 1 in 3500 males born each year in the United States. DMD is caused by mutation of the dystrophin gene (2.4 megabases—the largest known gene) that, together with its location at Xp21, provides a vulnerable target for new mutations. The cardiac and skeletal muscles of DMD patients are deficient in the dystrophin gene product, a 427-kD protein found primarily in the outer cell membrane in cardiac and skeletal muscle [1]. Without dystrophin in the outer membrane, the muscle fiber is particularly vulnerable to damage from normal daily activities [2]. As a result, damaged DMD muscle fibers eventually succumb to injury [3]. Normally, muscle damage is repaired by resident muscle stem cells (satellite cells) [4]. However, continuous cycles of damage eventually overwhelm the capacity for regeneration, potentially due to impaired ability of muscle satellite cells [5]. To address this progressive and ultimately fatal degeneration in DMD muscles, intense research efforts are aimed at tilting the balance in favor of regeneration. Stem cell transplantation therapy may offer one approach to enhance the regenerative ability of damaged and degenerating muscle cells in patients with DMD.

1.3   Can Human Cells Be Transplanted to Regenerate Damaged Muscles?

Diseased tissue may be regenerated in vivo by transplantation of healthy cells that are able to replicate extensively. Progenitor and stem cells have this intrinsic ability and are used in regenerative medicine to enhance or restore damaged tissue. According to data from the Centers for Disease Control, as many as one million Americans will die every year from disease that, in the future, may be treatable with tissues derived from stem cells [40]. Diseases that might benefit from stem cell-based therapies included diabetes, heart disease, cerebrovascular disease, liver and renal failure, spinal cord injuries, and Parkinson's disease. In the following section, classes of cells available for regenerative techniques and research advances using these cells in selected clinical conditions are discussed.

1.3.1   Adult Stem Cells

Adult stem cells are defined as any stem cell population that corresponds to a point in development subsequent to the inner cell mass (ICM), or possibly the slightly later epiblast, of embryos at the blastocyst stage, prior to gastrulation. Those from the ICM are called embryonic stem (ES) cells. Adult stem cells are relatively abundant during fetal development and persist in tissues throughout adult life. Much of our current understanding of stem and progenitor cell biology derives from studies of a particular class of adult stem cells, namely, those of the hematopoietic (blood forming) system [41]. In fact, the therapeutic use of adult stem cells dates back to the first bone marrow transplant in 1956 [42]. Early suggestions supporting the existence of cells that are capable of reconstituting the blood system came from experience with persons exposed to lethal doses of radiation during World War II. In the early 1960s, James Till and Ernest McCulloch in Toronto, Canada, found evidence for a specific subpopulation of cells in bone marrow that could restore hematopoiesis after transplantation into irradiated mice. In the spleens of recipient animals, they observed large colonies and showed that these arose as clones from individual stem cells in the donor marrow that both could replicate to generate more stem cells and could give rise to multiple types of mature, differentiated blood cells [43]. These two characteristics, termed self-renewal and multipotency, remain generally accepted as the defining features of stem cells.

1.3.2   Embryonic Stem Cells

In contrast to adult stem cells, embryonic stem (ES) cells show essentially unlimited capacity for self-renewal in culture and should be able to give rise to any of the more than 200 cell types in the body (with the exception of certain extraembryonic cell types). ES cells have great therapeutic potential, but there are also significant barriers that must be overcome for their implementation in the clinic. ES cells are pluripotent, that is, they have the ability to form tissue representing all three embryonic germ layers—the ectoderm, mesoderm, and endoderm. Thus, the ES cell has the intrinsic ability to give rise to daughter cells that can form virtually any tissue in the body. This was demonstrated convincingly by the finding that mouse ES cells injected into developing embryos can contribute to all adult cell types, including germ cells [44]. Human ES cells have been induced to generate representatives of all three germ layers in culture and are widely assumed to have the same degree of pluripotency as mouse ES cells.

The great plasticity of ES cells can also represent a drawback to their use, because it may prove more difficult to induce them exclusively to yield a single desired cell type than when starting with lineage-committed adult stem cells. However, significant progress has been made in the directed production of many specialized cells, or at least lineage-specific progenitors, from mouse and human ES cells [45] [46]. Examples include cells of neuronal, epidermal, cardiac myocytic, hematopoietic, endothelial, hepatic, endocrine pancreatic, and germ cell lineages. Another important problem that must be overcome is that undifferentiated ES cells form teratoma tumors [47, 48]. Therefore, for clinical use, it will be essential to differentiate the cells quantitatively before implantation and to rigorously exclude the presence of residual stem cells. Progress is being made in the derivation of human ES cells in the absence of xenogeneic feeder cells and animal proteins that may pose risks for clinical application.

The derivation and certain potential uses of human ES cell lines have engendered ethical debate, especially because embryos, though donated for research, are destroyed to isolate the inner cell mass. Technological advances may offer at least a partial resolution to the ethical controversy. For example, human ES lines can be developed from single-cell biopsy at the eight-cell stage of development, without compromising the viability of the embryo, similar to a widely used procedure for prenatal genetic diagnosis [49, 50]. The development of ES-like cell lines by reprogramming of normal adult cells ultimately may put the debate to rest by providing pluripotent cells entirely without the use of donated embryos or oocytes.

1.3.3   Cellular Reprogramming

Reprogramming is a technique that involves dedifferentiation of adult somatic cells to produce patient-specific pluripotent stem cells without the use of embryos. Cells generated by reprogramming are theoretically identical to somatic cells and would not be rejected by the donor. This method also avoids nuclear transfer into oocytes.

Takahashi and Yamanaka were the first to report that mouse embryonic fibroblasts (MEFs) and adult mouse fibroblasts can be reprogrammed into an induced pluripotent state (IPS) [51]. They identified four key genes required to bestow embryonic stem cell-like properties in fibroblasts. Mouse embryonic fibroblasts and adult fibroblasts were cotransduced with retroviral vectors, each carrying Oct3/4, Sox2, c-Myc, and Klf4. Reprogrammed cells were selected via drug resistance. In this case, a downstream gene of Oct4, Fbx15, was replaced with a drug resistance gene via homologous recombination. The resultant IPS cells possessed the immortal growth characteristics of self-renewing ES cells, expressed genes specific for ES cells, and generated embryoid bodies in vitro and teratomas in vivo. When the IPS cells were injected into mouse blastocysts, they contributed to a variety of diverse cell types, demonstrating their developmental potential. Although IPS cells selected by *Fbx15* were pluripotent, they were not identical to ES cells. Unlike ES cells, chimeras of IPS cells did not result in full-term pregnancies. Gene expression profiles of the IPS cells showed that they possessed a distinct gene expression signature compared to ES cells. The epigenetic state of the IPS cells was somewhere between their somatic origins and fully reprogrammed ES cells, suggesting that the reprogramming was incomplete.

These results were improved significantly by Wernig and Jaenisch [52]. Fibroblasts were infected with retroviral vectors and selected for the activation of endogenous *Oct4* or *Nanog* genes. Results from this study showed that DNA methylation, gene expression profiles, and chromatic state of the reprogrammed cells were similar to those of ES cells. Teratomas induced by these cells contained differentiated cell types representing all three embryonic germ layers. Most importantly, the reprogrammed cells from this experiment were able to form viable chimeras and contribute to the germ line-like ES cells, suggesting that these IPS cells were completely reprogrammed.

Reprogramming by transduction of defined factors can be done with human cells [53, 54] Yamanaka's group began by optimizing the transduction efficiencies of human dermal fibroblasts (HDF) and determined that the introduction of a mouse receptor for retroviruses into HDF cells using a lentivirus improved the transduction efficiency from 20 % to 60 %. Yamanaka then showed that retrovirus-mediated transfection of *OCT3/4*, *SOX2*, *KLF4*, and *c-MYC* generates human IPS cells that are similar to hES cells in terms of morphology, proliferation, gene expression, surface markers, and teratoma formation. In contrast, Thompson's group showed that retroviral transduction of *OCT4*, *SOX2*, *NANOG*, and *LIN28* could generate pluripotent stem cells without introducing any oncogenes (c-MYC). Both studies showed that human IPS cells were similar but not identical to hES cells. Another concern is that these IPS cells contain three to six retroviral integrations (one for each factor) that may increase the risk of tumorigenesis.

These studies used retroviral transduction to induce reprogramming of somatic cells into a pluripotent state. Okita et al. studied the tumor formation in chimeric mice generated from Nanog-IPS cells and found 20 % of the offspring developed tumors due to the retroviral expression of c-Myc (33). An alternative approach would be to use a transient expression method, such as adenovirus-mediated system, since both Meissner and Okita showed strong silencing of the viral-controlled transcripts in IPS cells (33, 34, 34). This indicates that they are only required for the induction, not the maintenance, of pluripotency. Another concern is the use of transgenic donor cells for reprogrammed cells in the mouse studies. In both mouse studies, IPS cells were isolated by selecting for the activation of a drug-resistant gene inserted into endogenous *Fbx15*, *Oct3/4*, or *Nanog*. The use of genetically modified donors hinders its clinical applicability for humans.

More recently, the development of direct reprogramming technology allows for the direct conversion from one cell type into another without reverting back into a stem cell state. Indeed, direct reprogramming has proven sufficient in yielding a diverse range of cell types from fibroblasts, including neurons, cardiomyocytes, endothelial cells, hematopoietic stem/progenitor cells, and hepatocytes [55]. Direct reprogramming studies [56–58] indicate that cell fate plasticity is much wider than previously anticipated and that direct reprogramming may offer a new system to study the mechanisms underlying cell fate decisions during development and also open up yet another possible regenerative medicine approach for muscle disease.

1.4 What Is Gene Therapy and When Will It Be Used in Clinical Practice?

Gene therapy—the process of introducing foreign genomic materials into host cells to elicit therapeutic benefit—became available for clinical practice on November 2, 2012, when Glybera (alipogene tiparvovec) became the first gene therapy in the Western world to receive market approval for patients with lipoprotein lipase (LPL) deficiency, a rare genetic disease previously without effective treatment [59, 60]. Since 1989, gene therapy clinical trials have been undertaken in 31 countries with more than 1800 human trials ongoing, completed, or approved worldwide [61]. Many of these trials target rare "orphan diseases." The Orphan Drug Act of 1983 defined an orphan product as a drug intended to treat a condition affecting fewer than 200,000 persons in the United States or a drug that would not be expected to be profitable within 7 years following FDA approval [62]. Orphan disease designation allows a sponsor to apply for market protection for the product following approval. 2700 orphan drug designations and more than 400 approvals associated with these designations were approved as of 2012 [62]. While most gene therapy trials have addressed cancer or cardiovascular disease, a significant number of gene therapy trials have targeted rare monogenic (single-gene) diseases (Table 1.2) [61]. These groundbreaking therapies involve the insertion of DNA sequences that encode functional, therapeutic genes into patients to replace mutated dysfunctional genes causing disease. In the case of Glybera, a DNA sequence encodes a therapeutic LPL gene packaged within a vector, in this case an adeno-associated virus (AAV). The recombinant AAV is injected into a patient harboring a mutant disease-causing LPL gene. The injected AAV is capable of shuttling the replacement DNA sequence from inside the vector to the cells of the targeted tissue. Once inside, the patient's

Table 1.2 Single-gene
disorders reported in gene
therapy clinical trials

Adrenoleukodystrophy
α-1 antitrypsin deficiency
Becker muscular dystrophy
β-thalassemia
Canavan disease
Chronic granulomatous disease
Cystic fibrosis
Duchenne muscular dystrophy
Fabry disease
Familial adenomatous polyposis
Familial hypercholesterolemia
Fanconi anemia
Galactosialidosis
Gaucher's disease
Gyrate atrophy
Hemophilia A and B
Hurler syndrome
Hunter syndrome
Huntington's chorea
Junctional epidermolysis bullosa
Late-infantile neuronal ceroid lipofuscinosis
Leukocyte adherence deficiency
Limb-girdle muscular dystrophy
Lipoprotein lipase deficiency[a]
Mucopolysaccharidosis type VII
Ornithine transcarbamylase deficiency
Pompe disease
Purine nucleoside phosphorylase deficiency
Recessive dystrophic epidermolysis bullosa
Sickle cell disease
Severe combined immunodeficiency
Tay–Sachs disease
Wiskott–Aldrich syndrome

[a]Glybera approved in Europe

own cellular machinery works to transcribe the new replacement DNA sequence to produce the therapeutic protein to treat the patient's disease. Besides the success of Glybera, other notable examples of gene therapy successes are highlighted in clinical trials undertaken in rare genetic childhood diseases such as X-linked severe combined immunodeficiency (SCID-X1) and Leber congenital amaurosis (LCA). In the first example, SCID-X1 results in recurrent and often fatal infections caused by genetic deficiency in cellular and humoral immunity. Long-term follow-up of nine SCID-X1 boys treated in a French adenovirus-mediated gene therapy trial reported eight survivors after nearly 10 years [63]. In the second example, LCA leads to congenital blindness caused by mutations in a retinal gene that causes progressive loss of vision; young patients become completely blind by adulthood. Three independent gene therapy trials for LCA patients have been initiated, and follow-up results indicate improvement in vision for up to 2 years with no serious adverse events [64–66]. These and other gene therapy successes have been offset by serious, and rarely fatal, adverse events that lead to early clinical trial "holds." The most famous case was the death of Jesse Gelsinger in 2000, who was the 19th patient enrolled in a gene therapy trial of a deficiency of ornithine transcarbamylase (OTCD) [67]. Despite these early setbacks, the field of gene therapy continues to move forward at an extraordinary pace.

1.5   Will Gene Replacement Therapy Be Useful for Inherited Muscle Diseases like Muscular Dystrophy?

Most gene therapy clinical trials have targeted genes involved in cancer [61] with fewer trials initiated in monogenic diseases, such as Duchenne muscular dystrophy (DMD). Muscular dystrophies make up only a small percentage of the single-gene, or monogenic, diseases (Table 1.2), notably Becker, Duchenne, and limb-girdle muscular dystrophy. Pompe disease, among others listed in Table 1.2, is not considered a muscular dystrophy even though skeletal muscles of affected patients grow progressively weaker due to accumulation of abnormal proteins within the muscle's contractile tissue. Therapeutic approaches at replacing defective genes in monogenic diseases of muscle, like DMD, include the use of recombinant viral vectors engineered to target specific tissues. In the 1960s, the discovery of naturally occurring adeno-associated virus (AAV) isolates led to the clinical application of recombinant AAV vectors with early successes in clinical trials [68]. AAV vectors are attractive for clinical use because AAVs are not associated with human disease [60]; the virus persists in the infected host for years, and a large "toolkit" of AAV vectors is available as clinical gene therapy delivery tools [69]. In addition to muscle diseases, AAV vectors have been used in clinical gene therapy trials targeted to the liver for the treatment of hemophilia B, to the lung for treatment of cystic fibrosis, to the brain for treatment of Parkinson's, Batten's, and Canavan's disease, to joints for treatment of rheumatoid arthritis, and to the eye for treatment of LCA [68].

In recent years, AAV-mediated gene replacement has rapidly moved from preclinical studies to clinical trials due to encouraging results from animal models. Major challenges surrounding this strategy include (1) effective delivery methods to target muscles throughout the body, including diaphragm and cardiac muscle, and (2) host immune responses to the therapeutic vector [70]. The latter challenge was encountered in a double-blinded, randomized, controlled phase I trial of limb-girdle muscular dystrophy [71, 72], AAV1-MCK. Human alpha-sarcoglycan (SGCA) was injected locally into the "extensor digitorum brevis" muscle in three patients, and sustained transgene expression was observed in two out three patients after 6 months. Humoral and cellular immune responses to the AAV capsid proteins were detected in the patient who failed to show expression. Another phase I trial on DMD from Mendell et al. [73] also raised the potential of cellular immune responses to either self or nonself dystrophin epitopes. In the study, AAV vectors carrying a truncated but functional dystrophin gene under the control of a CMV promoter were injected into the bicep of six DMD patients. None of the patients displayed transgene expression with four out of six patients showing detectable T cell responses against the transgene product. Two of the patients had dystrophin-specific T cell responses before the treatment. Taken together, these trials are informative to emphasize the importance of prescreening patients for preexisting immune responses to both AAV capsid proteins and transgene product and also to develop strategies to circumvent immune responses, such as using a transient course of immunosuppression, shown to be effective in a dog model of DMD [74, 75].

1.6 Can Genes Be Repaired?

While AAV-mediated gene therapy is regarded as a gene replacement strategy, another exciting development, termed exon skipping, focuses on gene repair. Approximately 70 % of all DMD mutations are due to single or multiple exon deletions. Such deletions disrupt the open reading frame of dystrophin and, hence, result in a premature truncated protein. In most cases, selective removal of specific exons can restore the reading frame and produce a partially functional dystrophin protein for clinical benefit [76–79]. Antisense oligonucleotides (AOs) targeting pre-mRNA to modulate splicing have been used to induce exon skipping, and the first human phase I trials [80, 81] focused on skipping exon 51, which, if successful, would be able to correct ~13 % of DMD patients with specific deletions within exons 42–50. A therapeutic exon skipping AO, 2-O-methyl AO termed "PRO051" [81], and AVI-4658, a morpholino-conjugated AO [80] that targets an internal sequence of exon 51, were injected intramuscularly into DMD patients. AOs used in these trials were well tolerated and demonstrated successful exon skipping, and all patients demonstrated dystrophin expression to levels between 3 and 12 % and 22 and 35 % by PRO051 and AVI-4658, respectively. Following these initial clinical trials of intramuscular administration, phase I/II trials using systemic delivery of the two drugs were tested for dose, safety, and efficacy [82, 83]. Dose-dependent restoration of dystrophin expression was observed following weekly administration of the experimental compounds through abdominal subcutaneous injections in two cohorts. While no significant differences were detected in patient's walking ability following a 12-week-long treatment [82], patients treated with weekly PMO-AO (AVI-4658) [83] for 48 weeks demonstrated improvement in stabilization of the muscle and in the 6-min walking distance test (6MWT) [84]. A larger confirmatory phase III trial is in the planning for 2014. Clinical trials for skipping other exons are also underway or in planning [76]. These "personalized gene therapy medicines" generate hope for DMD patients, with the caveat that effective therapy will need to restore dystrophin expression in both skeletal and cardiac muscle, and that treatment will need to be persistent. The future challenge for clinical development with AOs is that current forms have a short half-life, and more than 85 % of AOs are cleared from the circulation within 24 h. This short half-life requires weekly injections to maintain a therapeutic level. Long-term outcomes of these AOs are also unknown. Intramuscular administration of these agents may be limited in treating skeletal muscle, and treating the cardiac muscle will need to be addressed for this strategy to be an effective treatment for DMD-associated cardiomyopathy.

Nonsense stop codon read through is another method of gene editing and potentially benefits ~ 13 % of DMD boys with premature termination codon (PTC) mutations. Aminoglycoside antibiotics initially demonstrated the capacity to induce ribosomal read through of premature stop codon [85], but not efficient and too toxic to be used for long-term treatment. Ataluren (PTC124) was identified to be effective in this matter in *mdx* mice and subsequently tested in human DMD trials [86–88]. The drug was well tolerated in general, and the dystrophin protein was detected with treated patients showing improvement in the 6MWT. However, correlation between the level of dystrophin expression and the 6MWT remains unclear. More trials are currently under way, and the same treatment strategy could be potentially applied to other muscle diseases including spinal muscular atrophy. Progress in developing other approaches targeting repair of the muscle membrane due to lack of dystrophin is also under development for example, increasing the level of the compensatory protein utrophin [89, 90] and upregulation of glycosylation of a-dystroglycan to improve extracellular matrix attachment [91].

1.7 Can Gene Replacement Therapy Rescue Lethal Muscle Disease? Lessons Learned from Dogs

In 2008, a case report of a 5-month-old Labrador Retriever was published in a Canadian veterinary medical journal [92]. The dog presented with weakness, muscle atrophy, and histopathological changes in skeletal muscle consistent with a centronuclear myopathy. Because male littermates were similarly affected, the authors postulated that the disease was X-linked, giving rise to the possibility that the disorder could be analogous to X-linked myotubular myopathy (XLMTM). It was through the tireless and extraordinary efforts of Alison Rockett Frase that the author's research group was able to acquire a first-degree relative, a dog named "Nibs," a Labrador Retriever coming from a line of dogs with a history suspicious for XLMTM. We later discovered that Nibs harbored a canine MTM1 mutation(34), the same gene known to cause myotubular myopathy in patients [reviewed in "The Miracle of Nibs" [93]]. Mrs. Frase recalls the story of her odyssey locating and retrieving the founding carrier dog from a farmer in Canada. The following excerpt describes how a determined mother of an affected child can help shape the future of research for a disorder like myotubular myopathy:

"In the fall of 2008, a female Labrador Retriever was discovered to carry the same gene as I do for my son's muscle disorder called mytubular myopathy (MTM). To date, this was the first MTM large animal ever discovered by researchers anywhere in the world....Nibs was a beautiful Labrador Retriever that was instrumental in giving us puppies that carry the myotubular myopathy (MTM1) gene that affects our son Joshua. I am so grateful to Nibs for the initial 12 puppy litter...Our second litter of MTM pups has been born...Knowledge gained from these animals may 1 day lead to treatments not only for MTM, but other neuromuscular diseases. It will be a miracle for our son Josh and thousands of children like him if our goals are achieved."

Joshua Frase passed away on December 24, 2010, less than 2 years after this was written. His legacy, the *Joshua Frase Foundation*, set into motion research that will, hopefully, develop the first effective treatment for this devastating disorder.

XLMTM is an orphan disease, affecting 1/50,000 live male births worldwide [94] with only supportive, palliative care available for patients [95]. This inherited muscle disease results from loss-of-function mutations in the Myotubularin 1 gene (*MTM1*) [96] that encodes the founder of a family of 3-phosphoinositide phosphatases acting on the second messengers phosphatidylinositol 3-monophosphate [PI(3)P] and phosphatidylinositol 3,5-bisphosphate [PI(3,5)P$_2$] [97, 98]. Although myotubularin is expressed ubiquitously, loss of this enzyme profoundly affects skeletal muscles causing hypotrophic myofibers and structural abnormalities, with associated weakness [99]. No effective therapy exists for XLMTM. Management of the disease generally consists of mechanical ventilation, gastrostomy feeding tubes, antibiotics (for respiratory infections), orthotics to prevent skeletal limb contractures, and surgical treatment to alleviate severe spinal deformities. In spite of aggressive medical care, the average life expectancy is only about 2 years, and most who survive beyond this age require mechanical ventilation.

Animal models of the disease currently exist in zebra fish, mice, and notably, in dogs [99–101]. The murine phenotype resembles human XLMTM, with similar pathology and early mortality. In a mouse knockout model of XLMTM, local delivery of the wild-type myotubularin gene (*MTM1*) via an AAV vector reversed characteristic pathological features and rescued the function of injected limb muscles [102]. Buj-Bello et al. were the first to report a gene therapy success in the *Mtm1* knockout mouse in 2008. That same year, the *Joshua Frase Foundation* provided our research group access to a female Labrador Retriever harboring an *MTM1* gene mutation that was later proven by Beggs et al. to cause a canine version of the human disease [101]. From this single founding female, our group established a canine breeding colony to study effects of the disease in dogs. Initial data revealed that affected males display a phenotype directly analogous to human XLMTM: progressive and severe muscular weakness [103] leading to the inability to walk, weak ventilatory muscles leading to respiratory impairment [104], and early death. Based upon the early experience with gene replacement in the *Mtm1* knockout mouse [102], a similar gene replacement strategy was initiated in the XLMTM dog in 2011.

For eventual gene therapy of XLMTM patients, our goal was to use a predictive large animal model (the XLMTM dog) to refine the delivery system, to assess critical safety parameters such as the potential host immune response to vector and transgene, and to optimize efficacy measurements. In collaboration with the French nonprofit institute, Généthon [105], cohorts of *Mtm1* knockout mice were first tested for response to systemic *Mtm1* gene replacement via tail vein injection. Results indicated that a single systemic treatment with AAV-*Mtm1* sufficed for long-term (at least 1 year) survival and essentially complete amelioration of symptoms of mice with myotubularin-deficient muscles [106]. Using the same AAV vectors produced by Généthon scientists and tested in mice, our collaborative research group confirmed that local gene replacement therapy, delivered intramuscularly into the hind limb of young XLMTM dogs, reversed pathological changes in myotubularin-deficient skeletal muscles. Remarkably, the treated muscles also showed nearly normal strength at 6 weeks postinjection, compared to very weak muscles (only 20 % of normal strength) in saline-injected contralateral limbs. In subsequent experiments, intravascular administration of AAV8-*MTM1* at the same dose used in mice was well tolerated in dogs, rescued the skeletal muscle pathology and respiratory function, and prolonged life for over 1 year. Together, these initial studies demonstrated the feasibility, safety, and efficacy of gene therapy with AAV for long-term correction of muscle pathology and weakness observed in myotubularin-deficient mouse and dog models and support future clinical trials aimed at correcting this devastating disease in patients.

Conflict of Interest Statement MC is an inventor on patents related to recombinant AAV technology. MC owns stock options in a biotechnology company commercializing AAV for gene therapy applications. To the extent that the work in this manuscript increases the value of these commercial holdings, MC has a conflict of interest.

References

1. Regenerative Medicine. http://report.nih.gov/NIHfactsheets/ViewFactSheet.aspx?csid=62&key=R-R. Accessed 1 May 2014.
2. Terzic A, Nelson TJ. Regenerative medicine primer. Mayo Clin Proc. 2013;88(7):766–75.
3. Organ Procurement and Transplantation Network. http://optn.transplant.hrsa.gov/. Accessed 1 May 2014.
4. U.S. Government Information on Organ and Tissue Donation and Transplantation. http://www.organdonor.gov/about/data.html. Accessed 1 May 2014.
5. Colvin-Adams M, Smithy JM, Heubner BM, et al. OPTN/SRTR 2012 annual data report: heart. Am J Transplant. 2014;14 Suppl 1:113–38.
6. Smith JM, Skeans MA, Horslen SP, et al. OPTN/SRTR 2012 annual data report: intestine. Am J Transplant. 2014;14 Suppl 1:97–111.
7. Kim WR, Smith JM, Skeans MA, et al. OPTN/SRTR 2012 annual data report: liver. Am J Transplant. 2014;14 Suppl 1:69–96.
8. Matas AJ, Smith JM, Skeans MA, et al. OPTN/SRTR 2012 annual data report: kidney. Am J Transplant. 2014;14 Suppl 1:11–44.
9. Valapour M, Skeans MA, Heubner BM, et al. OPTN/SRTR 2012 annual data report: lung. Am J Transplant. 2014;14 Suppl 1:139–65.
10. Atala A, Bauer SB, Soker S, Yoo JJ, Retik AB. Tissue-engineered autologous bladders for patients needing cystoplasty. Lancet. 2006;367(9518):1241–6.
11. Atala A. Creation of bladder tissue in vitro and in vivo. A system for organ replacement. Adv Exp Med Biol. 1999;462:31–42.

12. Atala A. Tissue engineering for bladder substitution. World J Urol. 2000;18(5):364–70.
13. Atala A. Bladder regeneration by tissue engineering. BJU Int. 2001;88(7):765–70.
14. Atala A. Tissue engineering of human bladder. Br Med Bull. 2011;97:81–104.
15. Horst M, Madduri S, Gobet R, et al. Engineering functional bladder tissues. J Tissue Eng Regen Med. 2013;7(7):515–22.
16. Gunter CI, Machens HG. New strategies in clinical care of skin wound healing. Eur Surg Res. 2012;49(1):16–23.
17. Berg M, Ejnell H, Kovacs A, et al. Replacement of a tracheal stenosis with a tissue-engineered human trachea using autologous stem cells: a case report. Tissue Eng Part A. 2014;20(1–2): 389–97.
18. Wise J. Five year results show success of first tissue engineered trachea transplant. BMJ. 2013;347:f6365.
19. Pastides P, Chimutengwende-Gordon M, Maffulli N, Khan W. Stem cell therapy for human cartilage defects: a systematic review. Osteoarthritis Cartilage. 2013;21(5):646–54.
20. Kruse FE, Cursiefen C. Surgery of the cornea: corneal, limbal stem cell and amniotic membrane transplantation. Dev Ophthalmol. 2008;41:159–70.
21. Mase Jr VJ, Hsu JR, Wolf SE, et al. Clinical application of an acellular biologic scaffold for surgical repair of a large, traumatic quadriceps femoris muscle defect. Orthopedics. 2010; 33(7):511.
22. Carrier P, Deschambeault A, Audet C, et al. Impact of cell source on human cornea reconstructed by tissue engineering. Invest Ophthalmol Vis Sci. 2009;50(6):2645–52.
23. Proulx S, D'Arc Uwamaliya J, Carrier P, et al. Reconstruction of a human cornea by the self-assembly approach of tissue engineering using the three native cell types. Mol Vis. 2010;16:2192–201.
24. Cerqueira MT, Marques AP, Reis RL. Using stem cells in skin regeneration: possibilities and reality. Stem Cells Dev. 2012;21(8):1201–14.
25. Lo DD, Zimmermann AS, Nauta A, Longaker MT, Lorenz HP. Scarless fetal skin wound healing update. Birth Defects Res C Embryo Today. 2012;96(3):237–47.
26. Yildirimer L, Thanh NT, Seifalian AM. Skin regeneration scaffolds: a multimodal bottom-up approach. Trends Biotechnol. 2012;30(12):638–48.
27. Borestrom C, Simonsson S, Enochson L, et al. Footprint-free human induced pluripotent stem cells from articular cartilage with redifferentiation capacity: a first step toward a clinical-grade cell source. Stem Cells Transl Med. 2014;3(4):433–47.
28. Sato Y, Wakitani S, Takagi M. Xeno-free and shrinkage-free preparation of scaffold-free cartilage-like disc-shaped cell sheet using human bone marrow mesenchymal stem cells. J Biosci Bioeng. 2013;116(6):734–9.
29. Yoon HH, Bhang SH, Shin JY, Shin J, Kim BS. Enhanced cartilage formation via three-dimensional cell engineering of human adipose-derived stem cells. Tissue Eng Part A. 2012;18(19–20):1949–56.
30. Kobayashi S, Takebe T, Inui M, et al. Reconstruction of human elastic cartilage by a CD44+ CD90+ stem cell in the ear perichondrium. Proc Natl Acad Sci U S A. 2011;108(35): 14479–84.
31. Sicari BM, Dearth CL, Badylak SF. Tissue engineering and regenerative medicine approaches to enhance the functional response to skeletal muscle injury. Anat Rec (Hoboken). 2014; 297(1):51–64.
32. Turner NJ, Badylak SF. Regeneration of skeletal muscle. Cell Tissue Res. 2012;347(3): 759–74.
33. Chong JJ, Yang X, Don CW, et al. Human embryonic-stem-cell-derived cardiomyocytes regenerate non-human primate hearts. Nature. 2014;510(7504):273–7.
34. Rare diseases: facts and statistics, https://globalgenes.org/raredaily/rare-disease-facts-and-figures/.
35. Schakman O, Kalista S, Barbe C, Loumaye A, Thissen JP. Glucocorticoid-induced skeletal muscle atrophy. Int J Biochem Cell Biol. 2013;45(10):2163–72.

36. Mills GH, Kyroussis D, Jenkins P, et al. Respiratory muscle strength in Cushing's syndrome. Am J Respir Crit Care Med. 1999;160(5 Pt 1):1762–5.
37. Fournier M, Huang ZS, Li H, Da X, Cercek B, Lewis MI. Insulin-like growth factor I prevents corticosteroid-induced diaphragm muscle atrophy in emphysematous hamsters. Am J Physiol Regul Integr Comp Physiol. 2003;285(1):R34–43.
38. Short KR, Bigelow ML, Nair KS. Short-term prednisone use antagonizes insulin's anabolic effect on muscle protein and glucose metabolism in young healthy people. Am J Physiol Endocrinol Metab. 2009;297(6):E1260–8.
39. Hillel AT, Taube JM, Cornish TC, et al. Characterization of human mesenchymal stem cell-engineered cartilage: analysis of its ultrastructure, cell density and chondrocyte phenotype compared to native adult and fetal cartilage. Cells Tissues Organs. 2010;191(1):12–20.
40. Hipp J, Atala A. Sources of stem cells for regenerative medicine. Stem Cell Rev. 2008;4(1):3–11.
41. Ballas CB, Zielske SP, Gerson SL. Adult bone marrow stem cells for cell and gene therapies: implications for greater use. J Cell Biochem. 2002;38:20–8.
42. Thomas ED, Lochte Jr HL, Lu WC, Ferrebee JW. Intravenous infusion of bone marrow in patients receiving radiation and chemotherapy. N Engl J Med. 1957;257(11):491–6.
43. McCulloch EA, Till JE. Proliferation of hemopoietic colony-forming cells transplanted into irradiated mice. Radiat Res. 1964;22:383–97.
44. Bradley A, Evans M, Kaufman MH, Robertson E. Formation of germ-line chimaeras from embryo-derived teratocarcinoma cell lines. Nature. 1984;309(5965):255–6.
45. Odorico JS, Kaufman DS, Thomson JA. Multilineage differentiation from human embryonic stem cell lines. Stem Cells. 2001;19(3):193–204.
46. Keller G. Embryonic stem cell differentiation: emergence of a new era in biology and medicine. Genes Dev. 2005;19(10):1129–55.
47. Simerman AA, Dumesic DA, Chazenbalk GD. Pluripotent muse cells derived from human adipose tissue: a new perspective on regenerative medicine and cell therapy. Clin Transl Med. 2014;3:12.
48. Gokhale PJ, Andrews PW. The development of pluripotent stem cells. Curr Opin Genet Dev. 2012;22(5):403–8.
49. Klimanskaya I. Embryonic stem cells from blastomeres maintaining embryo viability. Semin Reprod Med. 2013;31(1):49–55.
50. Klimanskaya I, Chung Y, Becker S, Lu SJ, Lanza R. Human embryonic stem cell lines derived from single blastomeres. Nature. 2006;444(7118):481–5.
51. Takahashi K, Yamanaka S. Induction of pluripotent stem cells from mouse embryonic and adult fibroblast cultures by defined factors. Cell. 2006;126(4):663–76.
52. Wernig M, Meissner A, Foreman R, et al. In vitro reprogramming of fibroblasts into a pluripotent ES-cell-like state 1. Nature. 2007;448(7151):318–24.
53. Takahashi K, Tanabe K, Ohnuki M, et al. Induction of pluripotent stem cells from adult human fibroblasts by defined factors. Cell. 2007;131(5):861–72.
54. Yu J, Vodyanik MA, Smuga-Otto K, et al. Induced pluripotent stem cell lines derived from human somatic cells. Science. 2007;1151526.
55. Sadahiro T, Yamanaka S, Ieda M. Direct cardiac reprogramming: progress and challenges in basic biology and clinical applications. Circ Res. 2015;116(8):1378–91.
56. Kelaini S, Cochrane A, Margariti A. Direct reprogramming of adult cells: avoiding the pluripotent state. Stem Cells Cloning. 2014;7:19–29.
57. Budniatzky I, Gepstein L. Concise review: reprogramming strategies for cardiovascular regenerative medicine: from induced pluripotent stem cells to direct reprogramming. Stem Cells Transl Med. 2014;3(4):448–57.
58. Jung DW, Kim WH, Williams DR. Reprogram or reboot: small molecule approaches for the production of induced pluripotent stem cells and direct cell reprogramming. ACS Chem Biol. 2014;9(1):80–95.

59. Kastelein JJ, Ross CJ, Hayden MR. From mutation identification to therapy: discovery and origins of the first approved gene therapy in the Western world. Hum Gene Ther. 2013; 24(5):472–8.
60. Nayerossadat N, Maedeh T, Ali PA. Viral and nonviral delivery systems for gene delivery. Adv Biomed Res. 2012;1:27.
61. Ginn SL, Alexander IE, Edelstein ML, Abedi MR, Wixon J. Gene therapy clinical trials worldwide to 2012 - an update. J Gene Med. 2013;15(2):65–77.
62. Byrne BJ. Pathway for approval of a gene therapy orphan product: treading new ground. Mol Ther. 2013;21(8):1465–6.
63. Hacein-Bey-Abina S, Hauer J, Lim A, et al. Efficacy of gene therapy for X-linked severe combined immunodeficiency. N Engl J Med. 2010;363(4):355–64.
64. Simonelli F, Maguire AM, Testa F, et al. Gene therapy for Leber's congenital amaurosis is safe and effective through 1.5 years after vector administration. Mol Ther. 2010;18(3): 643–50.
65. Maguire AM, High KA, Auricchio A, et al. Age-dependent effects of RPE65 gene therapy for Leber's congenital amaurosis: a phase 1 dose-escalation trial. Lancet. 2009;374(9701): 1597–605.
66. Chung DC, Lee V, Maguire AM. Recent advances in ocular gene therapy. Curr Opin Ophthalmol. 2009;20(5):377–81.
67. Wilson JM. Lessons learned from the gene therapy trial for ornithine transcarbamylase deficiency. Mol Genet Metab. 2009;96(4):151–7.
68. Asokan A, Schaffer DV, Samulski RJ. The AAV vector toolkit: poised at the clinical crossroads. Mol Ther. 2012;20(4):699–708.
69. Choi VW, McCarty DM, Samulski RJ. AAV hybrid serotypes: improved vectors for gene delivery. Curr Gene Ther. 2005;5(3):299–310.
70. Wang Z, Tapscott SJ, Storb R. Local gene delivery and methods to control immune responses in muscles of normal and dystrophic dogs. Methods Mol Biol. 2011;709:265–75.
71. Mendell JR, Rodino-Klapac LR, Rosales XQ, et al. Sustained alpha-sarcoglycan gene expression after gene transfer in limb-girdle muscular dystrophy, type 2D. Ann Neurol. 2010;68(5): 629–38.
72. Mendell JR, Rodino-Klapac LR, Rosales-Quintero X, et al. Limb-girdle muscular dystrophy type 2D gene therapy restores alpha-sarcoglycan and associated proteins. Ann Neurol. 2009;66(3):290–7.
73. Mendell JR, Campbell K, Rodino-Klapac L, et al. Dystrophin immunity in Duchenne's muscular dystrophy. N Engl J Med. 2010;363(15):1429–37.
74. Wang Z, Kuhr CS, Allen JM, et al. Sustained AAV-mediated dystrophin expression in a canine model of Duchenne muscular dystrophy with a brief course of immunosuppression. Mol Ther. 2007;15(6):1160–6.
75. Wang Z, Storb R, Halbert CL, et al. Successful regional delivery and long-term expression of a dystrophin gene in canine muscular dystrophy: a preclinical model for human therapies. Mol Ther. 2012;20(8):1501–7.
76. Jarmin S, Kymalainen H, Popplewell L. Dickson G. Expert Opin Biol Ther: New developments in the use of gene therapy to treat Duchenne muscular dystrophy; 2013.
77. Fairclough RJ, Wood MJ, Davies KE. Therapy for Duchenne muscular dystrophy: renewed optimism from genetic approaches. Nat Rev Genet. 2013;14(6):373–8.
78. Hoffman EP, Bronson A, Levin AA, et al. Restoring dystrophin expression in duchenne muscular dystrophy muscle progress in exon skipping and stop codon read through. Am J Pathol. 2011;179(1):12–22.
79. Nelson SF, Crosbie RH, Miceli MC, Spencer MJ. Emerging genetic therapies to treat Duchenne muscular dystrophy. Curr Opin Neurol. 2009;22(5):532–8.
80. Kinali M, Arechavala-Gomeza V, Feng L, et al. Local restoration of dystrophin expression with the morpholino oligomer AVI-4658 in Duchenne muscular dystrophy: a single-blind, placebo-controlled, dose-escalation, proof-of-concept study. Lancet Neurol. 2009;8(10): 918–28.

81. van Deutekom JC, Janson AA, Ginjaar IB, et al. Local dystrophin restoration with antisense oligonucleotide PRO051. N Engl J Med. 2007;357(26):2677–86.
82. Goemans NM, Tulinius M, van den Akker JT, et al. Systemic administration of PRO051 in Duchenne's muscular dystrophy. N Engl J Med. 2011;364(16):1513–22.
83. Cirak S, Arechavala-Gomeza V, Guglieri M, et al. Exon skipping and dystrophin restoration in patients with Duchenne muscular dystrophy after systemic phosphorodiamidate morpholino oligomer treatment: an open-label, phase 2, dose-escalation study. Lancet. 2011; 378(9791):595–605.
84. McDonald CM, Henricson EK, Han JJ, et al. The 6-minute walk test as a new outcome measure in Duchenne muscular dystrophy. Muscle Nerve. 2010;41(4):500–10.
85. Barton-Davis ER, Cordier L, Shoturma DI, Leland SE, Sweeney HL. Aminoglycoside antibiotics restore dystrophin function to skeletal muscles of mdx mice. J Clin Invest. 1999;104(4):375–81.
86. Welch EM, Barton ER, Zhuo J, et al. PTC124 targets genetic disorders caused by nonsense mutations. Nature. 2007;447(7140):87–91.
87. Finkel RS, Flanigan KM, Wong B, et al. Phase 2a study of ataluren-mediated dystrophin production in patients with nonsense mutation duchenne muscular dystrophy. PLoS ONE. 2013;8(12), e81302.
88. Finkel RS. Read-through strategies for suppression of nonsense mutations in Duchenne/Becker muscular dystrophy: aminoglycosides and ataluren (PTC124). J Child Neurol. 2010;25(9):1158–64.
89. Sonnemann KJ, Heun-Johnson H, Turner AJ, Baltgalvis KA, Lowe DA, Ervasti JM. Functional substitution by TAT-utrophin in dystrophin-deficient mice. PLoS Med. 2009;6(5), e1000083.
90. Tinsley JM, Fairclough RJ, Storer R, et al. Daily treatment with SMTC1100, a novel small molecule utrophin upregulator, dramatically reduces the dystrophic symptoms in the mdx mouse. PLoS ONE. 2011;6(5), e19189.
91. Nguyen HH, Jayasinha V, Xia B, Hoyte K, Martin PT. Overexpression of the cytotoxic T cell GalNAc transferase in skeletal muscle inhibits muscular dystrophy in mdx mice. Proc Natl Acad Sci U S A. 2002;99(8):5616–21.
92. Cosford KL, Taylor SM, Thompson L, Shelton GD. A possible new inherited myopathy in a young Labrador retriever. Can Vet J. 2008;49(4):393–7.
93. Frase AR. The miracle of Nibs. http://www.joshuafrase.org/uploads/JFF-Thestory of Nibs. pdf. 2009.
94. Heckmatt JZ, Sewry CA, Hodes D, Dubowitz V. Congenital centronuclear (myotubular) myopathy. A clinical, pathological and genetic study in eight children. Brain. 1985;108(Pt 4):941–64.
95. Jungbluth H, Wallgren-Pettersson C, Laporte J. Centronuclear (myotubular) myopathy. Orphanet J Rare Dis. 2008;3:26.
96. Laporte J, Hu LJ, Kretz C, et al. A gene mutated in X-linked myotubular myopathy defines a new putative tyrosine phosphatase family conserved in yeast. Nat Genet. 1996;13(2): 175–82.
97. Laporte J, Blondeau F, Buj-Bello A, et al. Characterization of the myotubularin dual specificity phosphatase gene family from yeast to human. Hum Mol Genet. 1998;7(11):1703–12.
98. Cameron JM, Maj MC, Levandovskiy V, MacKay N, Shelton GD, Robinson BH. Identification of a canine model of pyruvate dehydrogenase phosphatase 1 deficiency. Mol Genet Metab. 2007;90(1):15–23.
99. Buj-Bello A, Laugel V, Messaddeq N, et al. The lipid phosphatase myotubularin is essential for skeletal muscle maintenance but not for myogenesis in mice. Proc Natl Acad Sci U S A. 2002;99(23):15060–5.
100. Dowling JJ, Vreede AP, Low SE, et al. Loss of myotubularin function results in T-tubule disorganization in zebrafish and human myotubular myopathy. PLoS Genet. 2009;5(2), e1000372.
101. Beggs AH, Bohm J, Snead E, et al. MTM1 mutation associated with X-linked myotubular myopathy in Labrador Retrievers. Proc Natl Acad Sci U S A. 2010;107(33):14697–702.

102. Buj-Bello A, Fougerousse F, Schwab Y, et al. AAV-mediated intramuscular delivery of myo-tubularin corrects the myotubular myopathy phenotype in targeted murine muscle and sug-gests a function in plasma membrane homeostasis. Hum Mol Genet. 2008;17(14):2132–43.
103. Grange RW. Muscle function in a canine model of X-Linked Myotubular Myopathy. Muscle Nerve. 2012.
104. Goddard MA, Mitchell EL, Smith BK, Childers MK. Establishing clinical end points of respiratory function in large animals for clinical translation. Phys Med Rehabil Clin N Am. 2012;23(1):75–94. xi.
105. Butler D. French move past Genethon to gene-therapy research. Nature. 1993;361(6414):671.
106. Childers MK, Joubert R, Poulard K, et al. Gene therapy prolongs survival and restores func-tion in murine and canine models of myotubular myopathy. Sci Transl Med. 2014;6(220), 220ra210.

Chapter 2
An Overview of rAAV Vector Product Development for Gene Therapy

Richard O. Snyder

2.1 Vector Design and Lead Identification

When designing a recombinant adeno-associated viral (rAAV) gene transfer vector, consideration is given to tissue-specific transgene expression control elements, the transgene (e.g., codon optimization) and size of the expression cassette, vector genome configurations (i.e., single-stranded (ss) vs. self-complementary (sc) [1]), along with the AAV capsid serotype that efficiently transduces the target tissue and cell type. To date, more than 100 serotypes of AAV have been identified [2] and others generated through directed evolution and other engineering methods ([3–7]), and gene transfer studies in animal models have shown dramatic differences in the transduction efficiency and cell specificity [8–14]. The different AAV serotypes exhibit profound differences in their ability to transduce diverse cell types in vivo and in vitro, with the basis for these differences in infection efficiency relating to the presence or absence of specific cell surface receptors [15] and intracellular factors [16, 17]. One or more of these elements incorporated into a particular vector may require the licensing of intellectual property. Once the candidate elements are procured and assembled, then screening different combinations for the desired product-specific attributes can proceed to in vitro and in vivo studies in animals (normal and disease models). When a lead construct is identified, then it is locked-in throughout preclinical and clinical trials, where significant financial resources will be consumed over several years.

R.O. Snyder, Ph.D. (✉)
Department of Molecular Genetics and Microbiology, University of Florida, College of Medicine, 1600 SW Archer Road, Gainesville, FL 32610-0266, USA

Atlantic Gene Therapies, INSERM UMR 1089, Université de Nantes, CHU de Nantes, Nantes 44007, France

Center of Excellence for Regenerative Health Biotechnology, University of Florida, Alachua, FL 32615, USA
e-mail: rsnyder@cerhb.ufl.edu

© Springer Science+Business Media New York 2016
M.K. Childers (ed.), *Regenerative Medicine for Degenerative Muscle Diseases*, Stem Cell Biology and Regenerative Medicine,
DOI 10.1007/978-1-4939-3228-3_2

Each therapeutic vector is engineered with unique attributes, thus dosing is vector specific, route and organ target specific, and disease specific, in the context of the vector's safety profile. Potency depends on the capsid serotype, promoter and other expression control elements, and the half-life of the transgene product needed to ameliorate disease. So, the required dose of different vectors will likely be unique to each vector, even for the same disease target and same organ target, and thus, choices made in the design of the product at this early stage will impact directly on the cost to manufacture the product and product differentiation for market positioning and market share.

2.2 Product Manufacturing

Recombinant adeno-associated virus (rAAV) vectors are manufactured in compliance with current good manufacturing practices (cGMPs) as outlined in the US, European, Japanese, or ICH regulations (21CFR210,211,610; EU guidelines on GMP Part I and II Annex 13; MHLW Ministerial Ordinance No. 179, and ICH Q7A) (http://www.fda.gov.cber/guidelines.htm, http://www.ich.org, http://ec.europa.eu/health/documents/eudralex/vol-4/index_en.htm, http://www.pmda.go.jp/english/index.html and http://www.mhlw.go.jp/english/policy/health-medical/pharmaceuticals/index.html). To meet these requirements, manufacturing controls and quality systems are established, including (1) adequate facilities and equipment, (2) personnel who have relevant education or experience and are trained for specific assigned duties, (3) raw materials that are qualified for use, and (4) a process (including production, purification, formulation, filling, storage, and shipping) that is controlled, aseptic, reliable, and consistent. Quality systems including quality control (QC) and quality assurance (QA) are also implemented. These manufacturing procedures and quality systems are designed to ensure that the product meets its release specifications so patients receive a safe, pure, potent, and stable investigational or commercial drug. It is important to keep in perspective that for cGMP manufacturing and testing, there is an expectation that systems and controls continuously improve as the product development activities transition through phases 1–3 towards licensure and commercial launch. Costs increase significantly as the project moves along this continuum. One solution to control manufacturing costs is to outsource the manufacturing and testing to a contract manufacturing organization (CMO), instead of maintaining a large staff and costly cGMP manufacturing facility.

2.2.1 Scale

Deciding on an appropriate manufacturing system depends on lead time, and pressures from investors and advocates to initiate clinical trials, as well as the scale of the clinical lot (e.g., dose, organ target, size of population), and access to manufacturing-related intellectual property. One misconception is that GMP compliance equates with the batch scale, but the two are independent parameters (e.g., a small batch will still need to be manufactured in compliance with the cGMP regulations for human use).

The rAAV vector amounts required in an animal study or a clinical trial need to be carefully calculated in the context of administration. For example, with frozen liquid formulations, if an animal dose-escalation study can accommodate administering vector to more than one animal in a few hour window, then one vial of vector could be thawed and used for multiple animals. However, for administration to humans, single-use vials are required. Therefore, the amount of vector needed for administration depends on the number of vials needed, the fill volume of each vial, and the vector concentration (as illustrated in Table 2.1). The fill volume needs to be compatible with the delivery device and the device's hold-up volume. Furthermore, approximately 20 % more vials are needed for QC product release and stability testing that runs concurrently with the animal study or human clinical trial. Additionally, separate product lot stability studies may be required for each volume of vector filled in a vial; therefore, filling vials at a single volume will reduce the amount of vector consumed in multiple stability studies, but this will necessitate combining multiple vials in the clinic for higher doses.

Manufacturing processes that are based on transient transfection [18] provide an ability to rapidly screen multiple candidate vector constructs in vitro and in vivo to settle on a lead product candidate. Transient transfection may also be suitable for clinical trials that require delivery of limited vector dose volumes (retina/brain) and small patient populations. Commercial services for establishing and qualifying *E. coli* master cell banks (MCBs) and manufacturing plasmid DNA are readily available. Transfection reagents can be made and qualified relatively quickly, and if a qualified 293 MCB is available, then preclinical and clinical lot manufacturing can be initiated on a relatively short timeline.

Other systems that are more scalable, such as baculovirus [19, 20], adenovirus [21, 22], or herpesvirus [23] production systems, require a longer lead time for banking the appropriate cell (MCB) and engineering and banking the production vectors used in manufacturing (i.e., master viral banks, MVBs). These systems are amenable to supplying the larger vector lots for the large doses needed for large organ targets as well as sizable patient populations. Once the qualified manufacturing reagents are established for these systems, then producing sequential large-scale lots is much more rapid than transient transfection.

2.2.2   Reagents and Materials

The highest-quality reagents and materials should be sourced (e.g., United States Pharmacopeia, National Formulary, American Chemical Society grades) or made for rAAV manufacturing, and a robust supply chain established. Depending on the product development plan and business plan, having maximum freedom to operate will be important, and this includes having ownership of MCB and MVBs when it is time to commercialize the product. Being beholden to another entity that owns an MCB (with

Table 2.1 Hypothetical lot size calculation

vg/kg	kg/pt	vg/pt	pts/cohort	vg/cohort	vg/vial*	vials/pt	vials/cohort	vg/cohort
1.00E+11	70	7.00E+12	5	3.50E+13	1.10E+13	1	5	5.50E+13
3.30E+11	70	2.31E+13	5	1.16E+14	1.10E+13	3	15	1.65E+14
1.00E+12	70	7.00E+13	5	3.50E+14	1.10E+13	7	35	3.85E+14
3.30E+12	70	2.31E+14	5	1.16E±15	1.10E+13	21	105	1.16E±15
vg needed for the clinic				**1.66E+15**				**1.76E+15**
vg needed for release and stability testing							40 vials	**0.44E+15**
Total vg needed							200 vials	**2.20E+15**

*filled 1.1ml at 1.00+13vg/ml

pt, patient
vg, vector genome
kg, kilogram

a finite number of vials), where a vial or two was purchased for a phase 1 trial, is not amenable to a full product development trajectory. Likewise, vector constructs, expression cassette elements, and vector configurations (ss vs. sc) may require obtaining licenses to technology, with payments synchronized to clinical and sales milestones.

2.2.3 Process Design and Development

The culture volume of producer cells needs to be increased to a scale that can support the preclinical study or clinical trial. Once the rAAV vector harvest is collected, purification of the vector away from process impurities is needed. For scale and capacity, column chromatography is a common method; however, separation of full (i.e., vector genome-containing) from empty capsids is not efficient, because there is usually overlap between the peaks resulting in loss of product yield. For some indications, the presence of empty capsids may be desired [24]. In other indications, removing the empty AAV capsids may be important in a situation where the highest potency is desired or lowest capsid load is warranted. In the latter case, density gradient ultracentrifugation is an efficient, albeit capacity constrained, method [18, 25, 26].

2.2.4 Upstream Processing

Production cells can be cultured on adherent plasticware (e.g., multilayer cell factories, roller bottles) or in suspension using rocker-type reactors or stirred tank reactors. All of these cell culture formats are available as disposables, with the latter two being highly scalable. Appropriate cell seed trains that precede the actual seeding of culture vessels and induction of production are also needed, as are appropriate sampling points for in-process and final product QC testing. For transient transfection, cells are co-transfected with plasmid DNA that supplies the cells with the AAV and helper genes necessary to replicate and package the rAAV genome. For the baculovirus, adenovirus, and HSV-based systems, cells are infected with recombinant production vectors to supply the cells with the AAV and helper genes necessary to replicate and package the rAAV genome.

2.2.5 Downstream Processing

To obtain the drug substance, the vector is separated from process impurities using a variety of filtration and chromatographic technologies with different standard chemistries and affinity molecules that are available in filter, bead, and monolith formats. These steps remove DNA, lipid, carbohydrate, RNA, and protein impurities that derive from the production cells, production vectors or plasmid DNA, and AAV components (e.g., unassembled capsids, unpackaged vector DNA, etc.). Here again, appropriate sampling is needed for in-process and final product release testing [27].

2.2.6   Formulation, Fill, Finish, and Product Configuration

For the drug product, rAAV vectors are most commonly formulated as liquids in simple salt recipes, but the formulation needs to be compatible not only with the vector itself – preserving its potency during storage – but also with the end-use (i.e., organ target, route of administration, device, and volumetric dose) and stability that may be needed in the clinic (i.e., the time from vial thaw to patient administration). As described above, the filling of vials with a single volume may be desired, so as to reduce the amount of vials/vector consumed in product lot stability testing. Appropriate labeling, packaging, and storage and shipping conditions need to be validated to ensure product potency along the supply chain.

2.3   Analytical Testing

2.3.1   Assay Design and Development

Although several assays are employed to evaluate AAV vectors used for research, additional assay development may be required to achieve appropriate specificity, accuracy, precision, reproducibility, sensitivity, range, and/or limits of detection and quantification. The QC assay development activity also involves generating standards and controls and writing test records and reagent preparation logs.

2.3.2   Assay Qualification

Once the methods are established, then the assays are qualified using calibrated equipment to demonstrate that they meet appropriate assay parameters (robustness, reproducibility, ruggedness, specificity, sensitivity, etc.) for eventual assay validation. Once the validity criteria are established, then the assays can be used for routine testing.

2.3.3   Sampling and Testing

Appropriate in-process and final product sampling and testing will depend on the specific manufacturing process, and some tests may need to be conducted repeatedly throughout the process [28]. Furthermore, the performance of the assays on samples in different matrices (e.g., final formulation buffer vs. a chromatography buffer with a different pH or salt concentration) may need to be evaluated during assay qualification.

2.3.4 Product Specifications Setting

Testing and characterization of each lot of the rAAV vector is performed under cGMPs using qualified or validated assays prior to product release. Release documentation (a Certificate of Analysis, COA) for each lot is submitted to the regulatory agency. Tests and specifications for product release are set prior to testing an actual lot of product. The development of product release specifications is an activity that relies on previous experience with the production of the batches of the specific vector manufactured for research or toxicology and is dependent on what is acceptable by regulatory authorities (e.g., passing a sterility test) and what the process is capable of generating. In some cases, no test specification is set so that data can be collected to monitor the manufacturing process or to establish a specification for products in later phases of development; these results are reported on the COA.

2.3.5 Stability Testing

A stability study should be designed to generate data for the purified rAAV viral vector drug product at the proper storage temperature, formulation, and fill volume and in the storage container used for patient doses. The study is designed to demonstrate genetic and physicochemical stability of the product and container integrity for at least the duration of the clinical trial. A typical testing schedule includes time points at 1, 3, 6, 9, 12, 18, and 24 months after filling or until the last patient is administered vector from the product lot.

During product development, an iterative process is undertaken as the manufacturing process becomes more refined and controlled, the analytical methods become more robust, and specifications become better defined with narrower ranges. As the analytics improve (e.g., become more sensitive), then the manufacturing process may need to be improved to better remove impurities previously undetected, and vice versa.

2.4 Toxicology Studies

Preclinical studies in animals address the efficacy of the vector in treating the targeted disease state and evaluate the toxicity profile and biodistribution of the vector. These studies are designed to assess the safety and efficacy of vector itself and identify a dose range for the route and method of delivery that mimics closely what is planned for the human clinical trial. For animal studies, the species-specific homologous transgene may be required to avoid complications of immune reactions. For toxicology, the vector should carry the human transgene (i.e., the exact same vector construct that will be used in humans). However, there are other considerations that make designing an animal toxicology study complex, such as human-specific genome editing or mRNA editing [29–31] where the genetic target is absent and the relevancy of off-targeting may be difficult to determine. Vector evaluated in these studies should be produced and purified using the same methods that will be used to manufacture vector for the human clinical studies. The vector does not need to be manufactured under cGMP; however, ideally for toxicology studies, the culture reagents should be the same as those used for the clinical lots (i.e., the same MCB/WCB and/or MVB/WVB), as the safety of the final vector lot and any impurities endemic in the production reagents (and their removal) are being evaluated in animals ahead of administration to humans. Nonclinical toxicology and biodistribution studies to support IND applications should be performed in compliance with good laboratory practices (GLPs) regulations (21CFR58, ICH S1-S8) where nonclinical studies are defined as in vivo or in vitro experiments in which the vector (test article) is evaluated prospectively in test systems (animal models, cell culture, etc.) under controlled laboratory conditions to determine the product's safety profile (e.g., toxicity, biodistribution, immune responses, dosing) [32, 33]. The design of the study includes parameters for route, dose, schedule of administration (e.g., single, repeat), animal species or model, sex and age of subjects, time dependence (long-term persistence and the time needed to achieve steady-state expression), and number of subjects. Qualified or validated testing methods should be employed to characterize the vector used in the toxicology studies as well as the tests used to evaluate toxicology and biodistribution. The GLP regulations state that the identity, purity, and composition of the vector (test article) used in a safety study must be known and documented. In addition, the stability of the vector preparation in the specific container used for the study must also be known prior to initiation of the study or acquired during the study itself.

Proper training is necessary to instruct the personnel on procedures for thawing and handling (including mixing) product vials, loading the administration device, and administering doses, as well as conducting the study, performing sampling and necropsy protocols, and conducting the analyses. These studies can be outsourced to a contract research organization (CRO), but direct involvement with the CRO, especially with novel agents like gene transfer vectors, can ensure the study is conducted smoothly.

2.5   Clinical Trials

Clinical trials are conducted in compliance with good clinical practices (GCPs) regulations (21 CFR 50, 56, and11 or ICH E6) to ensure appropriate conduct of the trial and patient protections. The design of the human clinical trial involves the route of administration and the device used for delivery, the number of patients in each cohort, the dose per patient in each cohort, the number of cohorts, inclusion/exclusion criteria, informed consent, patient safety monitoring, adverse event responses, stopping rules, and efficacy endpoints. For patient protections, there is on-site review in the form of Investigational Review Board (IRB) and Institutional Biosafety Committee (IBC) and external review from a Data Safety Monitoring Board (DSMB). Clinical trials can be outsourced to a CRO that specializes in clinical studies, but direct involvement with the CRO is important.

Phase 1 trials are designed to evaluate product safety, observe any side effects associated with increasing doses, gain early evidence of effectiveness, and evaluate dose increases with an observation time between cohorts (usually three to five patients each). Phase 2 trials are designed primarily to evaluate product safety, determine the long- and short-term side effects and risks, and evaluate the effectiveness of the drug for a particular indication in patients with the disease, along with dose ranging (phase 2a) and determination of potential efficacy endpoints (phase 2b). These studies usually involve cohorts of 5–10 patients each but can be much larger (on the order of hundreds of patients total) depending on the indication. Phase 3 trials are expanded randomized controlled and uncontrolled multicenter studies that are started after preliminary evidence suggesting effectiveness of the drug has been obtained. Product safety is evaluated and additional information is obtained to evaluate the overall benefit-risk relationship of the drug. The data provide an adequate basis for physician labeling for specified dose(s) in specified patient population(s). They include designated efficacy endpoints and comparison with current standard treatments and involve cohorts of 15–30 patients each but can be much larger (on the order of multiple hundreds of patients total) depending on the indication. Estimated costs and timelines (that can overlap) for a rAAV vector-based gene therapy product are listed in Table 2.2; the wide ranges reflect the spectrum of orphan to more common disease indications. Following licensure, Phase 4 postmarketing surveillance studies are conducted to delineate additional information from the larger number of patients receiving the prescription drug, including the drug's risks, benefits, and optimal use. These studies can be used to support extended claims or usage, suitability for a new population or indication, demonstrate manufacturing changes, validate surrogate clinical endpoints, and demonstrate long-term safety and efficacy.

Early input from the end user (i.e., clinician) is important as it will influence the specific attributes incorporated into the vector, the level of impurities that are acceptable for the specific route of delivery and organ target, the formulation of the drug product (i.e., concentration of the vector and excipients), the volume and container that the vector is dispensed into, the labeling of the vials, the carton and ancillary items included in the packaging, and the storage of the vector.

As is the case with the preclinical studies, proper training is necessary to instruct the clinical personnel (investigational pharmacist, study coordinator, nurses, doctors) on procedures for thawing and handling product vials, loading the device, and administering doses, as well as conducting the study, monitoring patients, and performing sampling and evaluation. In some cases, adding the vector product to a bag

Table 2.2 Major milestones for product development

Milestone	Activities	Duration	Costs
Vector design, POC studies, and lead identification	Depends on the number of variables to be screened (promoters, serotypes) and the number of vector preps to be produced, data collected and analyzed from proof-of-concept, and performance studies in vitro and in small animals. Long-term correction demonstration can be 6–18 months. The variables are compared and the lead vector configuration chosen based on desired performance metrics	0.5–2 years	$250K–1M
Process and analytical development and reagent preparation	Process development and evaluation is required to achieve larger scale, asepsis, automation, and reproducibility of yield and purity suitable to support human clinical trials. Transferring production and testing protocols and methods to the cGMP environment. Converting production and purification protocols and identifying raw materials and reagents suitable for human use, but having equivalent performance (i.e., generating the same vector yield, purity, potency, and stability) can take several months. Establishing and qualifying MCBs and/or MVBs, manufacturing plasmid DNA, etc., can take months. Once the actual methods, automated programs, and raw materials are established, then the documents (batch records, reagent preparation records, test records standard operating procedures, etc.) and specifications need to be written and personnel trained and shakedown runs need to be executed. In parallel, new and existing product-specific assays are developed/customized and qualified to then use for in-process and final product testing	0.5–1 year	$500K–$3M
Toxicology study	Preparing pre-IND document and FDA review. Manufacturing a suitable amount of vector followed by release testing, with some of the tests requiring 8 weeks. The actual study includes procuring and screening the animals (small vs. large), vector administration, time course, necropsy, and analyses, including biodistribution. Quality assurance audit and final report needs to be written	6–12 months	$250K–$2M
Phase I clinical trial	Preparing the IND document and FDA review. Manufacturing a suitable amount of vector followed by release testing, with some of the tests requiring 8 weeks. Enrolling patients, time between cohorts. Evaluations	1–3 years	$1.5–10M

(continued)

Table 2.2 (continued)

Milestone	Activities	Duration	Costs
Phase 2	Manufacturing a suitable amount of vector followed by release testing. Enrolling patients, time between cohorts. Evaluations	1–4 years	$3–20M
Phase 3	Manufacturing a suitable amount of vector followed by release testing. Enrolling patients, time between cohorts. Evaluations	2–4 years	$10–50M
Licensure	Preparing the BLA and FDA review and site inspections	8–12 months	$2–3M
Product launch	Building a manufacturing plant. Manufacturing a suitable amount of vector followed by release testing. Sales and marketing activities. Physician training. Phase 4 studies	8–36 months	$30–200M

of saline for infusion or other pharmacy manipulations (loading syringes) may be required prior to administration, and understanding the stability profile of the vector at the holding temperature in each format is necessary.

2.6   Regulatory

Ultimately, the goal of the research, preclinical, and clinical studies is to achieve marketing approval and commercial launch. The license application is reviewed by the US FDA, European Medicines Agency, or country-specific regulatory authorities. For clinical trials, meetings that may be held between sponsors and the US FDA include (a) pre-investigational new drug application (pre-IND), (b) end-of-phase 1 meeting, (c) end-of-phase 2 and pre-phase 3 meeting, and (d) pre-Biologics License Application (pre-BLA). These meetings may be requested by the sponsor to address outstanding questions and scientific issues that arise during the course of a clinical investigation, aid in the resolution of problems, and facilitate evaluation of drugs. The meetings, which often coincide with critical points in the drug development and/or regulatory process, are recommended, but are not mandatory. Additional meetings may be requested at other times, if warranted [34], and informal conversations may be possible. In addition, FDA drafts and updated guidance on the manufacture and testing of gene transfer vectors and evaluation of products in toxicology studies and human trials.

The formal pre-IND meeting is held to discuss the pre-IND document submitted in advance of the meeting that outlines the study objectives and design (scientific rational), animal data, toxicology study design, previous human experience, clinical protocol synopsis (intended patient population (risk/benefit, defined disease), patient accrual (inclusion/exclusion, numbers needed), proposed dose and escalation rationale, patient monitoring plan, safety evaluations, adverse events, statistical methods, potential efficacy endpoints, schedule of protocol events), manufacturing/characterization summary, and stopping rules (e.g., severe adverse events, germ line transmission, severe immune reaction). Following the incorporation of any changes that may have been requested at the pre-IND meeting, the toxicology study is conducted and manufacturing the clinical lot(s) can begin.

The Investigational New Drug Application (IND) covers general investigational plan; clinical protocol; chemistry, manufacturing, and control information (CMC); pharmacology and toxicology data; IRB approved consent form; and previous human experience. The IND automatically becomes effective 30 days after receipt by the FDA, unless the FDA places the clinical trial on hold within this 30-day time window. If a hold is placed on the trial, the sponsor and the FDA must resolve the issues before the clinical trial can start.

In addition to submission of an IND to the FDA, the protocol is also submitted to the National Institutes of Health (NIH) Office of Biotechnology Activities (OBA), in accordance with the NIH Guidelines for Research Involving Recombinant DNA. Compliance with these guidelines is required for institutions and their investigators who receive NIH funds for research involving recombinant DNA, as utilized in gene transfer vectors. Given the recent recommendations following review of the NIH OBA Recombinant DNA Advisory Committee's oversight of gene therapy protocols, review may be limited in the future to protocols involving (a) a new vector, genetic material, or delivery methodology which represents a first-in-human experience, thus presenting an unknown risk, (b) a protocol that relies on preclinical safety data obtained using a new preclinical model system of unknown and unconfirmed value, and (c) a proposed vector, gene construct, or method of delivery that is associated with possible toxicities that are not widely known which may render it difficult for local and federal regulatory bodies to rigorously evaluate the protocol [35].

Following the clinical trial phases, a Biologics License Application (BLA) is submitted to the FDA (in Europe, a Marketing Authorization Application is submitted to the EMA) that includes data from the manufacture and testing of the product, supporting laboratory and animal study results, and human trial results; composition and configuration of the product candidate; and information that may be included in the product label. Under the Prescription Drug User Fee Act (PDUFA), a user fee is paid with each BLA submission. For 2016, the user fee for a BLA requiring clinical data is $2,374,200, the annual product fee for biologics is $114,450, and an annual establishment fee of $585,200 is assessed on facilities manufacturing prescription biologics [36]. A waiver is granted for user fees assessed on BLAs for product candidates designated as orphan drugs (see below). The FDA will review the BLA to ensure all necessary information has been provided before accepting it for filing, and once filed, the BLA is reviewed in 10 months, but a priority review can be made in 6 months.

There is an expectation that compliance systems along with manufacturing and analytical procedures improve proportionally to each successive clinical phase. Manufacturing facilities are inspected and licensed by the FDA in advance of product licensure and product launch (i.e., a "prelicense inspection"). Manufacturing systems and quality systems are required to be sufficiently validated to support commercial manufacturing. FDA may also audit the nonclinical and clinical trial sites that generated the data in support of the BLA. In Europe, the manufacturing site is inspected and authorized, and the medicines regulatory authorities in Member States supervise these authorizations.

2.6.1   Expedited Programs

The FDA has programs that expedite the review process described above, including (a) fast-track designation, (b) breakthrough therapy designation, (c) accelerated approval, and (d) priority review. These programs are intended to help ensure that therapies for serious conditions are approved and available to patients as soon as it can be concluded that the benefits of the therapy justify its risks [37].

2.7   Commercial

Many of the rAAV-based gene transfer products are currently being developed to treat rare diseases. Under the Orphan Drug Act, the FDA may grant orphan designation to a biological product candidate intended to treat a rare disease or condition, generally a disease or condition that affects fewer than 200,000 individuals in the United States or which will not be profitable within 7 years following approval by the FDA. The request for orphan product designation is made prior to submitting a BLA. If a product candidate with orphan designation receives the first FDA approval for the specific disease or condition, the product is entitled to exclusivity, prohibiting the FDA from approving other applications to market the same product for the same indication for 7 years (10 years in Europe after obtaining the designation from EMA), except in limited situations (http://www.fda.gov/forindustry/developing-productsforrarediseasesconditions/howtoapplyfororphanproductdesignation/default.htm). Furthermore, the FDA provides grants (phase 1 trials of up to $200,000 per year for up to 3 years, and for phase 2 and 3 trials up to $400,000 per year for up to 4 years) for clinical studies on safety and/or effectiveness that will result in, or substantially contribute to, market approval of these products (http://www.fda.gov/ForIndustry/DevelopingProductsforRareDiseasesConditions/WhomtoContactaboutOrphanProductDevelopment/default.htm).   Other   benefits include a waiver of PDUFA fees for orphan product BLAs and a 50 % tax credit for clinical research and testing expenses [38].

Given the expected high price for gene therapy products, sales will depend on coverage by third-party payers [39]. In the United States, private and governmental payers determine the amount to which new biologics will be covered and reimbursed. Approval for coverage and reimbursement of rAAV-based products for the Medicare program will come from the US Health and Human Services Centers for Medicare & Medicaid Services and usually sets levels that are adopted by private payers. Coverage and reimbursement by a third-party payer is dependent on the effectiveness of the product, its price, and medical need. In some countries, a product candidate needs to be approved for reimbursement prior to being approved for sale. Models have been proposed for setting the price and negotiating the reimbursement of long-duration rAAV gene therapy products [40–44] that have the potential to ameliorate disease for a lifetime following a single administration.

For product launch, it is expensive to implement a suitable sales and marketing organization for patient and physician awareness, achieving market acceptance, and maximal product use. One approach is to form an alliance with another company with an established sales and marketing apparatus that is familiar with FDA advertising and promotion regulations pertaining to professional, direct-to-consumer (DTC) advertising and promotional labeling materials, off-label use, interactive and internet-based promotion, and drug product naming (http://www.fda.gov/BiologicsBloodVaccines/ DevelopmentApprovalProcess/AdvertisingLabelingPromotionalMaterials/ ucm164120.htm and [45]). Market positioning will be influenced by product efficacy and safety, competitive pricing, lower dose/higher potency, less invasive route of administration, fewer side effects (off-targeting), and competing technology. Furthermore, to support product launch, the manufacturing and distribution supply chain must be robust enough so as to avoid interruption of drug supply at this critical time and to support expanded market penetration.

As of this writing, there has been an impressive surge of investment of nearly $1B, new company formation, and entrance of large pharma and established biotechnology companies entering the cell and gene therapy sector. Much of this activity is driven by published reports of efficacy [46–50], and the recent market approval [51] and reimbursement pricing (€1.1 M) in Europe of the rAAV-based Glybera from UniQure [52], perhaps ushering in an era in the gene therapy industry beyond the Nascent stage [53].

Conflict of Interest Statement RS is an inventor on patents related to recombinant AAV technology. RS owns equity in a gene therapy company that is commercializing AAV for gene therapy applications. To the extent that the work in this manuscript increases the value of these commercial holdings, RS has a conflict of interest.

References

1. McCarty DM. Self-complementary AAV, vectors; advances and applications. Mol Ther. 2008;16(10):1648–56. Epub 2008/08/07.
2. Gao G, Vandenberghe LH, Alvira MR, Lu Y, Calcedo R, Zhou X, et al. Clades of adeno-associated viruses are widely disseminated in human tissues. J Virol. 2004;78(12):6381–8.
3. Kotterman MA, Schaffer DV. Engineering adeno-associated viruses for clinical gene therapy. Nat Rev Genet. 2014;15(7):445–51. Epub 2014/05/21.
4. Marsic D, Govindasamy L, Currlin S, Markusic DM, Tseng YS, Herzog RW, et al. Vector design tour de force: integrating combinatorial and rational approaches to derive novel adeno-associated virus variants. Mol Ther. 2014;22(11):1900–9. Epub 2014/07/23.

5. Rabinowitz JE, Bowles DE, Faust SM, Ledford JG, Cunningham SE, Samulski RJ. Cross-dressing the virion: the transcapsidation of adeno-associated virus serotypes functionally defines subgroups. J Virol. 2004;78(9):4421–32. Epub 2004/04/14.
6. Warrington Jr KH, Gorbatyuk OS, Harrison JK, Opie SR, Zolotukhin S, Muzyczka N. Adeno-associated virus type 2 VP2 capsid protein is nonessential and can tolerate large peptide insertions at its N terminus. J Virol. 2004;78(12):6595–609.
7. Gigout L, Rebollo P, Clement N, Warrington Jr KH, Muzyczka N, Linden RM, et al. Altering AAV tropism with mosaic viral capsids. Mol Ther. 2005;11(6):856–65. Epub 2005/06/01.
8. Burger C, Gorbatyuk OS, Velardo MJ, Peden CS, Williams P, Zolotukhin S, et al. Recombinant AAV viral vectors pseudotyped with viral capsids from serotypes 1, 2, and 5 display differential efficiency and cell tropism after delivery to different regions of the central nervous system. Mol Ther. 2004;10(2):302–17.
9. Chao H, Liu Y, Rabinowitz J, Li C, Samulski RJ, Walsh CE. Several log increase in therapeutic transgene delivery by distinct adeno-associated viral serotype vectors. Mol Ther. 2000;2(6):619–23.
10. Davidson BL, Chiorini JA. Recombinant adeno-associated viral vector types 4 and 5. Preparation and application for CNS gene transfer. Methods Mol Med. 2003;76:269–85.
11. Grimm D, Kay MA, Kleinschmidt JA. Helper virus-free, optically controllable, and two-plasmid-based production of adeno-associated virus vectors of serotypes 1 to 6. Mol Ther. 2003;7(6):839–50.
12. Rutledge EA, Halbert CL, Russell DW. Infectious clones and vectors derived from adeno-associated virus (AAV) serotypes other than AAV type 2. J Virol. 1998;72(1):309–19.
13. Xiao W, Chirmule N, Berta SC, McCullough B, Gao G, Wilson JM. Gene therapy vectors based on adeno-associated virus type 1. J Virol. 1999;73(5):3994–4003.
14. Gao GP, Alvira MR, Wang L, Calcedo R, Johnston J, Wilson JM. Novel adeno-associated viruses from rhesus monkeys as vectors for human gene therapy. Proc Natl Acad Sci U S A. 2002;99(18):11854–9.
15. Agbandje-McKenna M, Kleinschmidt J. AAV capsid structure and cell interactions. Methods Mol Biol. 2011;807:47–92. Epub 2011/10/29.
16. Schwartz RA, Palacios JA, Cassell GD, Adam S, Giacca M, Weitzman MD. The Mre11/Rad50/Nbs1 complex limits adeno-associated virus transduction and replication. J Virol. 2007;81(23):12936–45. Epub 2007/09/28.
17. Choi VW, McCarty DM, Samulski RJ. Host cell DNA repair pathways in adeno-associated viral genome processing. J Virol. 2006;80(21):10346–56.
18. Snyder RO, Xiao X, Samulski RJ, et al. Production of recombinant adeno-associated viral vectors. In: Dracopoli N, Haines J, Krof B, Moir D, Morton C, Seidman C, editors. Current protocols in human genetics. New York: Wiley; 1996. p. 12.1.1–24.
19. Mietzsch M, Grasse S, Zurawski C, Weger S, Bennett A, Agbandje-McKenna M, et al. OneBac: platform for scalable and high-titer production of adeno-associated virus serotype 1–12 vectors for gene therapy. Hum Gene Ther. 2014;25(3):212–22. Epub 2013/12/05.
20. Cecchini S, Virag T, Kotin RM. Reproducible high yields of recombinant adeno-associated virus produced using invertebrate cells in 0.02- to 200-liter cultures. Hum Gene Ther. 2011;22(8):1021–30. Epub 2011/03/09.
21. Gao GP, Qu G, Faust LZ, Engdahl RK, Xiao W, Hughes JV, et al. High-titer adeno-associated viral vectors from a Rep/Cap cell line and hybrid shuttle virus [In Process Citation]. Hum Gene Ther. 1998;9(16):2353–62.
22. Clark KR, Voulgaropoulou F, Fraley DM, Johnson PR. Cell lines for the production of recombinant adeno-associated virus. Hum Gene Ther. 1995;6(10):1329–41.
23. Thomas DL, Wang L, Niamke J, Liu J, Kang W, Scotti MM, et al. Scalable recombinant adeno-associated virus production using recombinant herpes simplex virus type 1 coinfection of suspension-adapted mammalian cells. Hum Gene Ther. 2009;20(8):861–70. Epub 2009/05/08.
24. Mingozzi F, Anguela XM, Pavani G, Chen Y, Davidson RJ, Hui DJ, et al. Overcoming preexisting humoral immunity to AAV using capsid decoys. Sci Transl Med. 2013;5(194), 194ra92. Epub 2013/07/19.

25. Ayuso E, Mingozzi F, Montane J, Leon X, Anguela XM, Haurigot V, et al. High AAV vector purity results in serotype- and tissue-independent enhancement of transduction efficiency. Gene Ther. 2010;17(4):503–10. Epub 2009/12/04.
26. Zolotukhin S, Byrne BJ, Mason E, Zolotukhin I, Potter M, Chesnut K, et al. Recombinant adeno-associated virus purification using novel methods improves infectious titer and yield. Gene Ther. 1999;6(6):973–85.
27. Snyder RO, Audit M, Francis JD. rAAV vector product characterization and stability studies. Methods Mol Biol. 2011;807:405–28. Epub 2011/10/29.
28. Snyder RO, Francis J. Adeno-associated viral vectors for clinical gene transfer studies. Curr Gene Ther. 2005;5(3):311–21. Epub 2005/06/25.
29. Senis E, Fatouros C, Grosse S, Wiedtke E, Niopek D, Mueller AK, et al. CRISPR/Cas9-mediated genome engineering: an adeno-associated viral (AAV) vector toolbox. Biotechnol J. 2014;9(11):1402–12. Epub 2014/09/05.
30. Li H, Haurigot V, Doyon Y, Li T, Wong SY, Bhagwat AS, et al. In vivo genome editing restores haemostasis in a mouse model of haemophilia. Nature. 2011;475(7355):217–21. Epub 2011/06/28.
31. Le Guiner C, Montus M, Servais L, Cherel Y, Francois V, Thibaud JL, et al. Forelimb treatment in a large cohort of dystrophic dogs supports delivery of a recombinant AAV for exon skipping in Duchenne patients. Mol Ther. 2014;22(11):1923–35. Epub 2014/09/10.
32. Bailey AM, Mendicino M, Au P. An FDA perspective on preclinical development of cell-based regenerative medicine products. Nat Biotechnol. 2014;32(8):721–3. Epub 2014/08/06.
33. MacLachlan TK, McIntyre M, Mitrophanous K, Miskin J, Jolly DJ, Cavagnaro JA. Not reinventing the wheel: applying the 3Rs concepts to viral vector gene therapy biodistribution studies. Hum Gene Ther Clin Dev. 2013;24(1):1–4. Epub 2013/05/23.
34. USFDA. Guidance for Industry, Formal meetings between the FDA and sponsors or applicants. 2009.
35. Lenzi RN, Altevogt BM, Gostin LO, editors. Oversight and review of clinical gene transfer protocols: assessing the role of the recombinant DNA advisory committee. Washington, DC; 2014.
36. USFDA. Prescription drug user fee rates for fiscal year 2016. Fed Regist. 2015;80(148): 46028–32.
37. USFDA. Guidance for Industry, expedited programs for serious conditions – drugs and biologics. 2014.
38. Field MJ, Boat TF, editors. Rare diseases and orphan products: accelerating research and development. Washington, DC; 2010.
39. Ridic G, Gleason S, Ridic O. Comparisons of health care systems in the United States, Germany and Canada. Mater Sociomed. 2012;24(2):112–20. Epub 2012/01/01.
40. Brennan TA, Wilson JM. The special case of gene therapy pricing. Nat Biotechnol. 2014;32(9):874–6. Epub 2014/09/10.
41. Philippidis A. Orphan drugs, big pharma. Hum Gene Ther. 2011;22(9):1035–8. Epub 2011/09/22.
42. Philippidis A. Crafting a robust business model for orphan drug development. Hum Gene Ther. 2011;22(7):781–3. Epub 2011/07/16.
43. Philippidis A. Developing a balanced business model for gene therapy. Hum Gene Ther. 2011;22(6):645–6. Epub 2011/05/19.
44. Abou-El-Enein M, Bauer G, Reinke P. The business case for cell and gene therapies. Nat Biotechnol. 2014;32(12):1192–3. Epub 2014/12/10.
45. USFDA. Guidance for Industry (Draft), Fulfilling regulatory requirements for postmarketing submissions of interactive promotional media for prescription human and animal drugs and biologics. 2014.
46. Nathwani AC, Reiss UM, Tuddenham EG, Rosales C, Chowdary P, McIntosh J, et al. Long-term safety and efficacy of factor IX gene therapy in hemophilia B. N Engl J Med. 2014;371(21):1994–2004. Epub 2014/11/20.

47. Carvalho LS, Vandenberghe LH. Promising and delivering gene therapies for vision loss. Vis Res. 2014. Epub 2014/08/06.
48. Hacein-Bey-Abina S, Pai SY, Gaspar HB, Armant M, Berry CC, Blanche S, et al. A modified gamma-retrovirus vector for X-linked severe combined immunodeficiency. N Engl J Med. 2014;371(15):1407–17. Epub 2014/10/09.
49. Gill S, June CH. Going viral: chimeric antigen receptor T-cell therapy for hematological malignancies. Immunol Rev. 2015;263(1):68–89. Epub 2014/12/17.
50. Cavazzana-Calvo M, Payen E, Negre O, Wang G, Hehir K, Fusil F, et al. Transfusion independence and HMGA2 activation after gene therapy of human beta-thalassaemia. Nature. 2010;467(7313):318–22. Epub 2010/09/17.
51. Kastelein JJ, Ross CJ, Hayden MR. From mutation identification to therapy: discovery and origins of the first approved gene therapy in the Western world. Hum Gene Ther. 2013;24(5):472–8. Epub 2013/04/13.
52. Burger L, HIirschler B. First gene therapy drug sets million-euro price record. Reuters. 2014.
53. Ledley FD, McNamee LM, Uzdil V, Morgan IW. Why commercialization of gene therapy stalled; examining the life cycles of gene therapy technologies. Gene Ther. 2014;21(2):188–94. Epub 2013/12/07.

Chapter 3
Gene Discovery in Congenital Myopathy

Laura L. Smith, Vandana A. Gupta, and Alan H. Beggs

Abbreviations

AD	Autosomal dominant
AR	Autosomal recessive
bp	Base pairs
CCD	Central core disease
CFTD	Congenital fiber-type disproportion
CM	Congenital myopathy
CNM	Centronuclear myopathy
COX	Cytochrome c oxidase
EM	Electron microscopy
EMG	Electromyography
FSD	Fiber size disproportion
H&E	Hematoxylin and eosin
KD	Knockdown
KI	Knock-in
KO	Knockout
MmD	Multiminicore disease
MRI	Magnetic resonance imaging
NADH-TR	Nicotinamide adenine dinucleotide–tetrazolium reductase
NGS	Next-generation sequencing
NM	Nemaline myopathy

L.L. Smith • V.A. Gupta • A.H. Beggs (✉)
Division of Genetics and Genomics, The Manton Center for Orphan
Disease Research, Boston Children's Hospital, Harvard Medical School,
300 Longwood Ave., Boston, MA 02115, USA
e-mail: llsmith@fas.harvard.edu; vgupta@enders.tch.harvard.edu;
beggs@enders.tch.harvard.edu

© Springer Science+Business Media New York 2016
M.K. Childers (ed.), *Regenerative Medicine for Degenerative Muscle Diseases*,
Stem Cell Biology and Regenerative Medicine,
DOI 10.1007/978-1-4939-3228-3_3

Tg	Transgenic
WBMRI	Whole body MRI
WES	Whole exome sequencing
WGS	Whole genome sequencing
XLMTM	X-linked myotubular myopathy

3.1 Congenital Myopathies

3.1.1 Overview

The congenital myopathies (CMs) are a heterogeneous group of inherited neuro-muscular disorders that manifest as skeletal muscle weakness at birth or early in life and are defined by the presence of specific morphological features on biopsy [1–3]. The most common forms of CM can be roughly subdivided into four categories based on the predominant pathological features observed under light and electron microscopy: (1) nemaline (or rod) myopathies, (2) core myopathies, (3) centronuclear myopathies, and (4) myopathies with congenital fiber-type disproportion. However, accurate diagnoses are often confounded due to broad variations in the clinical severity of each phenotype and to substantial histological overlap between the different forms of these disorders [4, 5]. Histological analysis is further complicated by biopsies lacking any defining features, typically placed into a catchall category known as "undefined" CM. CMs can also result from mutations in more than one gene (Table 3.1) with, causative genes associated with multiple pathologies (Table 3.2). Clinical features, such as the presentation of hypotonia during the newborn period, may be similar to features found in patients with congenital myasthenic syndromes, metabolic myopathies, spinal muscular atrophy, as well as muscular dystrophies. Thus, CMs are typically a diagnosis of exclusion [2, 47] and require detailed clinical data combined with electromyographic and histopathological findings to prioritize gene testing and establish a genetic basis.

More than twenty genes cause CM (Fig. 3.1) and their biological functions are widely studied in vitro and in vivo. Major pathophysiologic pathways responsible for weakness in CMs are hypothesized to result from either malformed contractile filaments, in the case of nemaline and other rod myopathies, or from disruptions in calcium homeostasis at the skeletal muscle triad, in the case of many centronuclear and core myopathies [1, 48]. However, it is becoming increasingly clear that each genetic condition likely involves multiple mechanisms, each of which provides potential therapeutic targets for development. In contrast to CMs, muscular dystrophies are characterized by skeletal muscle that retains the intrinsic ability to contract but becomes progressively weaker due to myofiber death and eventually fails to compensate through regeneration.

Today, investigators and clinicians are focused on understanding the biological basis underlying the CMs and on gene discovery, as the genetic cause in 30–40 %

Table 3.1 Genetics of congenital myopathies

Structural defect	Subtype	Gene	Chromosome location	Inheritance pattern	Method of gene discovery
Rods/protein accumulation	Nemaline myopathy	ACTA1	1q42.1	AD, AR	Candidate gene studies [6]
		CFL2	14q13.1	AR	Candidate gene studies [7]
		KBTBD13	15q22.31	AD	Linkage mapping [8]
		KLHL40	3p33.1	AR	WES [9, 10]
		KLHL41	2q31.1	AR	WES [11]
		LMOD3	3p14.1	AR	WES [12]
		NEB	2q23.3	AR	Linkage mapping [13]
		RYR1	19q13.2	AR	Candidate gene studies [14, 15]
		TNNT1	19q13.42	AR	Linkage mapping [16]
		TPM2	9p13.3	AD	Candidate gene studies [17]
		TPM3	1q21.3	AD, AR	Linkage mapping [18]
	Core–rod myopathy	ACTA1		AD	Candidate gene studies [19]
		CFL2		AR	Candidate gene studies [7]
		KBTBD13		AD	Linkage mapping [8]
		NEB		AR	Candidate gene studies [20]
		RYR1		AD, AR	Linkage mapping [21]
		TPM2		AD	Linkage mapping [22]
	Cap disease	ACTA1		AD	Candidate gene studies [23]
		TPM2		AD	Candidate gene studies [24]
		TPM3		AD	Candidate gene studies [25]
	Intranuclear rod myopathy	ACTA1		AD	Candidate gene studies [6]

(continued)

Table 3.1 (continued)

Structural defect	Subtype	Gene	Chromosome location	Inheritance pattern	Method of gene discovery
Cores	Central core disease	RYR1		AD, AR	Linkage mapping [26]
	Multiminicore disease	MEGF10	5q23.2	AR	WES [27]
		MYH7		AD	Candidate gene studies [28]
		RYR1		AR	Candidate gene studies [29]
		SEPN1	1p36.11	AR	Linkage mapping [30]
Central nuclei	Centronuclear myopathy	BIN1	2q14.3	AR	Candidate gene studies [31]
		CCDC78	16p13.3	AD	Linkage mapping; WES [32]
		DNM2	19p13.2	AD	Linkage mapping [33]
		MTM1	Xq28	XL	Linkage mapping [34]
		RYR1		AR	Candidate gene studies [35]
		SPEG	2q35	AR	WES [36]
		TTN	2q31.2	AR	WES [37]
Fiber size variation	Congenital fiber-type disproportion	ACTA1		AD	Candidate gene studies [38]
		MYH7		AD	Linkage mapping [39]
		RYR1		AR	Candidate gene studies [40]
		SEPN1		AR	Candidate gene studies [41]
		TPM2		AD	Candidate gene studies [42]
		TPM3		AD	Candidate gene studies [43, 44]

Table 3.2 Clinical and pathological features associated with particular congenital myopathy genes

Gene	Pathological features	Clinical features
Skeletal muscle alpha-actin (*ACTA1*)	• Rods in both fiber types (Gömöri trichrome) • Accumulation of actin: – Pale zones (H&E) – No ATPase or oxidative activity – Thin filament accumulations (EM) – Do not immunolabel with anti-myosin antibodies • Intranuclear rods • Zebra bodies (specificity unknown) • Fiber size disproportion (FSD) as the only feature	• Variable severity ranging from severe neonatal to adult onset • Neck flexor and ankle dorsiflexor weakness • Rare cases reported with cardiac involvement
Bridging integrator 1 (*BIN1*)	• Multiple central nuclei (H&E) • Dark perinuclear NADH-TR staining • Central membranous structures (EM)	• Full phenotype remains uncertain due to so few reported cases • Muscle MRI: no large series to date. Case report showed prominent involvement of soleus, tibialis anterior, peroneal, and extensor muscles, but sparing of the gastrocnemius
Dynamin 2 (*DNM2*)	• Central, internal, and sarcolemmal nuclei (H&E) • Radial sarcoplasmic strands on NADH-TR (may be age related) • Chains of internalized nuclei • Necklace-like fibers	• Range in severity from congenital to adult-onset weakness • Relative early weakness of neck flexors, external ophthalmoplegia, and ptosis • Facial weakness • Muscle MRI: progressive sequence with early involvement of the distal lower leg muscles and subsequent changes in the thigh • WBMRI: selective involvement of head (lateral pterygoid), forearm axis, and pelvic girdle muscles
Kelch-like family member 40 (*KLHL40*)	• Small numerous rods visible by EM • Myofibers composed of only rods and few myofibrils	• Severe congenital lethal; death in utero or in newborns is common • Fetal akinesia/hypokinesia • Contractures • Ophthalmoparesis • Respiratory failure and swallowing difficulties at birth

(continued)

Table 3.2 (continued)

Gene	Pathological features	Clinical features
Myotubularin (*MTM1*)	• Central nuclei in both fiber types spaced down the length of the fiber; few subsarcolemmal nuclei • Central areas of absent stains (i.e., no organelles) • Dark centers with subsarcolemmal peripheral halo with oxidative enzymes • Carriers and mild cases may show "necklace fibers"	• Severe perinatal onset in males • May have no voluntary movements at birth • Bilateral ptosis, facial diplegia, and limitation of eye movements • Thin ribs • Contractures of the hips and knees • Macrocephaly with or without hydrocephalus • Narrow elongated face • Long digits
Slow/beta-cardiac myosin heavy chain (*MYH7*)	• Hyaline bodies (H&E) • Prominent FSD (common)	• Scapuloperoneal or limb girdle patterns of weakness with foot drop, calf hypertrophy, scoliosis, and respiratory failure • Cardiomyopathy and arrhythmias in some cases
Nebulin (*NEB*)	• Rods in both fiber types • Rods as well as cores (uncommon) • No specific pathological markers	• Variable severity ranging from mild to severe, but more commonly associated with typical congenital NM • Facial weakness (particularly lower face) • Neck weakness and foot drop may be presenting signs • Pes cavus • Chest deformities and scoliosis • Muscle MRI: early involvement in lower leg, particularly the tibialis anterior, and relative sparing of the thigh (mild cases); diffuse involvement of lower limbs with sparing of gastrocnemii (severe cases) • WBMRI: selective involvement of lateral pterygoid muscles with sparing of tongue and other masticator muscles

| Ryanodine receptor (*RYR1*) | • Type I fiber predominance or uniform type I fibers
• Large, defined cores that develop with age
• Clear minicores or unevenness of oxidative enzyme stains (EM is recommended and shows areas without mitochondria)
• Multiple nuclei and/or central nuclei
• Focal dark centers with oxidative enzyme stains, in association with central nuclei
• Prominent connective and adipose tissues in the absence of fibers containing fetal myosin; variable number of small fibers (less than 5 μm) with fetal myosin
• Rods in the presence of cores
• Severe FSD as the only feature
• Variable recessive pathologies:
– Absence of cores
– NADH-TR irregularities
– Central nuclei
– Prominent fatty and fibrous tissue | • AD mutations associated with mild or moderate weakness; AR mutations have variable severity ranging from mild to severe congenital weakness
• Mild weakness
– Ptosis, but respiratory/bulbar/extraocular muscles usually spared
– Contractures and skeletal deformities may be prominent
• Moderate or severe weakness:
– Generalized hypotonia
– Congenital hip dysplasia
– External ophthalmoplegia
– Ptosis
– Amyotrophy and distal laxity
– Respiratory and bulbar muscle involvement
– Contractures, as well as knee and ankle dislocation
– Doughy skin and lax hands (observed in Samaritan myopathy) [45, 46]
• High risk of malignant hyperthermia unless in vitro contracture testing is normal
• Muscle MRI (CCD): marked involvement of vasti, sartorius, and adductor magnus muscles; relative sparing of rectus femoris, adductor longis, and hamstring muscles
• WBMRI: moderate involvement in biceps brachii, subscapularis, lumbar paravertebral, and glutei muscles; milder involvement of masticator, neck extensor, and forearm muscles |

(continued)

Table 3.2 (continued)

Gene	Pathological features	Clinical features
Selenoprotein N (*SEPN1*)	• Minicores in both fiber types with NADH-TR and COX stains • Wide variation in fiber size and increase in connective tissue (without necrosis) • Mild FSD as the only feature • Mallory body-like inclusions (H&E)	• Congenital axial weakness (with early head drop) and slim build (muscles and skeleton) • Scoliosis and respiratory failure due to diaphragmatic weakness by late childhood • Limb strength generally preserved • Most patients ambulant into adulthood • Muscle MRI: prominent involvement of the sartorius muscle within the thigh • WBMRI: atrophy of sternocleidomastoid muscle and fatty infiltration of the paravertebral muscles; tongue, shoulder, and arm muscles well preserved
Beta-tropomyosin (*TPM2*)	• Rods in both fiber types • Caps (Gömöri trichrome) • FSD as the only feature	• Similar to typical congenital NM caused by *NEB* and *TPM3* mutations in phenotype (e.g., neck weakness, foot drop) • Predisposition to distal arthrogryposis and/or congenital large joint contractures with pterygia (with or without typical congenital myopathy features) • Cardiomyopathy • Asymmetric limb weakness • WBMRI: affected masticator (temporal) muscles and distal lower leg (soleus, extensor, flexor) muscles, with no specific findings in the thigh
Alpha-tropomyosin (*TPM3*)	• Rods restricted to type I fibers • Caps • FSD as the only feature (common)	• Similar to typical congenital NM caused by *NEB* and *TPM2* mutations in phenotype (e.g., neck weakness, foot drop) • Variable severity • Marked discrepancy between upper and lower limb weakness

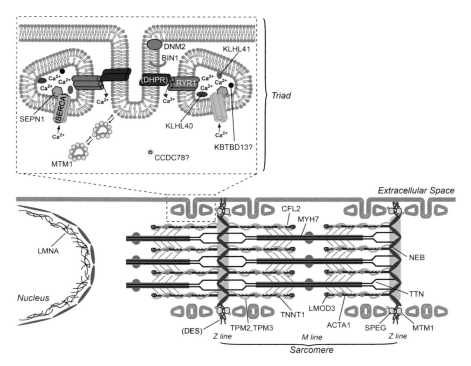

Fig. 3.1 Schematic diagram of human skeletal muscle illustrating many known CM genes and their encoded proteins. Disease genes discussed in the chapter text, the majority of which encode proteins of the myofilament or triad, are capitalized. Other important muscle proteins not currently known to be mutated in human CMs, but part of relevant complexes and responsible for other conditions, are selectively labeled in parentheses

of cases remains unknown. Due to the vast genetic and clinical heterogeneity of CMs, molecular diagnosis is important, both for disease management and for the development of gene-specific therapeutic strategies. Although certainly not an exhaustive list, the focus of this chapter will be limited to the four subtypes enumerated above.

3.1.2 Nemaline Myopathies

3.1.2.1 Clinical Features

Nemaline myopathy (NM) is histologically characterized by the presence of abnormal thread- or rod-like structures in muscle fibers (Fig. 3.2a). Clinically, NM phenotypes vary and are subclassified into different groups according to age of onset as well as severity of motor and respiratory involvement: (1) severe congenital NM, (2) intermediate congenital NM, (3) typical congenital NM, (4) childhood/juvenile-onset

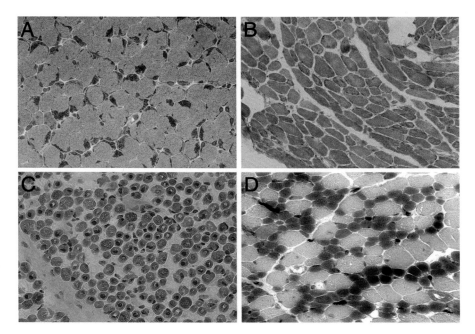

Fig. 3.2 Characteristic histological findings in representative congenital myopathies. (**a**) Gömöri trichrome stain of muscle from a 2 ½-year-old girl with nemaline myopathy due to a *NEB* mutation. Note dense subsarcolemmal accumulations of darkly staining rod bodies in many fibers. (**b**) NADH-TR staining reveals multiple minicores, appearing as patchy lightly stained regions within myofibers, in the muscle of a 4-year-old boy with MmD caused by a mutation of *SEPN1*. (**c**) H&E staining of muscle from a 1-month-old boy with XLMTM. An *MTM1* mutation is sufficient to reveal characteristic small myofibers with central nuclei. (**d**) Fiber typing by histochemical staining for myosin ATPase at pH 4.6 reveals uniform smallness of the population of darkly staining type I myofibers in a 1-year-old girl with CFTD (genotype unknown)

NM, (5) adult-onset NM, and (6) other forms with atypical clinical features such as cardiomyopathy and ophthalmoplegia [49]. Although these classifications have been well established and can be used to accurately predict prognosis, certain morphological and clinical features are common and shared between two or more groups.

Presentation of NM includes generalized muscle weakness and hypotonia, with the most pronounced severity in the face, neck flexors, and proximal limbs. In forms of NM with congenital onset, the face is often elongated and expressionless and the mouth tent-shaped with a high-arched palate. Body build is slender, especially in young children, but muscle bulk is not necessarily reduced. The spine is generally hyperlordotic, and chest deformities may be evident at birth. In terms of motor capacities, gross motor activity is slow, whereas fine motor activity is normal. Many patients show hypermobility of the joints and develop joint contractures. Gait is usually waddling [50].

Respiratory issues and bulbar muscle weakness are common in congenital NM, but do not always correlate with the severity of breathing difficulties. Asymptomatic patients may reveal a restrictive pattern on pulmonary function on testing. Sleep studies are an important part of NM management planning, as patients are at risk of insidious nocturnal hypoxia even in the absence of morning symptoms, and several patients have experienced sudden respiratory failure [5, 51–53]. Infants with congenital onset NM commonly have feeding difficulties, and older patients may experience trouble swallowing. Ophthalmoparesis is seen in a significant number of patients with *KLHL40* mutations. The central nervous system is typically unaffected in NM, and intelligence is normal. Cardiac contractility is also usually normal, although cardiac involvement, particularly dilated cardiomyopathy, does rarely occur [54].

Serum creatine kinase concentrations in NM are usually normal or only slightly elevated (up to five times normal levels). Electromyography (EMG) may be normal in young patients or in mild cases, but show "myopathic" features including polyphasic motor unit potentials with small amplitude and a full interference pattern during weak effort (early recruitment). Generally, ultrasonography reveals abnormally high echogenicity in affected muscles, computed tomography shows a low density of muscles with preservation of volume, and magnetic resonance imaging (MRI) demonstrates fatty infiltrations of the muscle tissue. Muscle MRI (particularly T1 imaging) is increasingly used as an adjunct diagnostic modality, and certain NM subtypes (particularly due to *NEB* mutations) have relatively specific patterns of selective muscle involvement [55–57].

3.1.2.2 Pathological Features

Nemaline bodies are the pathological and diagnostic hallmark of NM and by definition are shared among all genetic forms of this disorder. Rods appear as red or purple structures against a blue-green myofibrillar background upon modified Gömöri trichrome staining and show a tendency to cluster under the sarcolemma and around nuclei (Fig. 3.2a). Rods are considered to derive from the lateral expansion of the Z-line, based on their structural continuity with Z-lines, electron density, and crisscross pattern in electron micrographs. Additionally, rods stain positively for antibodies to alpha-actinin isoforms 2 and 3, the major components of the skeletal muscle Z-line [58–60]. With few exceptions (e.g., patients with *TPM3* mutations), rods are present in both type I (slow twitch) and type II (fast twitch) muscle fibers, although type I fiber predominance is a common feature of NM and fiber-type disproportions tend to become more prominent with age [61, 62]. The proportion of myofibers containing rods varies considerably between cases, however, and the rod size and count in a muscle specimen do not appear to correlate with disease severity [63]. Nemaline rods are typically cytoplasmic, although intranuclear rods are occasionally a prominent feature, particularly in severe cases [64].

3.1.2.3 Genetics

NMs are considered diseases in which mutations disrupt the ability of the myofiber to generate adequate force during contraction. To date, mutations in ten different genes have been identified in a subset of NM patients: alpha-skeletal muscle actin

(*ACTA1*) [6], slow alpha-tropomyosin (*TPM3*) [18, 65, 66], nebulin (*NEB*) [13], slow troponin T (*TNNT1*) [16, 67], beta-tropomyosin (*TPM2*) [17, 65], muscle-specific cofilin (*CFL2*) [7, 68, 69], leiomodin 3 (*LMOD3*) [12], kelch-like family members 40 (*KLHL40*) [9, 10] and 41 (*KLHL41*) [11], and kelch repeat and BTB domain containing 13 (*KBTBD13*) [8]. Seven of these ten genes encode protein components of the muscle fiber thin filament, while the other three likely participate as regulators of the thin filament degradation/turnover apparatus. Of these, the most common genetic cause of NM is autosomal recessive mutations in *NEB*, which may account for up to 50 % of cases, followed by mutations in *ACTA1* [47]. In practice, *ACTA1* is often the first gene to be sequenced due to its small size and relatively high mutation detection rate (up to 25 % of cases and over half of cases with severe congenital NM) [70]. Over 230 different *ACTA1* variants have been reported, the majority are de novo heterozygous missense mutations or show an autosomal dominant pattern of inheritance [71, 72]. *KLHL40, TPM3, TPM2*, and *LMOD3* mutations are the next most common of the remaining genes, whereas *CFL2, TNNT1, KLHL41*, and *KBTBD13* mutations are relatively rare [73].

As of yet, no definitive clinical or pathological markers for the various genetic forms of NM have been identified, although detailed pathologic studies may provide morphologic clues to guide mutation analysis in the future. *TPM3*, for example, is only expressed in type I fibers, and patients harboring mutations in this gene often experience fiber atrophy and nemaline bodies preferentially in this fiber type [43, 44, 63, 74, 75]. Similarly, abnormal accumulation of actin filaments and irregular distributions of actin staining have been observed in patients with *ACTA1* mutations and in extreme cases may be the predominant finding, leading to a histopathological diagnosis of "actin myopathy" [64, 76].

3.1.3   Core Myopathies

3.1.3.1   Clinical Features

Core myopathies are characterized by areas in the muscle fiber lacking oxidative and glycolytic enzymatic activity. Central cores run along the length of the myofiber, whereas minicores are short zones of myofibrillar disorganization that are wider than they are long on longitudinal section. Based on the presence of these abnormal features, patients with core myopathies are traditionally subclassified as having either central core disease (CCD) or multiminicore disease (MmD) [77]. This subsection will address the clinical features of these two "pure" forms of core myopathy, although it is important to note that cores reported in association with other characteristic findings, such as nemaline rods, can also result in hybrid diagnoses referred to as core–rod myopathies.

CCD was the first CM defined on the basis of specific morphological changes in skeletal muscle. In 1956, Magee and Shy described the first patient with centrally placed cores [78], and the term "central core disease" was introduced soon afterwards to reflect the absence of oxidative enzymes, phosphorylase, and glycogen in the core area due to mitochondrial depletion [79]. CCD typically presents in infancy with hypotonia or in early childhood with delays in motor development. Weakness is mild and symmetrical, preferentially affecting proximal musculature, including hip girdle and axial muscles [80]. Extraocular muscles are usually spared. Significant respiratory insufficiency and cardiac abnormalities are unusual, and intellectual capacities are not impaired. Orthopedic complications such as congenital dislocation of the hips [81], kyphoscoliosis [82], and foot deformities (e.g., talipes equinovarus, pes planus) occur, but contractures other than at the ankle are rare [83]. Almost all CCD patients achieve the ability to walk independently, with the exception of cases with debilitating hip dislocations or severe cases presenting with neonatal weakness, arthrogryposis, and respiratory failure [84–87].

Serum creatine kinase levels are normal or slightly raised (up to six to fourteen times higher than normal) in CCD patients, and the EMG is normal or myopathic, with short-duration, small amplitude polyphasic motor unit potentials. Muscle ultrasound often detects increased echogenicity associated with a primary myopathic process [88]. Importantly, a characteristic pattern of selective involvement on muscle MRI has been reported in CCD patients [56, 57] and is distinct from that observed in other CMs [56, 57]. Muscle MRI is therefore useful for aiding genetic diagnosis in cases with mixed pathologies featuring both cores and rods [89, 90].

3.1.3.2 Pathological Features

The cardinal diagnostic feature of patients with CCD is the presence of single, well-circumscribed circular regions in the center of most type I fibers. These "central cores" are devoid of mitochondria and deficient in oxidative enzymes, phosphorylase activity, and glycogen. Cores are best observed on sections stained for oxidative enzyme activity (e.g., succinate dehydrogenase [SDH], cytochrome c oxidase [COX], or nicotinamide adenine dinucleotide [NADH] dehydrogenase-reacted sections) and may be overlooked on routine sections with hematoxylin and eosin (H&E) staining. Cores examined using electron microscopy contain densely packed and disorganized myofibrils and have been divided into two types on the basis of whether myofibrillar organization is maintained. Structured cores preserve basic sarcomeric architecture, although sarcomeres may be out of register with adjacent fibrils as well as with each other, whereas unstructured cores contain large areas of Z-line streaming [91].

Fifteen years after the initial description of CCD, Engel and colleagues reported a family with two affected siblings exhibiting multiple small cores within muscle fibers or "minicores." Patients with the classic form of MmD typically present in infancy or childhood with pronounced hypotonia and proximal weakness, although select cases of prenatal or adult onset have been recognized [92–96]. Axial muscle weakness, particularly affecting the neck and trunk flexors, is a prominent feature of MmD, and failure to acquire head control is an early clinical sign. Spinal rigidity and scoliosis are also common. The clinical course of MmD is static for the majority of patients, although some experience cardiac involvement secondary to marked decline in respiratory function during adolescence or young adulthood [96].

MmD is diagnosed on muscle biopsy by the presence of multifocal, well-circumscribed areas in the muscle fiber with reduced oxidative staining and low myofibrillar ATPase activity (Fig. 3.2b) [97]. In contrast to central cores, minicores are typically unstructured, extend for only a short distance along the longitudinal axis of the myofiber, and may affect both type I and type II fibers [98]. Minicores in electron micrographs appear as regions of myofibrillar disruption lacking mitochondria, with sarcomere degeneration and structural abnormalities of the triad. Muscle MRI can be used to complement histological assessments and aid in the choice of genetic testing. Distinct patterns of muscle involvement can distinguish between different genetic forms of MmD, as well as between MmD and other forms of CM [56, 57, 85].

3.1.3.3   Genetics

The vast majority of CCD patients (>90 %) have mutations in the skeletal muscle isoform of the ryanodine receptor (*RYR1*), a calcium release channel of the sarcoplasmic reticulum involved in excitation–contraction coupling [21, 86, 89, 99]. These are most commonly autosomal dominant or de novo dominant mutations, with occasional autosomal recessive cases observed. Of note, mutations causing the classical CCD phenotype most commonly occur in the C-terminal transmembrane pore-forming domain of *RYR1*, although mutations in other domains may account for patients with atypical clinical or histopathological presentations [4, 100, 101].

Only occasionally will autosomal recessive *RYR1* mutations yield CCD pathology and can instead lead to a wide range of findings on muscle biopsy, including characteristic features of MmD [102], CNM [35], congenital fiber-type disproportion (CFTD) [40, 103], and muscular dystrophy [104]. Clinically, patients with recessive *RYR1* mutations usually have a more severe presentation than those with dominant mutations [105]. In addition to weakness, ophthalmoplegia is commonly observed and is a distinguishing feature compared to CCD and *SEPN1*-related MmD [106]. Patients typically have two missense mutations or one missense mutation and one hypomorphic allele (i.e., frameshift, splice site mutation), and the presence of a hypomorphic allele is correlated with more severe clinical symptoms [107, 108].

Rarely, CCD patients have also been observed with mutations in the gene encoding beta-myosin heavy chain (*MYH7*). Weakness in distal muscles is often a clinical clue to *MYH7* mutations, particularly a "dropped" big toe, and muscle pathology tends to show more minicores or atypical cores in these individuals [109]. Lastly, there are cases of CCD without *RYR1* or *MYH7* mutations, but the causative gene have not yet been identified.

MmD core myopathy is most commonly caused by recessive mutations in the selenoprotein N (*SEPN1*) gene [94, 95]. Selenoproteins are characterized by inclusion of one or more selenocysteine residues, which contain the biological form of selenium and are evidenced to be catalytically active in cellular redox processes. Apart from MmD, recessive mutations in *SEPN1* are associated with rigid spine muscular dystrophy with or without cores [30], desmin-related myopathy with Mallory body-like inclusions [110], as well as CFTD [41]. Although these four myopathies are pathologically distinguishable, they are relatively homogenous in clinical presentation and as a result are now collectively referred to as SEPN1-related myopathy. Additional cases of MmD have been described with *MYH7* mutations, as well as mutations in the giant sarcomeric protein titin (*TTN*) [28, 111]. In both instances, cardiomyopathy was reported as a complicating feature.

3.1.4 Centronuclear Myopathies

3.1.4.1 Clinical Features

Centronuclear myopathies (CNM) are classically defined by a high incidence of muscle fibers containing central nuclei, in the absence of other diagnostic abnormalities such as rods and cores [112]. Several genetically distinct forms of CNM have been described based on age of onset, severity of symptoms, and mode of inheritance [113, 114]. Clinically, these can be categorized as the severe X-linked recessive form with prenatal or neonatal onset, the autosomal recessive form with onset in infancy or childhood, and the autosomal dominant form, typically mild with late onset. Despite wide variability in the clinical features among these groups, emerging evidence suggests that defective excitation–contraction coupling at the level of the triad may be a unifying pathophysiological feature in CNM.

X-linked centronuclear myopathy, also referred to as myotubular myopathy or X-linked myotubular myopathy (XLMTM), frequently presents with polyhydramnios and reduced fetal movements in utero. Affected newborn boys present at birth with severe hypotonia, generalized weakness, and muscle wasting. Respiratory insufficiency requires immediate ventilator support, and swallowing difficulties necessitate a feeding tube to prevent death by aspiration. Additional features include thin ribs, ophthalmoplegia, ptosis, pyloric stenosis, and contractures of the hips and knees. XLMTM carries a poor long-term prognosis, with death due to respiratory failure occurring within the first year of life in 20–25 % of cases [115, 116]. Although a small proportion of boys may be less affected in the neonatal period and survive into childhood or even adulthood, most remain severely impaired and require permanent ventilation.

Autosomal recessive centronuclear myopathy is a relatively rare form of CNM and generally presents in infancy or early childhood with diffuse muscle weakness and respiratory distress. Facial diplegia, ptosis, and varying degrees of ophthalmoplegia are common features. The clinical course of the disease is marked by slowly progressive weakness, development of scoliosis or kyphosis, and delays in motor milestones such as walking, running, and stair climbing. By adolescence, most patients are wheelchair dependent.

Autosomal dominant CNM showing late adult onset with slowly progressive weakness, together with de novo forms of CNM with earlier onset, has a broader range of clinical presentation than the X-linked or recessive forms [45, 46, 117, 118]. Generally, limb girdle, trunk, and neck muscles are involved, but distal muscles may be affected [119]. Ptosis and limitation of eye movements often parallel age of onset, with complete ophthalmoplegia associated with early onset cases. While most patients are ventilator independent, phases of respiratory decline may require noninvasive ventilation but rarely necessitate invasive support. Achilles tendon contractures, reduced jaw opening, and atrophy of the calves are common, with scoliosis often becoming progressive in adolescence [59, 60, 120–122].

3.1.4.2 Pathological Features

The principal pathological feature of CNM is abnormal centralization of nuclei in >25 % of muscle fibers [112], although there can be considerable variability both in the number of myofibers with central nuclei and in the number of central nuclei within a single myofiber. Unlike other CMs, particularly CCD and NM, diagnosis of CNM can be achieved with H&E staining of formalin-fixed, paraffin-embedded tissue (Fig. 3.2c). There is typically a predominance of type I fibers [123], with most fibers smaller and rounder than those of age-matched control subjects. On electron microscopy, mitochondrial aggregates, glycogen granules, and a reduction in myofilaments often surround centrally placed nuclei. The remainder of the muscle fiber is structurally normal, with mild increases in perimysial fibrous connective tissue in select cases.

3.1.4.3 Genetics

Mutations in five different genes have been associated with CNM to date. Of these, the most common and extensively studied are mutations in the myotubularin gene (*MTM1*), which lead to the X-linked form of the disease [34]. Over 500 mutations have now been identified in *MTM1* [124, 125]. While most are believed to result in loss of protein expression, there are a few recurring missense mutations, such as p.R69C, which can cause either the classic severe phenotype or a milder presentation [126]. Mutations in the large GTPase dynamin 2 (*DNM2*) are the second most common cause of CNM and display dominant inheritance [33, 45, 46]. Variants located in the middle domain of *DNM2* generally correlate with milder symptoms, while variants in the pleckstrin homology domain are associated with more severe presentations [45, 46].

Mutations in *RYR1*, *TTN*, and bridging integrator 1 (*BIN1*) have been predominantly identified in recessive forms of CNM, with *RYR1* likely the most common of these three [31, 35, 37]. *RYR1*-related CNM is typically severe, but highly variable in presentation [35, 127]. Although internalized nuclei are the most prominent histological feature in these patients, abnormalities observed with oxidative staining are also common and suggest that *RYR1*-related CNM is part of a continuous spectrum that overlaps with *RYR1*-related core myopathies [128]. Homozygous recessive *BIN1* mutations are rare and have been reported in fewer than 20 families where they lead to early onset of clinical symptoms [129–133]. Only recently have heterozygous *BIN1* mutations been identified in nine cases of adult-onset autosomal dominant CNM [134]. *TTN* mutations are associated with milder symptoms in skeletal muscle as well as cardiomyopathy and are probably underappreciated due to the diagnostic difficulties associated with sequencing and with interpretation of variants in the enormous *TTN* gene [37]. Of note, sequence variants of *MTMR14*, a gene functionally related to *MTM1*, have also been reported in two individuals with CNM [135]. However, it is unclear whether these variants are a true cause of disease or instead act as modifiers of a second site gene mutation. There are clearly additional genetic loci that cause CNM. A recent study using whole exome sequencing on a cohort of 29 individuals revealed that only a fraction of subjects had a mutation in one of the abovementioned genes [37].

3.1.5 Congenital Fiber-Type Disproportion

3.1.5.1 Clinical and Pathological Features

Congenital fiber-type disproportion (CFTD) is a histological diagnosis with multiple etiologies [103]. Brooke and Engel first used the term in 1973 in a large morphology study of childrens' biopsies to describe a group of 14 patients who all had clinical features of a CM and whose predominant abnormality on muscle biopsy was a discrepancy in muscle fiber size (Fig. 3.2d) [136]. It is now the consensus that the diagnosis of CFTD can be made when a mutation in a CM gene has been identified and type I (slow twitch) fibers are consistently smaller than type II (fast twitch) fibers by at least 35–40 % [137]. However, type I fiber hypotrophy is also observed in a variety of metabolic myopathies and central nervous system malformations, as well as the severe neonatal form of myotonic dystrophy [138–142]. Some experts therefore also include other histological features in the definition of CFTD, such as type I fiber predominance (>55 % type I fibers) or a paucity (<5 %) of type 2B fibers [137]. If a biopsy has several atypical features, such as type II fiber predominance and/or frequent type 2B fibers, pathologic processes other than CFTD should be considered. In general, patients with CFTD mimic the clinical course of other forms of CM that share the same genetic cause, although early respiratory failure is a common feature in CFTD and nocturnal hypoventilation should be monitored even in ambulant patients [3].

3.1.5.2 Genetics

Most cases of CFTD are associated with mutations in the *TPM3* gene, encoding the type I fiber-specific protein slow alpha-tropomyosin [43, 44, 143]. *TPM3* mutations are hypothesized to alter the interaction between tropomyosin and actin [43], and some data suggest that clinical weakness may arise from misregulated actin–myosin interactions [144]. *RYR1* mutations are less commonly seen [40] and only rarely have mutations in *ACTA1* [38], *MYH7* [39, 145], *SEPN1* [41], *TPM2* [146], as well as an X-linked form [147] been reported. Most recently, mutations of the *LMNA* gene, coding for lamin A/C, were identified in several Japanese patients with CFTD [148]. Since *LMNA* defects are known to cause a variety of different muscular dystrophies and related cardiomyopathies [149], these patients may represent a subset of CFTD cases at risk for cardiac disease. As *TPM3* is expressed only in type I fibers, it is readily apparent why its mutation preferentially leads to type I fiber defects. Conversely, the other aforementioned genes are expressed across all fiber types. The appearance of CFTD in a tiny subset of patients with these gene mutations may therefore reflect unusual environmental, epigenetic, or sampling effects, making it uncertain whether they should be considered true forms of CFTD in the same sense as *TPM3* mutations. Although data are lacking from large cohort studies, known genes are predicted to only account for 50–70 % of families with CFTD, and other genetic causes of CFTD remain to be identified [103].

3.1.6   Undefined Congenital Myopathies

The term "undefined" CM was created to describe patients for whom a distinct CM diagnosis cannot be determined due to a lack of specific pathological features on muscle biopsy. While this catchall diagnosis may result from mutations in known genes, it is likely that the discovery of novel CM genes will have significant impact on how these unsolved cases are classified in the future.

3.2   Gene Discovery in Congenital Myopathies

3.2.1   Diagnostic Challenges

Over the past 15 years, spectacular progress has been made in identifying the genetic basis of many of the CMs. These same forward strides, however, have also blurred the classical nosological boundaries between different forms of CM and between CM and other neuromuscular diseases. As the list of causative genes grows, the relationship between each CM, defined by characteristic morphological features observed on muscle biopsy, and its genetic cause becomes considerably more complex.

3.2.1.1   Genetic Heterogeneity

There is significant genetic heterogeneity within the main groups of CMs, meaning that one clinicopathological diagnosis can be caused by mutations in more than one gene. This suggests that mutations in different genes can trigger the same pathogenetic mechanism and lead to similar structural changes in the muscle. The concept of genetic heterogeneity is perhaps most clearly exemplified in NM, where ten different genetic loci have already been associated with rod formation secondary to abnormalities of thin filament proteins or proteins thought to be related to thin filament protein stabilization or turnover, as discussed in greater detail above. Nevertheless, the genetic cause of NM remains unknown in a significant proportion of cases, emphasizing the vast opportunities that remain for novel gene discovery.

3.2.1.2 Clinical Heterogeneity

CMs also display considerable clinical heterogeneity, the term applied when mutations in a single gene cause diverse clinical phenotypes and different muscle pathologies. Mutations in *ACTA1*, for example, may lead to rods in the sarcoplasm of the muscle fiber, as frequently observed in classic forms of NM, or lead to intranuclear nemaline bodies, or even to CFTD [38, 64, 150–153]. Similarly, mutations in *SEPN1* can result in MmD, rigid spine muscular dystrophy, desmin-related myopathy with Mallory body-like inclusions, and CFTD. These examples suggest that different mutations in the same gene can affect different pathogenetic pathways within the muscle sarcomere.

Clinical heterogeneity also arises when the same genetic mutation leads to different pathological features in members of the same family or in the same individual over time. This has been notably demonstrated for mutations in the skeletal muscle ryanodine receptor gene (*RYR1*), which form a continuous spectrum of pathological features spanning malignant hyperthermia [154], CCD [154], MmD [29], CM with cores and rods [21, 90], CNM [35, 128], and CFTD [40]. Furthermore, mutations in *TPM3* have been reported in both CFTD and NM in members of the same family. Age, genetic modifiers, environmental factors, and/or the site of the muscle biopsy may therefore also influence the pathological appearance of skeletal muscle [155].

3.2.1.3 Lack of Specific Biomarkers

Gene discovery in the CMs is further challenged by the lack of specific biomarkers for use in clinical diagnoses. In recent years, muscle ultrasound and MRI have become increasingly popular as a means to differentiate between different forms of CM. Muscle ultrasound is a practical way to image muscle that can be performed bedside and does not require general anesthetic [156]. Additionally, it can be helpful for identifying possible neurogenic changes and in selecting an appropriate muscle for biopsy. Its utility, however, is highly dependent on the expertise of the ultrasonographer. Similarly, selective muscle involvement on MRI can suggest a disease gene, but its specificity is variable and most often needs to be interpreted in conjunction with the clinical report as well as with biopsy results [2]. Alternative investigations such as serum creatine kinase, EMG, and nerve conduction studies are widely used to exclude other potential diagnoses, but are rarely diagnostic in cases of CM. Serum creatine kinase, for example, is normal or mildly elevated in CM, and only if levels are dramatically raised should muscular dystrophy be considered. Nerve conduction studies in CMs are also usually normal and used to exclude congenital myasthenic syndromes, yet some CMs can be associated with neuromuscular junction abnormalities [143].

Lastly, there can be significant clinical overlap between CM and other neuromuscular disorders. Congenital myotonic dystrophy, congenital myasthenic syndromes, metabolic myopathies such as Pompe disease, spinal muscular atrophy (SMA), as well as Prader–Willi syndrome, all present in newborns with significant weakness and/or hypotonia. In the absence of specific biomarkers, muscle biopsy, coupled with analysis of muscle histology, immunohistochemistry, and ultrastructure by light and electron microscopy, has therefore been the crucial aspect of establishing the diagnosis of a specific form of CM. Historically, the presence of dystrophic features excluded a diagnosis of CM, but mutations in *MTM1*, *DNM2*, *RYR1*, and *ACTA1* may also result in endomysial fibrosis, fiber size variation, as well as fat infiltration, and can thus sometimes mimic a dystrophic pattern [4, 122, 157]. Furthermore, *SEPN1* and *TTN* mutations can lead to both myopathic and dystrophic phenotypes [94, 95].

3.2.2   Genetic Approaches

Since the launch of the Human Genome Project in 1990, development of genetic maps, physical maps, and technologies for gene identification has had a remarkable impact on gene discovery in Mendelian disorders [158]. For CM patients whose diagnoses have traditionally relied on clinicopathological criteria, these genetic advances have created a second level of diagnosis, and defining the specific genetic cause of a presenting phenotype is now the gold standard. Genetic insight helps the clinician to predict prognosis and monitor health issues and will become increasingly important as gene-specific therapies are developed. This subsection will introduce current methods used in gene discovery and how these approaches have specifically enhanced our molecular understanding of the CMs.

3.2.2.1   Linkage Mapping and Positional Cloning

Linkage mapping, which may lead to positional cloning, is the process of scanning genomes using regularly spaced, highly variable DNA segments of known position [159, 160]. This approach, first successfully applied to a human disease gene in 1983 to map the gene for Huntington's disease, begins with the collection of pedigrees in which the disease gene is segregating [161]. Affected families are then studied with polymorphic markers until genetic regions are associated or "linked" with the disease, by observing that affected individuals co-inherit certain marker variants (i.e., alleles) located within those regions more frequently than would be expected by chance alone

[162]. The number of informative meioses available in the pedigree limits the span of these candidate regions. Finally, transcripts in this candidate region can be isolated (i.e., cloned) for further analysis and full characterization of the responsible gene. Transcripts may be prioritized for analysis if additional information is known, such as tissue expression patterns of the genes within in the candidate region [163]. These methods were first successfully applied to the identification and cloning of genes causing chronic granulomatous disease and Duchenne muscular dystrophy [164–166]. The primary advantage of the positional cloning approach is that no prior knowledge of the underlying physiology or biology of the disease of interest is required.

Linkage mapping and positional cloning techniques have been important in identifying several CM-related genes. The locus responsible for XLMTM, for example, was linkage mapped to Xq28 in the early 1990s [167–169]. The candidate region was then refined to 600 kilobases (kb) by characterizing an activating deletion found on an X-chromosome of a young female patient with a mild form of the disorder [170]. By combining linkage analyses in families with informative recombination events with a study of two male patients carrying deletions [171, 172], Laporte and colleagues were able to isolate the *MTM1* gene in 1996 [34]. Positional cloning strategies were similarly used to identify *SEPN1* [30], as well as a number of other genes as detailed in Table 3.1.

3.2.2.2 Homozygosity Mapping

Homozygosity mapping is an alternative method used to narrow down the location of a disease gene but is a more powerful and effective approach than classical linkage mapping when studying recessive disorders in inbred populations and in consanguineous families [173, 174]. Briefly, the observation that autosomal recessive pathogenic alleles are introduced at a rate proportionate to the human mutation rate of 1.2 in 10^8 per nucleotide per generation predicts that unaffected individuals carry a number of disease-causing recessive alleles in the heterozygous state [175]. This strategy takes advantage of the fact that when a population traces its origin to a few founders, these alleles will assume a high frequency such that two individuals taken at random from the reproductive pool will likely be carriers themselves and are at risk of having clinically affected offspring. Since small chromosomal regions tend to be transmitted whole, not only will the affected descendant have two identical copies of the ancestral allele, but the surrounding DNA segment will also be homozygous. Therefore, homozygosity across a stretch of DNA can serve as a clue to robustly map the pathogenic allele [174].

Homozygosity mapping led to the identification of the 3-hydroxyacyl-CoA dehydratase 1 (*HACD1*) as a causal gene of CFTD in a highly inbred family of Bedouin ancestry [176]. This strategy also helped to map a genetic locus of Native American myopathy (NAM) finally identified as *STAC3*, which is predominantly characterized by malignant hyperthermia, muscle wasting, and kyphoscoliosis [177].

3.2.2.3   Candidate Gene Studies

In contrast to linkage mapping, which is an unbiased search of the entire genome without any preconceptions about the functional role of a certain gene, the candidate gene approach allows research scientists to investigate the validity of a hypothesized genetic basis of a disorder [162, 178]. This strategy involves examining the relationship between a gene that may be causing a disease, the "candidate gene," and the disease itself. One drawback of this technique is that selecting a particular candidate requires basic understanding of the disease pathophysiology. However, candidate gene studies do not require large families with both affected and unaffected members. They instead can be performed with small families (e.g., a proband and parents) or with unrelated cases and control subjects [179], as is often the case with CMs.

The candidate gene approach is well illustrated in the discovery of *BIN1*, the gene responsible for autosomal recessive CNM. In 2007, Nicot and colleagues considered candidate genes encoding proteins with cellular functions related to membrane trafficking, similar to *MTM1* in XLMTM and *DNM2* in autosomal dominant CNM. Since *BIN1* is regulated by phosphoinositides and is involved in membrane remodeling, the gene was prioritized and directly sequenced in 55 affected individuals using Sanger-based methods. Three homozygous variants in *BIN1* were identified in four individuals from three consanguineous families. These variants were absent in 280 genomic DNA samples collected from unaffected controls [31]. Similarly, several NM genes have been identified by candidate gene approaches on the basis of their encoding components of the thin filament (Table 3.1) [7, 17].

3.2.2.4   Next-Generation Sequencing

Increased demand for improved technologies to sequence large numbers of human genomes has caused a fundamental shift away from "first-generation" automated Sanger methods [180]. Next-generation sequencing (NGS), often referred to as massively parallel sequencing, refers to a collective group of newer methods in which numerous sequencing reactions take place simultaneously. In contrast to Sanger sequencing, which generates single long reads of several hundreds of base pairs (bp) using dye terminator chemistry, NGS methods produce massive numbers of short reads (50–250 bp in length) in parallel using sequencing chemistries that can be performed at a fraction of the cost per base. These short reads are first aligned and mapped to a reference genome, and then variants are called from the mapped data. NGS platforms have become widely available since 2005 and currently include the Illumina HiSeq/MiSeq, Life Technologies SOLiD and Ion Torrent/Ion Proton, and Roche 454. Together, these serve as today's gold standard for genetic analysis [181, 182].

NGS and high-throughput sequence capture methods may be used to generate whole genome or whole exome (all protein-coding sequences in the genome) data or to specifically target genes or loci of interest [180]. Whole genome sequencing

(WGS) does not involve any specific sequence capture steps, providing the most comprehensive and unbiased look at the majority of the genome. Sequence data include genes as well as many regulatory elements and structural motifs between the genes. Coverage across the genome is generally even. However, the cost and time required for WGS are significantly greater than for targeted approaches. The volume of data and necessary bioinformatic analyses are also considerably greater. Furthermore, at the present time, the many variants in nonprotein-coding regions are currently difficult to interpret due to limited knowledge of their biological function. For the moment, whole exome sequencing (WES) achieves a better balance between scope and the desired depth of sequence coverage when the disease gene is not already known. This hypothesis-independent approach is particularly suited to accelerating gene discovery in the CMs, as it is amendable to rare Mendelian disorders with (1) only a few available cases, (2) unrelated cases from different families, (3) sporadic cases due to de novo variants, and (4) genetic and/or phenotypic heterogeneity. It also has considerable advantages for sequencing select genes implicated in CM, which are some of the largest in the human genome: *RYR1* (106 exons), *NEB* (183 exons), or *TTN* (363 exons). Briefly, exons are enriched through a sequence capture step by hybridization of the mechanically fragmented DNA sample to biotinylated complementary bait sequences. The captured sequences are then subjected to NGS, and reads are computationally aligned to the known reference genome. Each nucleotide of the target sequence is represented in a large number of reads, defined as the depth of sequencing and typically averages at least 30X coverage for the whole exome [183]. Coverage may be increased to 1000× or higher when specific genes or regions of interest are targeted.

One challenge regarding the application of WES has been how best to define the exome. Whereas initial efforts conservatively targeted only high-confidence protein coding genes identified by the Consensus Coding Sequence (CCDS) Project, commercial kits now target, at minimum, the entire RefSeq collection as well as an increasingly large set of hypothetical proteins. Nevertheless, WES is not without limitations. Capture probes, for example, only target exons in the human genome that have been identified, and because probe efficiency varies dramatically, some exons fail to be experimentally isolated at all [184]. It is also important to note that commercially available sequence capture methods have focused only on exons and splice sites, leaving other important regulatory sequences such as promoters, enhancers, microRNAs, and other evolutionarily conserved noncoding sequences with limited or no coverage. Finally, the identification of large copy number variations and chromosomal rearrangements is not yet reliable, although improvements in both hardware and bioinformatics analytical methods to address this deficiency are under development [185].

Even with these caveats, WES is rapidly being implemented as a powerful new strategy to identify the genetic basis of both known and novel forms of CM. Examples of recent successes include the discoveries of *KLHL40*, *KLHL41*, and *LMOD3* mutations in NM [9–12], *TTN* and *SPEG* mutations in CNM [36, 37], and *CCDC78* mutations in a myopathy with prominent internal nuclei and atypical cores [32].

3.2.3   Clinical Implications of NGS

In the CMs, as in other inherited disorders, the aim of genetic testing in most patients is to achieve a definitive molecular diagnosis. Knowledge of the specific genetic defect can then be used by the patient and their family to optimize disease management and reproductive counseling and ultimately, in the selection of targeted therapies. Proceeding directly to genetic analysis has typically not been practical in the CMs, however, due to the significant degree of clinical heterogeneity. Instead, genetic testing has always been prioritized based on a combination of clinical clues, muscle imaging, and histological features [2].

Although sequential Sanger sequencing of candidate genes has been the cornerstone of genetic diagnosis and still remains a popular and reasonably successful method of molecular testing for CM, it is no longer the most efficient or cost-effective approach, even when the patient has been carefully assessed in clinic. Moreover, in cases where genes are particularly large (e.g., *NEB*, *RYR1*), Sanger sequencing often covers only a portion of the gene. With more than twenty CM genes now described, it is difficult to keep abreast of each new reported gene and to maintain a working knowledge of their respective phenotypes. Furthermore, it is now clear that a single gene can cause several phenotypes and exhibit different modes of inheritance (Tables 3.1 and 3.2). Thanks to recent advances in NGS technology, multiple genes can be tested in parallel and done so that the read depth and sequencing coverage are sufficient for use in clinical practice.

3.2.3.1   NGS Gene Panels

NGS-based diagnostics has the potential to detect a full range of genetic variation in an affected individual, including pathogenic variants in previously unrecognized disease genes as well as incidental findings of carrier status or disease predisposition for conditions unrelated to the patient's personal complaint. Although valuable applications, most diagnostic laboratories are currently focused on finding mutations in known genes for which the results will be interpretable and meaningful to a specific patient in a clinical setting. The "medical exome" is newly defined as the coding regions of four to five thousand genes known to be associated with a particular disease. To speed up the sequencing of known genes and to ensure that the tests examine only disease genes relevant to the patient's disease, customized gene panels are now being designed to sequence only the coding regions of the exome that contain known CM genes [180, 186].

For gene panels, NGS is typically used to cover the full coding regions of known causal genes, plus approximately 20 bp of noncoding DNA flanking each exon. Sanger sequencing is then used to confirm regions with insufficient NGS coverage, as well as all rare, undocumented variants. One of the biggest challenges at present is to identify which sequence changes are pathogenic and to distinguish these from rare nonpathogenic variants, although several control population databases now exist to assist in this process. The Leiden Open Variation Database, for example, is an excellent open access resource that offers a list of DNA sequence variants in specific genes and their associated phenotypes (http://www.lovd.nl) and has recently been used to characterize novel variants in the CMs [125].

The cost of a panel depends on its sensitivity as well as the number of genes included, keeping in mind that its clinical utility will decrease as new pathogenic genes are discovered. One option now being explored is to sequence an entire genome, but limit the bioinformatic analysis to just the "panel genes" and retain the underlying data for future reanalysis as new genes are found. Another is to develop smaller, highly focused panels that only cover a particular phenotype—for example, NM. These "phenotype-specific" panels are most closely aligned to the traditional method of genetic testing for a single gene in the CMs, using clinical features and histological markers as a first line of investigation. Moreover, the number of irrelevant variants likely to be discovered in phenotype-specific panels is reduced compared to complete CM panels. Phenotype-specific panels may be cheaper and need to be updated less often [187]. Regardless of scope, disease-specific panels can be designed with better read depth and sequencing coverage than WES or WGS and, until this practical aspect of WES and WGS improves, will remain the most economical tool for targeted but comprehensive simultaneous sequencing of the known CM genes.

3.2.3.2 Structural Variation

Sequencing and alignment of large repeats >300 bp and identification of large genomic structural rearrangements, insertions, and deletions are not yet reliable with NGS. Today, the detection of structural DNA variations (e.g., translocations, inversions, large insertions/deletions, copy number variations) is accomplished using a variety of other methods, each of which has played a substantial role in the diagnosis of Mendelian disorders. Conventional cytogenetics, in which metaphase chromosomes are stained and morphologically evaluated, is among the oldest of these techniques, but often suffers from limited resolution and sensitivity. More popular strategies include multiple ligation-dependent probe amplification (MLPA), array comparative genomic hybridization (CGH), fluorescent in situ hybridization, and DNA single nucleotide polymorphism arrays.

The overwhelming majority of pathogenic mutations currently associated with different forms of CM are single nucleotide variants or small insertions/deletions. XLMTM is a good example, where 93.3 % of known sequence variants described in the *MTM1* gene are small missense and splicing mutations and only 6.3 % are large intragenic deletions [107, 108, 125]. This disproportion may at least partially be attributed to the inherent bias of standard detection methods, as Sanger candidate gene sequencing does not allow for the identification of complex rearrangements and often fails to detect mosaics. MLPA analysis, a variation of the multiplex polymerase chain reaction that permits multiple targets to be amplified with only a single primer pair, has been used to identify the only two large duplications reported in *MTM1* patients to date [125, 188].

Similarly, more than 140 different NM-causing *NEB* mutations have been found in 125 families, and most are intragenic point mutations (55 %) or small deletions of 2–19 bp (31 %) [189]. While the majority of these variants have been successfully detected using denaturing high-performance liquid chromatography (dHPLC) followed by sequencing [190], this method cannot be used to identify large deletion/duplications, intronic mutations affecting splicing, or mutations in the *NEB* promoter. Structural variation analysis in *NEB* has been made even more difficult since there is no commercial MLPA kit available for its 183 exons. Prior to 2013, only one large mutation, the 2.5 kb deletion of exon 55, had been reported in *NEB* [191, 192]. Since it is now possible to detect large copy number variations throughout in the genome within a single experiment, a targeted custom CGH microarray more recently identified two novel deletions in *NEB* inherited from a healthy parent, as well as copy number variants in the triplicate region of *NEB* [189]. Clinical diagnostic laboratories are increasingly offering array CGH-based testing for insertions and deletions, but in the future, it is likely that low-coverage genome sequencing by NGS will serve as an additional technique to detect structural rearrangements in the CMs and will be offered at a considerably lower cost than conventional workups [193].

3.3 Development of Targeted Therapies

Treatment for the CMs has traditionally been limited to supportive care, including respiratory, nutritional, and orthopedic interventions. With the identification of causative genes for many of these conditions, the prospect for development of targeted therapies becomes viable. Substantial advances in the creation of animal models for the CMs (Table 3.3) have fostered ideas for novel therapeutic strategies and set the stage for preclinical testing.

Zebrafish models have been particularly important in broadening our understanding of the molecular pathophysiology of CMs due to their embryonic transparency, genetic tractability, and rapid development [9–11, 205, 211, 212, 220, 225, 230, 236, 240, 242]. Zebrafish genes share a high degree of synteny and 60–80 % amino acid identity with most human orthologues [243]. The benefits of zebrafish research become even more explicit when modeling congenital diseases. In contrast to placental animals, zebrafish embryos develop ex utero, and their cardiovascular function is not essential during early embryogenesis. Zebrafish can therefore bypass most secondary defects that cause embryonic lethality and be used to investigate the primary cause of CMs at the cellular and molecular level [244]. Zebrafish models of BIN1 and MTM1 deficiency, for example, suggest that defects in either protein may promote skeletal muscle weakness in human CNM by disrupting the machinery necessary for excitation–contraction coupling [220]. Zebrafish are also the only vertebrates suited for drug discovery and efficacy assessment in a high-throughput, whole-organism context. Large-scale chemical screens have successfully identified

Table 3.3 Animal models of congenital myopathies

Gene	Invertebrate	Zebrafish	Mouse	Large vertebrate
ACTA1	Fly (D. melanogaster): hemizygous loss-of-function KM88 line is flightless and has defective myofibrils; Act88F mutants display a highly variable antimorphic phenotype, ranging from mild to severe alterations in skeletal muscle, but no electron-dense nemaline bodies have been reported [194, 195]	None to date	• Homozygous Acta1−/− KO mice die by postnatal day 9 [196, 197] • Transgenic (ACTA1)D286G mice have a normal lifespan, but produce less specific force and are less active than wild-type mice [198, 199] • Crosses of transgenic (ACTA1)D286G mice to hemizygous Acta1 mice yield progeny with progressive phenotype; paralyzed hindlimbs and reduced musculature in these extremities present between postnatal day 10–17 [198, 199] • Transgenic (ACTA1)D286G-EGFP model reproduces structural lesions observed in human patients and exhibits muscle weakness [198, 199] • KI (Acta1)H40Y mice display actin aggregates and cytoplasmic as well as intranuclear rods in skeletal muscle; female mice have normal lifespan, while male mice die by either 9 weeks or 5 months of age [200]	Nemaline rods have been identified in biopsies from a Border Collie with early onset muscle atrophy and weakness, a Schipperke with stiff gait and difficulty jumping [201], and in a family of cats [202]; specific mutations have not yet been determined
BIN1	• Fly (D. melanogaster): mutations in BIN1 orthologue (amph) result in severe disorganization and reduction in T-tubules; flightless, sluggish flies [203] • Worm (C. elegans): BIN1 orthologue (AMPH-1) deletion mutants show defects in membrane recycling and remodeling [204]	Morpholino KD leads to delays in motor functions, triad defects, and disrupted calcium signaling [205]	• KO is perinatally lethal (within 24 h of birth); exhibits cardiac structural defects [206]	Great Danes with altered splicing of muscle-specific exon 11 demonstrates highly progressive form of centronuclear myopathy [130]

(continued)

Table 3.3 (continued)

Gene	Invertebrate	Zebrafish	Mouse	Large vertebrate
DNM2	Fly (*D. melanogaster*): mutations in *shibirets1* (*shi*) gene, the only orthologue of human dynamins, prevent GTP hydrolysis and abolish vesicle translocation at nonpermissive temperatures [207–209]	• Transient overexpression of p.S619L-DNM2 causes severe weakness and motor deficits [210–212] • Morpholino KD results in delays in motor development and smaller myofibers with abnormal membrane accumulations [211, 212]	• KO die prior to embryonic day 10 [213] • Homozygous KI(*Dnm2*)R465W is perinatally lethal (within 24 h of birth), potentially due to defects in clathrin-mediated endocytosis [214] • Heterozygous KI(*Dnm2*)R465W is viable with no reduced lifespan; mild myopathic phenotype [214] • AAV injections of p.R465W-DNM2 into adult skeletal muscle lead to weakness associated with disruptions in mitochondrial and T-tubule networks [215]	None to date
KLHL40	None to date	Morpholino KD of *klhl40a* and *klhl40b* isoforms results in disruption of skeletal muscle structure as well as loss of movement [211, 212]	Homozygous *Klhl40*−/− KO mice are smaller than wild-type littermates and die between 7 days and 3 weeks of age; skeletal muscles is disorganized with widened Z-disks, nemaline bodies, and Z-line streaming [216]	None to date
KLHL41	None to date	Morpholino KD of *klhl41* leads to highly diminished motor functions and myofibrillar disorganization with nemaline body formation [11]	None to date	None to date

| *MTM1* | • Fly (*D. melanogaster*): muscle-specific RNAi KD of *MTM1* orthologue (*mtm*) in larvae shows developmental delays and T-tubule disorganization [217, 218]
• Worm (*C. elegans*): RNAi knockdown showed that *MTM1* orthologue (Y110A7A.5) negatively regulates phosphatidylinositol 3-phosphate levels [219] | Morpholino KD exhibits abnormally located nuclei and structural disruptions at the triad; externally, impaired motor behaviors and reduced skeletal muscle birefringence [220, 221] | • KO (CMV-Cre) is postnatally lethal (week 6–14), most likely due to cachexia and respiratory insufficiency [222]
• Muscle-specific KO (HSA-Cre) shows similar viability to ubiquitous KO [222]
• AAV-mediated deletion of *Mtm1* gene in adult mouse myofibers results in atrophy and weakness, abnormal positioning of organelles, and dysregulated satellite cell numbers as well as autophagy markers [223] | Labrador Retrievers with *MTM1* missense variant (p.N155K-MTM1) are nonambulatory by week 3–4; muscle biopsies show "necklace fibers" and disorganized T-tubules [224] |
| *NEB* | None to date | Recessive mutation in *neb* results in impaired force generation, thin filament alterations, and the presence of nemaline bodies [225] | • Homozygous *Neb*$^{-/-}$ KO mice have normal myofibrillogenesis but exhibit sarcomeric disorganization after muscle usage as well as reductions in thin filament length; die by postnatal day 8–20 [226, 227]
• Deletion of exon 55 (NebΔExon55) have developmental delays, reduced thin filament lengths, and decreased calcium sensitivity to force generation [228]
• Mice with deletion of C-terminal SH3 domain (NebΔSH3) do not show structural or histological abnormalities in skeletal muscle but are more susceptible to eccentric contraction-induced injury [229] | None to date |

(continued)

Table 3.3 (continued)

Gene	Invertebrate	Zebrafish	Mouse	Large vertebrate
RYR1	None to date	Spontaneous intron insertion and aberrant splicing cause small, amorphous cores and reduced calcium transients in muscle fibers [230]	• Homozygous *Ryr1^-/-* KO mice die perinatally with gross abnormalities in skeletal muscle; contractile response to electrical stimulation absent under physiological conditions [231] • Homozygous KI(*Ryr1*)Y522S mice exhibit skeletal muscle defects and die early in development [232] • Heterozygous KI(*Ryr1*)Y522S mice are susceptible to malignant hyperthermia; skeletal muscles display caffeine- and heat-induced contractures in vitro and have increased calcium release from leaky Ryr1 channels under oxidative stress [232, 233] • Heterozygous KI(*Ryr1*)R163C mice undergo full malignant hyperthermia episodes upon exposure to halothane or heat and increased resting intracellular calcium concentrations [234] • Heterozygous KI(*Ryr1*)I4895T mice exhibit a slowly progressive myopathy with neonatal respiratory stress and histological formation of minicores, cores, and nemaline rods [235]	None to date

SEPN1	None to date	Morpholino KD results in reduced motility, as well as disruptions in sarcomeric organization, myofiber attachment, and myoseptum integrity [236, 237]	Homozygous *Sepn1*−/− KO mice have normal growth and are macroscopically indistinguishable from wild-type mice except for abnormalities in lung development; body rigidity, impaired motor behaviors, and subtle core lesions are observed only after stress induction [238, 239]	None to date
TPM2, TPM3	None to date	Transient overexpression of p.K7del-Tpm2 shows that mutant protein fails to localize properly within the thin filament compartment and alters sarcomere length [240]	Transgenic (*TPM3*)[HSA.M9R] mice have reduced sensitivity to calcium activation in skeletal muscles and limited type I fiber predominance [241]	None to date
TTN	None to date	Genomic duplications of *ttn* locus and morpholino KD both lead to collapsed myofibrils, disrupted expression of sarcomeric proteins, decreased skeletal muscle birefringence, and reduced mobility [242]	None to date	None to date

phosphodiesterase inhibitors as a viable treatment strategy for Duchenne muscular dystrophy [245]. Targeted approaches with candidate compounds have also shown that the antioxidant N-acetylcysteine can ameliorate muscle pathologies by reducing oxidative stress in the zebrafish model of *RYR1*-related core myopathy [246].

To date, perhaps the most promising observation has been that either gene or protein replacement therapies can successfully ameliorate the disease process in XLMTM. Recombinant adeno-associated viruses are powerful tools to express transgenes in vivo, and several serotypes transduce skeletal muscle with very high efficiency [247]. Local studies in *Mtm1* knockout mice and in dogs carrying a naturally occurring *MTM1* gene mutation have demonstrated the efficacy of adeno-associated virus-mediated myotubularin delivery. Only a single intravascular injection was required to improve strength and force generation, as well as correct muscle pathology, in treated mouse and dog muscles [248, 249]. In dogs, treatment prolonged survival without any observed toxicity or a clinical immune response. At the protein level, a myotubularin replacement strategy using a single-chain fragment derived from the mouse monoclonal antibody 3E10 (3E10Fv) has been published. Short-term exogenous myotubularin supplementation with a 3E10Fv-MTM1 fusion protein improves both muscle pathology and contractile function [250]. *Mtm1* knockout mice have additionally been treated via injections of a soluble activin-receptor type IIB fusion protein, which binds to TGF-beta family members and increases muscle fiber growth in vivo [251].

Alternative approaches have also been explored to rescue loss-of-function CM phenotypes. Reduction of *DNM2* expression in *Mtm1* knockout mice, for example, is sufficient to rescue the early lethality and most pathological features of the X-linked disease [252]. This finding also supports the hypothesis that MTM1 and DNM2 function in a common molecular pathway, with MTM1 acting as a negative regulator of DNM2 function in skeletal muscle. In NM, cardiac alpha-actin (the predominant actin isoform in fetal muscle) has been shown to successfully substitute for skeletal muscle alpha-actin and prevent the early postnatal death of *Acta1* knockout mice [197]. Transgenic overexpression of cardiac alpha-actin also ameliorates the phenotype and increases survival of mice with dominant *ACTA1* p.D286G mutations [9, 10]. The discovery that SEPN1 is a redox regulator of the Ca^{2+}-ATPase SERCA2 pump, acting antagonistically with the ER oxidoreductase, ERO1, raises the possibility that ERO1 inhibitors may provide protection against pathology resulting from inhibition of SERCA channel activity in the absence of SEPN1 [253]. Finally, ex vivo treatment with N-acetylcysteine significantly improves cell survival and decreases protein oxidation levels in cultured SEPN1-deficient myoblasts [254]. As a reduced thiol donor, N-acetylcysteine is hypothesized to act by partially replacing the thiol activity of absent SEPN1 protein. Alleviating oxidative and nitrosative stress in human *SEPN1*-related myopathy patients is currently being pursued as a therapeutic strategy to treat this debilitating disease.

3.4   Conclusion

Gene discovery in the CMs has accelerated in recent years, thanks to the advent of several improved technologies. A number of causative genes have now been identified that were previously inaccessible using traditional approaches. However, the genetic causes remain unknown for approximately 30–40 % of CM patients due to heterogeneous clinical presentations as well as technical limitations in current gene discovery methods. These cases are further complicated by the possibilities of modifier genes, multifactorial inheritance, and/or somatic mutations. The imminent optimization of technologies such as global RNA sequencing (RNA-Seq) and deep genomic sequencing, as well as efforts to create and expand large-scale control population databases, will likely help address existing limitations in genetic diagnoses and offer insights into the mechanistic basis for these devastating disorders.

References

1. Nance JR, Dowling JJ, Gibbs EM, Bonnemann CG. Congenital myopathies: an update. Curr Neurol Neurosci Rep. 2012;12(2):165–74.
2. North KN, Wang CH, Clarke N, Jungbluth H, Vainzof M, Dowling JJ, et al. Approach to the diagnosis of congenital myopathies. Neuromuscul Disord. 2014;24(2):97–116.
3. Romero NB, Clarke NF. Congenital myopathies. Handb Clin Neurol. 2013;113:1321–36.
4. Romero NB, Monnier N, Viollet L, Cortey A, Chevallay M, Leroy JP, et al. Dominant and recessive central core disease associated with *RYR1* mutations and fetal akinesia. Brain. 2003;126(Pt 11):2341–9.
5. Ryan MM, Schnell C, Strickland CD, Shield LK, Morgan G, Iannaccone ST, et al. Nemaline myopathy: a clinical study of 143 cases. Ann Neurol. 2001;50(3):312–20.
6. Nowak KJ, Wattanasirichaigoon D, Goebel HH, Wilce M, Pelin K, Donner K, et al. Mutations in the skeletal muscle alpha-actin gene in patients with actin myopathy and nemaline myopathy. Nat Genet. 1999;23(2):208–12.
7. Agrawal PB, Greenleaf RS, Tomczak KK, Lehtokari VL, Wallgren-Pettersson C, Wallefeld W, et al. Nemaline myopathy with minicores caused by mutation of the *CFL2* gene encoding the skeletal muscle actin-binding protein, cofilin-2. Am J Hum Genet. 2007;80(1):162–7.
8. Sambuughin N, Yau KS, Olive M, Duff RM, Bayarsaikhan M, Lu S, et al. Dominant mutations in *KBTBD13*, a member of the BTB/Kelch family, cause nemaline myopathy with cores. Am J Hum Genet. 2010;87(6):842–7.
9. Ravenscroft G, McNamara E, Griffiths LM, Papadimitriou JM, Hardeman EC, Bakker AJ, et al. Cardiac alpha-actin over-expression therapy in dominant *ACTA1* disease. Hum Mol Genet. 2013;22(19):3987–97.
10. Ravenscroft G, Miyatake S, Lehtokari VL, Todd EJ, Vornanen P, Yau KS, et al. Mutations in *KLHL40* are a frequent cause of severe autosomal-recessive nemaline myopathy. Am J Hum Genet. 2013;93(1):6–18.
11. Gupta VA, Ravenscroft G, Shaheen R, Todd EJ, Swanson LC, Shiina M, et al. Identification of *KLHL41* mutations implicates BTB-Kelch-mediated ubiquitination as an alternate pathway to myofibrillar disruption in nemaline myopathy. Am J Hum Genet. 2013;93(6):1108–17.
12. Yuen M, Sandaradura SA, Dowling JJ, Kostyukova AS, Moroz N, Quinlan KG, et al. Leiomodin-3 dysfunction results in thin filament disorganization and nemaline myopathy. J Clin Invest. 2014;24.

13. Pelin K, Hilpela P, Donner K, Sewry C, Akkari PA, Wilton SD, et al. Mutations in the nebulin gene associated with autosomal recessive nemaline myopathy. Proc Natl Acad Sci U S A. 1999;96(5):2305–10.
14. Quane KA, Healy JM, Keating KE, Manning BM, Couch FJ, Palmucci LM, et al. Mutations in the ryanodine receptor gene in central core disease and malignant hyperthermia. Nat Genet. 1993;5(1):51–5.
15. Zhang Y, Chen HS, Khanna VK, De Leon S, Phillips MS, Schappert K, et al. A mutation in the human ryanodine receptor gene associated with central core disease. Nat Genet. 1993; 5(1):46–50.
16. Johnston JJ, Kelley RI, Crawford TO, Morton DH, Agarwala R, Koch T, et al. A novel nemaline myopathy in the Amish caused by a mutation in troponin T1. Am J Hum Genet. 2000; 67(4):814–21.
17. Donner K, Ollikainen M, Ridanpaa M, Christen HJ, Goebel HH, de Visser M, et al. Mutations in the beta-tropomyosin (*TPM2*) gene--a rare cause of nemaline myopathy. Neuromuscul Disord. 2002;12(2):151–8.
18. Laing NG, Wilton SD, Akkari PA, Dorosz S, Boundy K, Kneebone C, et al. A mutation in the alpha tropomyosin gene *TPM3* associated with autosomal dominant nemaline myopathy. Nat Genet. 1995;9(1):75–9.
19. Jungbluth H, Sewry CA, Brown SC, Nowak KJ, Laing NG, Wallgren-Pettersson C, et al. Mild phenotype of nemaline myopathy with sleep hypoventilation due to a mutation in the skeletal muscle alpha-actin (*ACTA1*) gene. Neuromuscul Disord. 2001;11(1):35–40.
20. Romero NB, Lehtokari VL, Quijano-Roy S, Monnier N, Claeys KG, Carlier RY, et al. Core-rod myopathy caused by mutations in the nebulin gene. Neurology. 2009;73(14):1159–61.
21. Monnier N, Romero NB, Lerale J, Nivoche Y, Qi D, MacLennan DH, et al. An autosomal dominant congenital myopathy with cores and rods is associated with a neomutation in the *RYR1* gene encoding the skeletal muscle ryanodine receptor. Hum Mol Genet. 2000;9(18): 2599–608.
22. Gommans IM, Davis M, Saar K, Lammens M, Mastaglia F, Lamont P, et al. A locus on chromosome 15q for a dominantly inherited nemaline myopathy with core-like lesions. Brain. 2003;126(Pt 7):1545–51.
23. Hung RM, Yoon G, Hawkins CE, Halliday W, Biggar D, Vajsar J. Cap myopathy caused by a mutation of the skeletal alpha-actin gene *ACTA1*. Neuromuscul Disord. 2010;20(4): 238–40.
24. Lehtokari VL, Ceuterick-de Groote C, de Jonghe P, Marttila M, Laing NG, Pelin K, et al. Cap disease caused by heterozygous deletion of the beta-tropomyosin gene *TPM2*. Neuromuscul Disord. 2007;17(6):433–42.
25. De Paula AM, Franques J, Fernandez C, Monnier N, Lunardi J, Pellissier JF, et al. A *TPM3* mutation causing cap myopathy. Neuromuscul Disord. 2009;19(10):685–8.
26. Haan EA, Freemantle CJ, McCure JA, Friend KL, Mulley JC. Assignment of the gene for central core disease to chromosome 19. Hum Genet. 1990;86(2):187–90.
27. Boyden SE, Mahoney LJ, Kawahara G, Myers JA, Mitsuhashi S, Estrella EA, et al. Mutations in the satellite cell gene *MEGF10* cause a recessive congenital myopathy with minicores. Neurogenetics. 2012;13(2):115–24.
28. Cullup T, Lamont PJ, Cirak S, Damian MS, Wallefeld W, Gooding R, et al. Mutations in *MYH7* cause multi-minicore disease (MmD) with variable cardiac involvement. Neuromuscul Disord. 2012;22(12):1096–104.
29. Monnier N, Ferreiro A, Marty I, Labarre-Vila A, Mezin P, Lunardi J. A homozygous splicing mutation causing a depletion of skeletal muscle *RYR1* is associated with multi-minicore disease congenital myopathy with ophthalmoplegia. Hum Mol Genet. 2003;12(10): 1171–8.
30. Moghadaszadeh B, Petit N, Jaillard C, Brockington M, Quijano Roy S, Merlini L, et al. Mutations in *SEPN1* cause congenital muscular dystrophy with spinal rigidity and restrictive respiratory syndrome. Nat Genet. 2001;29(1):17–8.

31. Nicot AS, Toussaint A, Tosch V, Kretz C, Wallgren-Pettersson C, Iwarsson E, et al. Mutations in amphiphysin 2 (*BIN1*) disrupt interaction with dynamin 2 and cause autosomal recessive centronuclear myopathy. Nat Genet. 2007;39(9):1134–9.
32. Majczenko K, Davidson AE, Camelo-Piragua S, Agrawal PB, Manfready RA, Li X, et al. Dominant mutation of *CCDC78* in a unique congenital myopathy with prominent internal nuclei and atypical cores. Am J Hum Genet. 2012;91(2):365–71.
33. Bitoun M, Maugenre S, Jeannet PY, Lacene E, Ferrer X, Laforet P, et al. Mutations in dynamin 2 cause dominant centronuclear myopathy. Nat Genet. 2005;37(11):1207–9.
34. Laporte J, Hu LJ, Kretz C, Mandel JL, Kioschis P, Coy JF, et al. A gene mutated in X-linked myotubular myopathy defines a new putative tyrosine phosphatase family conserved in yeast. Nat Genet. 1996;13(2):175–82.
35. Wilmshurst JM, Lillis S, Zhou H, Pillay K, Henderson H, Kress W, et al. *RYR1* mutations are a common cause of congenital myopathies with central nuclei. Ann Neurol. 2010;68(5): 717–26.
36. Agrawal PB, Pierson CR, Joshi M, Liu X, Ravenscroft G, Moghadaszadeh B, et al. SPEG interacts with myotubularin, and its deficiency causes centronuclear myopathy with dilated cardiomyopathy. Am J Hum Genet. 2014;95(2):218–26.
37. Ceyhan-Birsoy O, Agrawal PB, Hidalgo C, Schmitz-Abe K, DeChene ET, Swanson LC, et al. Recessive truncating titin gene, *TTN*, mutations presenting as centronuclear myopathy. Neurology. 2013;81(14):1205–14.
38. Laing NG, Clarke NF, Dye DE, Liyanage K, Walker KR, Kobayashi Y, et al. Actin mutations are one cause of congenital fibre type disproportion. Ann Neurol. 2004;56(5):689–94.
39. Ortolano S, Tarrio R, Blanco-Arias P, Teijeira S, Rodriguez-Trelles F, Garcia-Murias M, et al. A novel *MYH7* mutation links congenital fiber type disproportion and myosin storage myopathy. Neuromuscul Disord. 2011;21(4):254–62.
40. Clarke NF, Waddell LB, Cooper ST, Perry M, Smith RL, Kornberg AJ, et al. Recessive mutations in *RYR1* are a common cause of congenital fiber type disproportion. Hum Mutat. 2010;31(7):E1544–50.
41. Clarke NF, Kidson W, Quijano-Roy S, Estournet B, Ferreiro A, Guicheney P, et al. *SEPN1*: associated with congenital fiber-type disproportion and insulin resistance. Ann Neurol. 2006;59(3):546–52.
42. Clarke NF, Waddell LB, Sie LT, van Bon BW, McLean C, Clark D, et al. Mutations in *TPM2* and congenital fibre type disproportion. Neuromuscul Disord. 2012;22(11):955–8.
43. Clarke NF, Kolski H, Dye DE, Lim E, Smith RL, Patel R, et al. Mutations in *TPM3* are a common cause of congenital fiber type disproportion. Ann Neurol. 2008;63(3):329–37.
44. Lawlor MW, Dechene ET, Roumm E, Geggel AS, Moghadaszadeh B, Beggs AH. Mutations of tropomyosin 3 (*TPM3*) are common and associated with type 1 myofiber hypotrophy in congenital fiber type disproportion. Hum Mutat. 2010;31(2):176–83.
45. Bohm J, Leshinsky-Silver E, Vassilopoulos S, Le Gras S, Lerman-Sagie T, Ginzberg M, et al. Samaritan myopathy, an ultimately benign congenital myopathy, is caused by a *RYR1* mutation. Acta Neuropathol. 2012;124(4):575–81.
46. Bohm J, Biancalana V, Dechene ET, Bitoun M, Pierson CR, Schaefer E, et al. Mutation spectrum in the large GTPase dynamin 2, and genotype-phenotype correlation in autosomal dominant centronuclear myopathy. Hum Mutat. 2012;33(6):949–59.
47. North KN. Clinical approach to the diagnosis of congenital myopathies. Semin Pediatr Neurol. 2011;18(4):216–20.
48. Al-Qusairi L, Laporte J. T-tubule biogenesis and triad formation in skeletal muscle and implication in human diseases. Skelet Muscle. 2011;1(1):26.
49. Wallgren-Pettersson C, Laing NG. Report of the 70th ENMC international workshop: nemaline myopathy, 11–13 June 1999, Naarden, The Netherlands. Neuromuscul Disord. 2000; 10(4–5):299–306.
50. North KN, Laing NG, Wallgren-Pettersson C. Nemaline myopathy: current concepts. The ENMC international consortium and nemaline myopathy. J Med Genet. 1997;34(9):705–13.

51. Howard RS, Wiles CM, Hirsch NP, Spencer GT. Respiratory involvement in primary muscle disorders: assessment and management. Q J Med. 1993;86(3):175–89.
52. Sasaki M, Yoneyama H, Nonaka I. Respiratory muscle involvement in nemaline myopathy. Pediatr Neurol. 1990;6(6):425–7.
53. Wallgren-Pettersson C. Congenital nemaline myopathy. A clinical follow-up of twelve patients. J Neurol Sci. 1989;89(1):1–14.
54. D'Amico A, Graziano C, Pacileo G, Petrini S, Nowak KJ, Boldrini R, et al. Fatal hypertrophic cardiomyopathy and nemaline myopathy associated with ACTA1 K336E mutation. Neuromuscul Disord. 2006;16(9–10):548–52.
55. Jarraya M, Quijano-Roy S, Monnier N, Behin A, Avila-Smirnov D, Romero NB, et al. Whole-Body muscle MRI in a series of patients with congenital myopathy related to TPM2 gene mutations. Neuromuscul Disord. 2012;22 Suppl 2:S137–47.
56. Jungbluth H, Sewry CA, Counsell S, Allsop J, Chattopadhyay A, Mercuri E, et al. Magnetic resonance imaging of muscle in nemaline myopathy. Neuromuscul Disord. 2004;14(12):779–84.
57. Jungbluth H, Davis MR, Muller C, Counsell S, Allsop J, Chattopadhyay A, et al. Magnetic resonance imaging of muscle in congenital myopathies associated with RYR1 mutations. Neuromuscul Disord. 2004;14(12):785–90.
58. Jockusch BM, Veldman H, Griffiths GW, van Oost BA, Jennekens FG. Immunofluorescence microscopy of a myopathy. alpha-Actinin is a major constituent of nemaline rods. Exp Cell Res. 1980;127(2):409–20.
59. Wallgren-Pettersson C, Clarke A, Samson F, Fardeau M, Dubowitz V, Moser H, et al. The myotubular myopathies: differential diagnosis of the X linked recessive, autosomal dominant, and autosomal recessive forms and present state of DNA studies. J Med Genet. 1995;32(9):673–9.
60. Wallgren-Pettersson C, Jasani B, Newman GR, Morris GE, Jones S, Singhrao S, et al. Alpha-actinin in nemaline bodies in congenital nemaline myopathy: immunological confirmation by light and electron microscopy. Neuromuscul Disord. 1995;5(2):93–104.
61. Miike T, Ohtani Y, Tamari H, Ishitsu T, Une Y. Muscle fiber type transformation in nemaline myopathy and congenital fiber type disproportion. Brain Dev. 1986;8(5):526–32.
62. Volpe P, Damiani E, Margreth A, Pellegrini G, Scarlato G. Fast to slow change of myosin in nemaline myopathy: electrophoretic and immunologic evidence. Neurology. 1982;32(1):37–41.
63. Ryan MM, Ilkovski B, Strickland CD, Schnell C, Sanoudou D, Midgett C, et al. Clinical course correlates poorly with muscle pathology in nemaline myopathy. Neurology. 2003;60(4):665–73.
64. Goebel HH, Warlo I. Nemaline myopathy with intranuclear rods--intranuclear rod myopathy. Neuromuscul Disord. 1997;7(1):13–9.
65. Marttila M, Lehtokari VL, Marston S, Nyman TA, Barnerias C, Beggs AH, et al. Mutation update and genotype-phenotype correlations of novel and previously described mutations in TPM2 and TPM3 causing congenital myopathies. Hum Mutat. 2014;35(7):779–90.
66. Wattanasirichaigoon D, Swoboda KJ, Takada F, Tong HQ, Lip V, Iannaccone ST, et al. Mutations of the slow muscle alpha-tropomyosin gene, TPM3, are a rare cause of nemaline myopathy. Neurology. 2002;59(4):613–7.
67. van der Pol WL, Leijenaar JF, Spliet WG, Lavrijsen SW, Jansen NJ, Braun KP, et al. Nemaline myopathy caused by TNNT1 mutations in a Dutch pedigree. Mol Genet Genomic Med. 2014;2(2):134–7.
68. Ockeloen CW, Gilhuis HJ, Pfundt R, Kamsteeg EJ, Agrawal PB, Beggs AH, et al. Congenital myopathy caused by a novel missense mutation in the CFL2 gene. Neuromuscul Disord. 2012;22(7):632–9.
69. Ong RW, AlSaman A, Selcen D, Arabshahi A, Yau KS, Ravenscroft G, et al. Novel cofilin-2 (CFL2) four base pair deletion causing nemaline myopathy. J Neurol Neurosurg Psychiatry. 2014;85(9):1058–60.

70. Agrawal PB, Strickland CD, Midgett C, Morales A, Newburger DE, Poulos MA, et al. Heterogeneity of nemaline myopathy cases with skeletal muscle alpha-actin gene mutations. Ann Neurol. 2004;56(1):86–96.
71. Laing NG, Dye DE, Wallgren-Pettersson C, Richard G, Monnier N, Lillis S, et al. Mutations and polymorphisms of the skeletal muscle alpha-actin gene (*ACTA1*). Hum Mutat. 2009;30(9): 1267–77.
72. Nowak KJ, Ravenscroft G, Laing NG. Skeletal muscle alpha-actin diseases (actinopathies): pathology and mechanisms. Acta Neuropathol. 2013;125(1):19–32.
73. Citirak G, Witting N, Duno M, Werlauff U, Petri H, Vissing J. Frequency and phenotype of patients carrying *TPM2* and *TPM3* gene mutations in a cohort of 94 patients with congenital myopathy. Neuromuscul Disord. 2014;24(4):325–30.
74. Tan P, Briner J, Boltshauser E, Davis MR, Wilton SD, North K, et al. Homozygosity for a nonsense mutation in the alpha-tropomyosin slow gene *TPM3* in a patient with severe infantile nemaline myopathy. Neuromuscul Disord. 1999;9(8):573–9.
75. Wallgren-Pettersson C. Genetics of the nemaline myopathies and the myotubular myopathies. Neuromuscul Disord. 1998;8(6):401–4.
76. Ilkovski B, Cooper ST, Nowak K, Ryan MM, Yang N, Schnell C, et al. Nemaline myopathy caused by mutations in the muscle alpha-skeletal-actin gene. Am J Hum Genet. 2001;68(6):1333–43.
77. Jungbluth H, Sewry CA, Muntoni F. Core myopathies. Semin Pediatr Neurol. 2011;18(4): 239–49.
78. Magee KR, Shy GM. A new congenital non-progressive myopathy. Brain. 1956;79(4): 610–21.
79. Greenfield JG, Cornman T, Shy GM. The prognostic value of the muscle biopsy in the floppy infant. Brain. 1958;81(4):461–84.
80. Dubowitz V. Muscle disorders in childhood. 2nd ed. London: Elsevier Health Sciences; 1995.
81. Ramsey PL, Hensinger RN. Congenital dislocation of the hip associated with central core disease. J Bone Joint Surg Am. 1975;57(5):648–51.
82. Merlini L, Mattutini P, Bonfiglioli S, Granata C. Non-progressive central core disease with severe congenital scoliosis: a case report. Dev Med Child Neurol. 1987;29(1):106–9.
83. Gamble JG, Rinsky LA, Lee JH. Orthopaedic aspects of central core disease. J Bone Joint Surg Am. 1988;70(7):1061–6.
84. Jungbluth H. Central core disease. Orphanet J Rare Dis. 2007;2:25.
85. Jungbluth H, Muller CR, Halliger-Keller B, Brockington M, Brown SC, Feng L, et al. Autosomal recessive inheritance of *RYR1* mutations in a congenital myopathy with cores. Neurology. 2002;59(2):284–7.
86. Klein A, Lillis S, Munteanu I, Scoto M, Zhou H, Quinlivan R, et al. Clinical and genetic findings in a large cohort of patients with ryanodine receptor 1 gene-associated myopathies. Hum Mutat. 2012;33(6):981–8.
87. Manzur AY, Sewry CA, Ziprin J, Dubowitz V, Muntoni F. A severe clinical and pathological variant of central core disease with possible autosomal recessive inheritance. Neuromuscul Disord. 1998;8(7):467–73.
88. Heckmatt JZ, Dubowitz V. Ultrasound imaging and directed needle biopsy in the diagnosis of selective involvement in muscle disease. J Child Neurol. 1987;2(3):205–13.
89. Lynch PJ, Tong J, Lehane M, Mallet A, Giblin L, Heffron JJ, et al. A mutation in the transmembrane/luminal domain of the ryanodine receptor is associated with abnormal Ca2+ release channel function and severe central core disease. Proc Natl Acad Sci U S A. 1999; 96(7):4164–9.
90. Scacheri PC, Hoffman EP, Fratkin JD, Semino-Mora C, Senchak A, Davis MR, et al. A novel ryanodine receptor gene mutation causing both cores and rods in congenital myopathy. Neurology. 2000;55(11):1689–96.
91. Hayashi K, Miller RG, Brownell AK. Central core disease: ultrastructure of the sarcoplasmic reticulum and T-tubules. Muscle Nerve. 1989;12(2):95–102.

92. Ferreiro A, Fardeau M. 80th ENMC international workshop on multi-minicore disease: 1st international MmD workshop. 12-13th May, 2000, Soestduinen, The Netherlands. Neuromuscul Disord. 2002;12(1):60–8.
93. Ferreiro A, Estournet B, Chateau D, Romero NB, Laroche C, Odent S, et al. Multi-minicore disease--searching for boundaries: phenotype analysis of 38 cases. Ann Neurol. 2000;48(5): 745–57.
94. Ferreiro A, Monnier N, Romero NB, Leroy JP, Bonnemann C, Haenggeli CA, et al. A recessive form of central core disease, transiently presenting as multi-minicore disease, is associated with a homozygous mutation in the ryanodine receptor type 1 gene. Ann Neurol. 2002; 51(6):750–9.
95. Ferreiro A, Quijano-Roy S, Pichereau C, Moghadaszadeh B, Goemans N, Bonnemann C, et al. Mutations of the selenoprotein N gene, which is implicated in rigid spine muscular dystrophy, cause the classical phenotype of multiminicore disease: reassessing the nosology of early-onset myopathies. Am J Hum Genet. 2002;71(4):739–49.
96. Jungbluth H, Sewry C, Brown SC, Manzur AY, Mercuri E, Bushby K, et al. Minicore myopathy in children: a clinical and histopathological study of 19 cases. Neuromuscul Disord. 2000;10(4–5):264–73.
97. Engel AG, Gomez MR, Groover RV. Multicore disease. A recently recognized congenital myopathy associated with multifocal degeneration of muscle fibers. Mayo Clin Proc. 1971; 46(10):666–81.
98. Dubowitz V, Sewry C. Muscle biopsy: a practical approach. 3rd ed. Oxford: Elsevier - Health Sciences Division; 2006.
99. Davis MR, Haan E, Jungbluth H, Sewry C, North K, Muntoni F, et al. Principal mutation hotspot for central core disease and related myopathies in the C-terminal transmembrane region of the *RYR1* gene. Neuromuscul Disord. 2003;13(2):151–7.
100. Monnier N, Romero NB, Lerale J, Landrieu P, Nivoche Y, Fardeau M, et al. Familial and sporadic forms of central core disease are associated with mutations in the C-terminal domain of the skeletal muscle ryanodine receptor. Hum Mol Genet. 2001;10(22):2581–92.
101. Monnier N, Marty I, Faure J, Castiglioni C, Desnuelle C, Sacconi S, et al. Null mutations causing depletion of the type 1 ryanodine receptor (*RYR1*) are commonly associated with recessive structural congenital myopathies with cores. Hum Mutat. 2008;29(5):670–8.
102. Zhou H, Lillis S, Loy RE, Ghassemi F, Rose MR, Norwood F, et al. Multi-minicore disease and atypical periodic paralysis associated with novel mutations in the skeletal muscle ryanodine receptor (*RYR1*) gene. Neuromuscul Disord. 2010;20(3):166–73.
103. DeChene ET, Kang PB, Beggs AH. Congenital fiber-type disproportion. In: Pagon RA, Adam MP, Ardinger HH, Bird TD, Dolan CR, Fong CT, et al., editors. GeneReviews®[Internet]. Seattle: University of Washington; 1993.
104. Bharucha-Goebel DX, Santi M, Medne L, Zukosky K, Dastgir J, Shieh PB, et al. Severe congenital *RYR1*-associated myopathy: the expanding clinicopathologic and genetic spectrum. Neurology. 2013;80(17):1584–9.
105. Zhou H, Yamaguchi N, Xu L, Wang Y, Sewry C, Jungbluth H, et al. Characterization of recessive *RYR1* mutations in core myopathies. Hum Mol Genet. 2006;15(18):2791–803.
106. Jungbluth H, Zhou H, Hartley L, Halliger-Keller B, Messina S, Longman C, et al. Minicore myopathy with ophthalmoplegia caused by mutations in the ryanodine receptor type 1 gene. Neurology. 2005;65(12):1930–5.
107. Amburgey K, Lawlor MW, Del Gaudio D, Cheng YW, Fitzpatrick C, Minor A, et al. Large duplication in *MTM1* associated with myotubular myopathy. Neuromuscul Disord. 2013; 23(3):214–8.
108. Amburgey K, Bailey A, Hwang JH, Tarnopolsky MA, Bonnemann CG, Medne L, et al. Genotype-phenotype correlations in recessive *RYR1*-related myopathies. Orphanet J Rare Dis. 2013;8:117.
109. Lamont PJ, Wallefeld W, Hilton-Jones D, Udd B, Argov Z, Barboi AC, et al. Novel mutations widen the phenotypic spectrum of slow skeletal/beta-cardiac myosin (*MYH7*) distal myopathy. Hum Mutat. 2014;35(7):868–79.

110. Ferreiro A, Ceuterick-de Groote C, Marks JJ, Goemans N, Schreiber G, Hanefeld F, et al. Desmin-related myopathy with Mallory body-like inclusions is caused by mutations of the selenoprotein N gene. Ann Neurol. 2004;55(5):676–86.

111. Carmignac V, Salih MA, Quijano-Roy S, Marchand S, Al Rayess MM, Mukhtar MM, et al. C-terminal titin deletions cause a novel early-onset myopathy with fatal cardiomyopathy. Ann Neurol. 2007;61(4):340–51.

112. Pierson CR, Tomczak K, Agrawal P, Moghadaszadeh B, Beggs AH. X-linked myotubular and centronuclear myopathies. J Neuropathol Exp Neurol. 2005;64(7):555–64.

113. Biancalana V, Beggs AH, Das S, Jungbluth H, Kress W, Nishino I, et al. Clinical utility gene card for: centronuclear and myotubular myopathies. Eur J Hum Genet. 2012;20(10). doi:10.1038/ejhg.2012.91

114. Romero NB, Bitoun M. Centronuclear myopathies. Semin Pediatr Neurol. 2011;18(4):250–6.

115. Herman GE, Finegold M, Zhao W, de Gouyon B, Metzenberg A. Medical complications in long-term survivors with X-linked myotubular myopathy. J Pediatr. 1999;134(2):206–14.

116. McEntagart M, Parsons G, Buj-Bello A, Biancalana V, Fenton I, Little M, et al. Genotype-phenotype correlations in X-linked myotubular myopathy. Neuromuscul Disord. 2002;12(10):939–46.

117. Hanisch F, Muller T, Dietz A, Bitoun M, Kress W, Weis J, et al. Phenotype variability and histopathological findings in centronuclear myopathy due to *DNM2* mutations. J Neurol. 2011;258(6):1085–90.

118. Jeannet PY, Bassez G, Eymard B, Laforet P, Urtizberea JA, Rouche A, et al. Clinical and histologic findings in autosomal centronuclear myopathy. Neurology. 2004;62(9):1484–90.

119. Fischer D, Herasse M, Bitoun M, Barragan-Campos HM, Chiras J, Laforet P, et al. Characterization of the muscle involvement in dynamin 2-related centronuclear myopathy. Brain. 2006;129(Pt 6):1463–9.

120. Bitoun M, Bevilacqua JA, Prudhon B, Maugenre S, Taratuto AL, Monges S, et al. Dynamin 2 mutations cause sporadic centronuclear myopathy with neonatal onset. Ann Neurol. 2007;62(6):666–70.

121. McLeod JG, Baker Wde C, Lethlean AK, Shorey CD. Centronuclear myopathy with autosomal dominant inheritance. J Neurol Sci. 1972;15(4):375–87.

122. Susman RD, Quijano-Roy S, Yang N, Webster R, Clarke NF, Dowling J, et al. Expanding the clinical, pathological and MRI phenotype of *DNM2*-related centronuclear myopathy. Neuromuscul Disord. 2010;20(4):229–37.

123. Romero NB. Centronuclear myopathies: a widening concept. Neuromuscul Disord. 2010;20(4):223–8.

124. Biancalana V, Caron O, Gallati S, Baas F, Kress W, Novelli G, et al. Characterisation of mutations in 77 patients with X-linked myotubular myopathy, including a family with a very mild phenotype. Hum Genet. 2003;112(2):135–42.

125. Oliveira J, Oliveira ME, Kress W, Taipa R, Pires MM, Hilbert P, et al. Expanding the *MTM1* mutational spectrum: novel variants including the first multi-exonic duplication and development of a locus-specific database. Eur J Hum Genet. 2013;21(5):540–9.

126. Pierson CR, Agrawal PB, Blasko J, Beggs AH. Myofiber size correlates with *MTM1* mutation type and outcome in X-linked myotubular myopathy. Neuromuscul Disord. 2007;17(7):562–8.

127. Bevilacqua JA, Monnier N, Bitoun M, Eymard B, Ferreiro A, Monges S, et al. Recessive *RYR1* mutations cause unusual congenital myopathy with prominent nuclear internalization and large areas of myofibrillar disorganization. Neuropathol Appl Neurobiol. 2011;37(3):271–84.

128. Jungbluth H, Zhou H, Sewry CA, Robb S, Treves S, Bitoun M, et al. Centronuclear myopathy due to a de novo dominant mutation in the skeletal muscle ryanodine receptor (*RYR1*) gene. Neuromuscul Disord. 2007;17(4):338–45.

129. Bohm J, Yis U, Ortac R, Cakmakci H, Kurul SH, Dirik E, et al. Case report of intrafamilial variability in autosomal recessive centronuclear myopathy associated to a novel *BIN1* stop mutation. Orphanet J Rare Dis. 2010;5:35.

130. Bohm J, Vasli N, Maurer M, Cowling B, Shelton GD, Kress W, et al. Altered splicing of the *BIN1* muscle-specific exon in humans and dogs with highly progressive centronuclear myopathy. PLoS Genet. 2013;9(6), e1003430.

131. Claeys KG, Maisonobe T, Bohm J, Laporte J, Hezode M, Romero NB, et al. Phenotype of a patient with recessive centronuclear myopathy and a novel *BIN1* mutation. Neurology. 2010; 74(6):519–21.

132. Mejaddam AY, Nennesmo I, Sejersen T. Severe phenotype of a patient with autosomal recessive centronuclear myopathy due to a *BIN1* mutation. Acta Myol. 2009;28(3):91–3.

133. Toussaint A, Cowling BS, Hnia K, Mohr M, Oldfors A, Schwab Y, et al. Defects in amphiphysin 2 (*BIN1*) and triads in several forms of centronuclear myopathies. Acta Neuropathol. 2011;121(2):253–66.

134. Bohm J, Biancalana V, Malfatti E, Dondaine N, Koch C, Vasli N, et al. Adult-onset autosomal dominant centronuclear myopathy due to *BIN1* mutations. Brain. 2014;25.

135. Tosch V, Rohde HM, Tronchere H, Zanoteli E, Monroy N, Kretz C, et al. A novel PtdIns3P and PtdIns(3,5)P2 phosphatase with an inactivating variant in centronuclear myopathy. Hum Mol Genet. 2006;15(21):3098–106.

136. Brooke MH, Engel WK. The histographic analysis of human muscle biopsies with regard to fiber types. 4. Children's biopsies. Neurology. 1969;19(6):591–605.

137. Clarke NF. Congenital fiber-type disproportion. Semin Pediatr Neurol. 2011;18(4):264–71.

138. Clarke NF, North KN. Congenital fiber type disproportion--30 years on. J Neuropathol Exp Neurol. 2003;62(10):977–89.

139. Dehkharghani F, Sarnat HB, Brewster MA, Roth SI. Congenital muscle fiber-type disproportion in Krabbe's leukodystrophy. Arch Neurol. 1981;38(9):585–7.

140. Del Bigio MR, Chudley AE, Sarnat HB, Campbell C, Goobie S, Chodirker BN, et al. Infantile muscular dystrophy in Canadian aboriginals is an alphaB-crystallinopathy. Ann Neurol. 2011;69(5):866–71.

141. Sarnat HB, Silbert SW. Maturational arrest of fetal muscle in neonatal myotonic dystrophy. A pathologic study of four cases. Arch Neurol. 1976;33(7):466–74.

142. Sarnat HB, Roth SI, Jimenez JF. Neonatal myotubular myopathy: neuropathy and failure of postnatal maturation of fetal muscle. Can J Neurol Sci. 1981;8(4):313–20.

143. Munot P, Lashley D, Jungbluth H, Feng L, Pitt M, Robb SA, et al. Congenital fibre type disproportion associated with mutations in the tropomyosin 3 (*TPM3*) gene mimicking congenital myasthenia. Neuromuscul Disord. 2010;20(12):796–800.

144. Ottenheijm CA, Lawlor MW, Stienen GJ, Granzier H, Beggs AH. Changes in cross-bridge cycling underlie muscle weakness in patients with tropomyosin 3-based myopathy. Hum Mol Genet. 2011;20(10):2015–25.

145. Sobrido MJ, Fernandez JM, Fontoira E, Perez-Sousa C, Cabello A, Castro M, et al. Autosomal dominant congenital fibre type disproportion: a clinicopathological and imaging study of a large family. Brain. 2005;128(Pt 7):1716–27.

146. Brandis A, Aronica E, Goebel HH. *TPM2* mutation. Neuromuscul Disord. 2008;18(12):1005.

147. Clarke NF, Smith RL, Bahlo M, North KN. A novel X-linked form of congenital fiber-type disproportion. Ann Neurol. 2005;58(5):767–72.

148. Kajino S, Ishihara K, Goto K, Ishigaki K, Noguchi S, Nonaka I, et al. Congenital fiber type disproportion myopathy caused by *LMNA* mutations. J Neurol Sci. 2014;340(1–2):94–8.

149. Benedetti S, Menditto I, Degano M, Rodolico C, Merlini L, D'Amico A, et al. Phenotypic clustering of lamin A/C mutations in neuromuscular patients. Neurology. 2007;69(12):1285–92.

150. Barohn RJ, Jackson CE, Kagan-Hallet KS. Neonatal nemaline myopathy with abundant intranuclear rods. Neuromuscul Disord. 1994;4(5–6):513–20.

151. Goebel HH, Piirsoo A, Warlo I, Schofer O, Kehr S, Gaude M. Infantile intranuclear rod myopathy. J Child Neurol. 1997;12(1):22–30.

152. Goebel HH, Brockman K, Bonnemann CG, Warlo IA, Hanefeld F, Labeit S, et al. Patient with actin aggregate myopathy and not formerly identified *ACTA1* mutation is heterozygous for the Gly15Arg mutation of *ACTA1*, which has previously been associated with actinopathy. J Child Neurol. 2006;21(6):545.

153. Rifai Z, Kazee AM, Kamp C, Griggs RC. Intranuclear rods in severe congenital nemaline myopathy. Neurology. 1993;43(11):2372–7.
154. Robinson R, Carpenter D, Shaw MA, Halsall J, Hopkins P. Mutations in *RYR1* in malignant hyperthermia and central core disease. Hum Mutat. 2006;27(10):977–89.
155. North KN, Laing NG. Skeletal muscle alpha-actin diseases. Adv Exp Med Biol. 2008;642: 15–27.
156. Heckmatt JZ, Leeman S, Dubowitz V. Ultrasound imaging in the diagnosis of muscle disease. J Pediatr. 1982;101(5):656–60.
157. Wallefeld W, Krause S, Nowak KJ, Dye D, Horvath R, Molnar Z, et al. Severe nemaline myopathy caused by mutations of the stop codon of the skeletal muscle alpha actin gene (*ACTA1*). Neuromuscul Disord. 2006;16(9–10):541–7.
158. Collins FS. Sequencing the human genome. Hosp Pract (1995). 1997;32(1):35–43. 6–9, 53–4.
159. Altshuler D, Daly MJ, Lander ES. Genetic mapping in human disease. Science. 2008; 322(5903):881–8.
160. Botstein D, White RL, Skolnick M, Davis RW. Construction of a genetic linkage map in man using restriction fragment length polymorphisms. Am J Hum Genet. 1980;32(3):314–31.
161. Gusella JF, Wexler NS, Conneally PM, Naylor SL, Anderson MA, Tanzi RE, et al. A polymorphic DNA marker genetically linked to Huntington's disease. Nature. 1983;306(5940):234–8.
162. Kwon JM, Goate AM. The candidate gene approach. Alcohol Res Health. 2000;24(3): 164–8.
163. Collins FS. Positional cloning: let's not call it reverse anymore. Nat Genet. 1992;1(1):3–6.
164. Kunkel LM, Hejtmancik JF, Caskey CT, Speer A, Monaco AP, Middlesworth W, et al. Analysis of deletions in DNA from patients with Becker and Duchenne muscular dystrophy. Nature. 1986;322(6074):73–7.
165. Monaco AP, Neve RL, Colletti-Feener C, Bertelson CJ, Kurnit DM, Kunkel LM. Isolation of candidate cDNAs for portions of the Duchenne muscular dystrophy gene. Nature. 1986; 323(6089):646–50.
166. Royer-Pokora B, Kunkel LM, Monaco AP, Goff SC, Newburger PE, Baehner RL, et al. Cloning the gene for the inherited disorder chronic granulomatous disease on the basis of its chromosomal location. Cold Spring Harb Symp Quant Biol. 1986;51(Pt 1):177–83.
167. Darnfors C, Larsson HE, Oldfors A, Kyllerman M, Gustavson KH, Bjursell G, et al. X-linked myotubular myopathy: a linkage study. Clin Genet. 1990;37(5):335–40.
168. Liechti-Gallati S, Muller B, Grimm T, Kress W, Muller C, Boltshauser E, et al. X-linked centronuclear myopathy: mapping the gene to Xq28. Neuromuscul Disord. 1991;1(4):239–45.
169. Thomas NS, Williams H, Cole G, Roberts K, Clarke A, Liechti-Gallati S, et al. X linked neonatal centronuclear/myotubular myopathy: evidence for linkage to Xq28 DNA marker loci. J Med Genet. 1990;27(5):284–7.
170. Dahl N, Samson F, Thomas NS, Hu LJ, Gong W, Herman G, et al. X linked myotubular myopathy (*MTM1*) maps between DXS304 and DXS305, closely linked to the DXS455 VNTR and a new, highly informative microsatellite marker (DXS1684). J Med Genet. 1994; 31(12):922–4.
171. Hu LJ, Laporte J, Kress W, Kioschis P, Siebenhaar R, Poustka A, et al. Deletions in Xq28 in two boys with myotubular myopathy and abnormal genital development define a new contiguous gene syndrome in a 430 kb region. Hum Mol Genet. 1996;5(1):139–43.
172. Smolenicka Z, Laporte J, Hu L, Dahl N, Fitzpatrick J, Kress W, et al. X-linked myotubular myopathy: refinement of the critical gene region. Neuromuscul Disord. 1996;6(4):275–81.
173. Alkuraya FS. Discovery of rare homozygous mutations from studies of consanguineous pedigrees. Curr Protoc Hum Genet;2012. Chapter 6:Unit6 12.
174. Alkuraya FS. Impact of new genomic tools on the practice of clinical genetics in consanguineous populations: the Saudi experience. Clin Genet. 2013;84(3):203–8.
175. Kong A, Frigge ML, Masson G, Besenbacher S, Sulem P, Magnusson G, et al. Rate of de novo mutations and the importance of father's age to disease risk. Nature. 2012;488(7412):471–5.

176. Muhammad E, Reish O, Ohno Y, Scheetz T, Deluca A, Searby C, et al. Congenital myopathy is caused by mutation of *HACD1*. Hum Mol Genet. 2013;22(25):5229–36.
177. Stamm DS, Aylsworth AS, Stajich JM, Kahler SG, Thorne LB, Speer MC, et al. Native American myopathy: congenital myopathy with cleft palate, skeletal anomalies, and susceptibility to malignant hyperthermia. Am J Med Genet A. 2008;146A(14):1832–41.
178. Zhu M, Zhao S. Candidate gene identification approach: progress and challenges. Int J Biol Sci. 2007;3(7):420–7.
179. Collins FS, Guyer MS, Charkravarti A. Variations on a theme: cataloging human DNA sequence variation. Science. 1997;278(5343):1580–1.
180. Xue Y, Ankala A, Wilcox WR, Hegde MR. Solving the molecular diagnostic testing conundrum for Mendelian disorders in the era of next-generation sequencing: single-gene, gene panel, or exome/genome sequencing. Genet Med. 2014;17:444–51.
181. Abel HJ, Al-Kateb H, Cottrell CE, Bredemeyer AJ, Pritchard CC, Grossmann AH, et al. Detection of gene rearrangements in targeted clinical next-generation sequencing. J Mol Diagn. 2014;16(4):405–17.
182. Metzker ML. Sequencing technologies - the next generation. Nat Rev Genet. 2010;11(1):31–46.
183. Ku CS, Cooper DN, Polychronakos C, Naidoo N, Wu M, Soong R. Exome sequencing: dual role as a discovery and diagnostic tool. Ann Neurol. 2012;71(1):5–14.
184. Ng SB, Turner EH, Robertson PD, Flygare SD, Bigham AW, Lee C, et al. Targeted capture and massively parallel sequencing of 12 human exomes. Nature. 2009;461(7261):272–6.
185. Tan R, Wang Y, Kleinstein SE, Liu Y, Zhu X, Guo H, et al. An evaluation of copy number variation detection tools from whole-exome sequencing data. Hum Mutat. 2014;35(7):899–907.
186. Mook OR, Haagmans MA, Soucy JF, van de Meerakker JB, Baas F, Jakobs ME, et al. Targeted sequence capture and GS-FLX titanium sequencing of 23 hypertrophic and dilated cardiomyopathy genes: implementation into diagnostics. J Med Genet. 2013;50(9):614–26.
187. Rossor AM, Polke JM, Houlden H, Reilly MM. Clinical implications of genetic advances in Charcot-Marie-Tooth disease. Nat Rev Neurol. 2013;9(10):562–71.
188. Trump N, Cullup T, Verheij JB, Manzur A, Muntoni F, Abbs S, et al. X-linked myotubular myopathy due to a complex rearrangement involving a duplication of *MTM1* exon 10. Neuromuscul Disord. 2012;22(5):384–8.
189. Kiiski K, Laari L, Lehtokari VL, Lunkka-Hytonen M, Angelini C, Petty R, et al. Targeted array comparative genomic hybridization--a new diagnostic tool for the detection of large copy number variations in nemaline myopathy-causing genes. Neuromuscul Disord. 2013;23(1):56–65.
190. Lehtokari VL, Pelin K, Sandbacka M, Ranta S, Donner K, Muntoni F, et al. Identification of 45 novel mutations in the nebulin gene associated with autosomal recessive nemaline myopathy. Hum Mutat. 2006;27(9):946–56.
191. Anderson SL, Ekstein J, Donnelly MC, Keefe EM, Toto NR, LeVoci LA, et al. Nemaline myopathy in the Ashkenazi Jewish population is caused by a deletion in the nebulin gene. Hum Genet. 2004;115(3):185–90.
192. Lehtokari VL, Greenleaf RS, DeChene ET, Kellinsalmi M, Pelin K, Laing NG, et al. The exon 55 deletion in the nebulin gene--one single founder mutation with world-wide occurrence. Neuromuscul Disord. 2009;19(3):179–81.
193. Abel HJ, Duncavage EJ. Detection of structural DNA variation from next generation sequencing data: a review of informatic approaches. Cancer Genet. 2014;206:432–40.
194. Beall CJ, Sepanski MA, Fyrberg EA. Genetic dissection of Drosophila myofibril formation: effects of actin and myosin heavy chain null alleles. Genes Dev. 1989;3(2):131–40.
195. Haigh SE, Salvi SS, Sevdali M, Stark M, Goulding D, Clayton JD, et al. Drosophila indirect flight muscle specific Act88F actin mutants as a model system for studying congenital myopathies of the human *ACTA1* skeletal muscle actin gene. Neuromuscul Disord. 2010;20(6):363–74.
196. Crawford K, Flick R, Close L, Shelly D, Paul R, Bove K, et al. Mice lacking skeletal muscle actin show reduced muscle strength and growth deficits and die during the neonatal period. Mol Cell Biol. 2002;22(16):5887–96.

197. Nowak KJ, Ravenscroft G, Jackaman C, Filipovska A, Davies SM, Lim EM, et al. Rescue of skeletal muscle alpha-actin-null mice by cardiac (fetal) alpha-actin. J Cell Biol. 2009;185(5): 903–15.
198. Ravenscroft G, Jackaman C, Bringans S, Papadimitriou JM, Griffiths LM, McNamara E, et al. Mouse models of dominant *ACTA1* disease recapitulate human disease and provide insight into therapies. Brain. 2011;134(Pt 4):1101–15.
199. Ravenscroft G, Jackaman C, Sewry CA, McNamara E, Squire SE, Potter AC, et al. Actin nemaline myopathy mouse reproduces disease, suggests other actin disease phenotypes and provides cautionary note on muscle transgene expression. PLoS ONE. 2011;6(12), e28699.
200. Nguyen MA, Joya JE, Kee AJ, Domazetovska A, Yang N, Hook JW, et al. Hypertrophy and dietary tyrosine ameliorate the phenotypes of a mouse model of severe nemaline myopathy. Brain. 2011;134(Pt 12):3516–29.
201. Delauche AJ, Cuddon PA, Podell M, Devoe K, Powell HC, Shelton GD. Nemaline rods in canine myopathies: 4 case reports and literature review. J Vet Intern Med. 1998;12(6):424–30.
202. Cooper BJ, De Lahunta A, Gallagher EA, Valentine BA. Nemaline myopathy of cats. Muscle Nerve. 1986;9(7):618–25.
203. Razzaq A, Robinson IM, McMahon HT, Skepper JN, Su Y, Zelhof AC, Jackson AP, Gay NJ, O'Kane CJ. Amphiphysin is necessary for organization of the excitation-contraction coupling machinery of muscles, but not for synaptic vesicle endocytosis in Drosophila. Genes Dev. 2001;15(22):2967–79.
204. Pant S, Sharma M, Patel K, Caplan S, Carr CM, Grant BD. AMPH-1/Amphiphysin/Bin1 functions with RME-1/Ehd1 in endocytic recycling. Nat Cell Biol. 2009;11(12):1399–410.
205. Smith LL, Gupta VA, Beggs AH. Bridging integrator 1 (Bin1) deficiency in zebrafish results in centronuclear myopathy. Hum Mol Genet. 2014;23(13):3566–78.
206. Muller AJ, Baker JF, DuHadaway JB, Ge K, Farmer G, Donover PS, et al. Targeted disruption of the murine bin1/amphiphysin II gene does not disable endocytosis but results in embryonic cardiomyopathy with aberrant myofibril formation. Mol Cell Biol. 2003;23(12):4295–306.
207. Estes PS, Roos J, van der Bliek A, Kelly RB, Krishnan KS, Ramaswami M. Traffic of dynamin within individual Drosophila synaptic boutons relative to compartment-specific markers. J Neurosci. 1996;16(17):5443–56.
208. Kosaka T, Ikeda K. Reversible blockage of membrane retrieval and endocytosis in the garland cell of the temperature-sensitive mutant of Drosophila melanogaster, shibirets1. J Cell Biol. 1983;97(2):499–507.
209. Ramaswami M, Krishnan KS, Kelly RB. Intermediates in synaptic vesicle recycling revealed by optical imaging of Drosophila neuromuscular junctions. Neuron. 1994;13(2):363–75.
210. Gibbs EM, Davidson AE, Telfer WR, Feldman EL, Dowling JJ. The myopathy-causing mutation *DNM2*-S619L leads to defective tubulation in vitro and in developing zebrafish. Dis Model Mech. 2014;7(1):157–61.
211. Gibbs EM, Clarke NF, Rose K, Oates EC, Webster R, Feldman EL, et al. Neuromuscular junction abnormalities in *DNM2*-related centronuclear myopathy. J Mol Med (Berl). 2013; 91(6):727–37.
212. Gibbs EM, Davidson AE, Trickey-Glassman A, Backus C, Hong Y, Sakowski SA, et al. Two dynamin-2 genes are required for normal zebrafish development. PLoS ONE. 2013;8(2), e55888.
213. Ferguson SM, Raimondi A, Paradise S, Shen H, Mesaki K, Ferguson A, et al. Coordinated actions of actin and BAR proteins upstream of dynamin at endocytic clathrin-coated pits. Dev Cell. 2009;17(6):811–22.
214. Durieux AC, Vignaud A, Prudhon B, Viou MT, Beuvin M, Vassilopoulos S, et al. A centro-nuclear myopathy-dynamin 2 mutation impairs skeletal muscle structure and function in mice. Hum Mol Genet. 2010;19(24):4820–36.
215. Cowling BS, Toussaint A, Amoasii L, Koebel P, Ferry A, Davignon L, et al. Increased expression of wild-type or a centronuclear myopathy mutant of dynamin 2 in skeletal muscle of adult mice leads to structural defects and muscle weakness. Am J Pathol. 2011;178(5): 2224–35.

216. Garg A, O'Rourke J, Long C, Doering J, Ravenscroft G, Bezprozvannaya S, et al. KLHL40 deficiency destabilizes thin filament proteins and promotes nemaline myopathy. J Clin Invest. 2014;124(8):3529–39.
217. Ribeiro I, Yuan L, Tanentzapf G, Dowling JJ, Kiger A. Phosphoinositide regulation of integrin trafficking required for muscle attachment and maintenance. PLoS Genet. 2011;7(2), e1001295.
218. Velichkova M, Juan J, Kadandale P, Jean S, Ribeiro I, Raman V, et al. Drosophila Mtm and class II PI3K coregulate a PI(3)P pool with cortical and endolysosomal functions. J Cell Biol. 2010;190(3):407–25.
219. Xue Y, Fares H, Grant B, Li Z, Rose AM, Clark SG, et al. Genetic analysis of the myotubularin family of phosphatases in Caenorhabditis elegans. J Biol Chem. 2003;278(36):34380–6.
220. Dowling JJ, Vreede AP, Low SE, Gibbs EM, Kuwada JY, Bonnemann CG, et al. Loss of myotubularin function results in T-tubule disorganization in zebrafish and human myotubular myopathy. PLoS Genet. 2009;5(2), e1000372.
221. Smith LL, Beggs AH, Gupta VA. Analysis of skeletal muscle defects in larval zebrafish by birefringence and touch-evoke escape response assays. J Vis Exp. 2013;82:e50925.
222. Buj-Bello A, Laugel V, Messaddeq N, Zahreddine H, Laporte J, Pellissier JF, et al. The lipid phosphatase myotubularin is essential for skeletal muscle maintenance but not for myogenesis in mice. Proc Natl Acad Sci U S A. 2002;99(23):15060–5.
223. Joubert R, Vignaud A, Le M, Moal C, Messaddeq N, Buj-Bello A. Site-specific *Mtm1* mutagenesis by an AAV-Cre vector reveals that myotubularin is essential in adult muscle. Hum Mol Genet. 2013;22(9):1856–66.
224. Beggs AH, Böhm J, Snead E, Kozlowski M, Maurer M, Minor K, Childers MK, Taylor SM, Hitte C, Mickelson JR, Guo LT, Mizisin AP, Buj-Bello A, Tiret L, Laporte J, Shelton GD. MTM1 mutation associated with X-linked myotubular myopathy in Labrador Retrievers. Proc Natl Acad Sci USA. 2010;107(33):14697–702.
225. Telfer WR, Nelson DD, Waugh T, Brooks SV, Dowling JJ. Neb: a zebrafish model of nemaline myopathy due to nebulin mutation. Dis Model Mech. 2012;5(3):389–96.
226. Bang ML, Li X, Littlefield R, Bremner S, Thor A, Knowlton KU, et al. Nebulin-deficient mice exhibit shorter thin filament lengths and reduced contractile function in skeletal muscle. J Cell Biol. 2006;173(6):905–16.
227. Witt CC, Burkart C, Labeit D, McNabb M, Wu Y, Granzier H, et al. Nebulin regulates thin filament length, contractility, and Z-disk structure in vivo. EMBO J. 2006;25(16):3843–55.
228. Ottenheijm CA, Buck D, de Winter JM, Ferrara C, Piroddi N, Tesi C, et al. Deleting exon 55 from the nebulin gene induces severe muscle weakness in a mouse model for nemaline myopathy. Brain. 2013;136(Pt 6):1718–31.
229. Yamamoto DL, Vitiello C, Zhang J, Gokhin DS, Castaldi A, Coulis G, et al. The nebulin SH3 domain is dispensable for normal skeletal muscle structure but is required for effective active load bearing in mouse. J Cell Sci. 2013;126(Pt 23):5477–89.
230. Hirata H, Watanabe T, Hatakeyama J, Sprague SM, Saint-Amant L, Nagashima A, et al. Zebrafish relatively relaxed mutants have a ryanodine receptor defect, show slow swimming and provide a model of multi-minicore disease. Development. 2007;134(15):2771–81.
231. Takeshima H, Iino M, Takekura H, Nishi M, Kuno J, Minowa O, et al. Excitation-contraction uncoupling and muscular degeneration in mice lacking functional skeletal muscle ryanodine-receptor gene. Nature. 1994;369(6481):556–9.
232. Chelu MG, Goonasekera SA, Durham WJ, Tang W, Lueck JD, Riehl J, et al. Heat- and anesthesia-induced malignant hyperthermia in an RyR1 knock-in mouse. FASEB J. 2006; 20(2):329–30.
233. Durham WJ, Aracena-Parks P, Long C, Rossi AE, Goonasekera SA, Boncompagni S, et al. RyR1 S-nitrosylation underlies environmental heat stroke and sudden death in Y522S RyR1 knockin mice. Cell. 2008;133(1):53–65.
234. Yang T, Riehl J, Esteve E, Matthaei KI, Goth S, Allen PD, et al. Pharmacologic and functional characterization of malignant hyperthermia in the R163C RyR1 knock-in mouse. Anesthesiology. 2006;105(6):1164–75.

235. Zvaritch E, Kraeva N, Bombardier E, McCloy RA, Depreux F, Holmyard D, et al. Ca2+ dysregulation in *Ryr1*(I4895T/wt) mice causes congenital myopathy with progressive formation of minicores, cores, and nemaline rods. Proc Natl Acad Sci U S A. 2009;106(51): 21813–8.
236. Deniziak M, Thisse C, Rederstorff M, Hindelang C, Thisse B, Lescure A. Loss of selenoprotein N function causes disruption of muscle architecture in the zebrafish embryo. Exp Cell Res. 2007;313(1):156–67.
237. Jurynec MJ, Xia R, Mackrill JJ, Gunther D, Crawford T, Flanigan KM, et al. Selenoprotein N is required for ryanodine receptor calcium release channel activity in human and zebrafish muscle. Proc Natl Acad Sci U S A. 2008;105(34):12485–90.
238. Moghadaszadeh B, Rider BE, Lawlor MW, Childers MK, Grange RW, Gupta K, et al. Selenoprotein N deficiency in mice is associated with abnormal lung development. FASEB J. 2013;27(4):1585–99.
239. Rederstorff M, Castets P, Arbogast S, Laine J, Vassilopoulos S, Beuvin M, et al. Increased muscle stress-sensitivity induced by selenoprotein N inactivation in mouse: a mammalian model for *SEPN1*-related myopathy. PLoS ONE. 2011;6(8), e23094.
240. Davidson AE, Siddiqui FM, Lopez MA, Lunt P, Carlson HA, Moore BE, et al. Novel deletion of lysine 7 expands the clinical, histopathological and genetic spectrum of *TPM2*-related myopathies. Brain. 2013;136(Pt 2):508–21.
241. Corbett MA, Akkari PA, Domazetovska A, Cooper ST, North KN, Laing NG, et al. An alpha Tropomyosin mutation alters dimer preference in nemaline myopathy. Ann Neurol. 2005; 57(1):42–9.
242. Steffen LS, Guyon JR, Vogel ED, Howell MH, Zhou Y, Weber GJ, et al. The zebrafish runzel muscular dystrophy is linked to the titin gene. Dev Biol. 2007;309(2):180–92.
243. Barbazuk WB, Korf I, Kadavi C, Heyen J, Tate S, Wun E, et al. The syntenic relationship of the zebrafish and human genomes. Genome Res. 2000;10(9):1351–8.
244. Lin YY. Muscle diseases in the zebrafish. Neuromuscul Disord. 2012;22(8):673–84.
245. Kawahara G, Karpf JA, Myers JA, Alexander MS, Guyon JR, Kunkel LM. Drug screening in a zebrafish model of Duchenne muscular dystrophy. Proc Natl Acad Sci U S A. 2011;108(13): 5331–6.
246. Dowling JJ, Arbogast S, Hur J, Nelson DD, McEvoy A, Waugh T, et al. Oxidative stress and successful antioxidant treatment in models of *RYR1*-related myopathy. Brain. 2012;135(Pt 4):1115–27.
247. Wu Z, Asokan A, Samulski RJ. Adeno-associated virus serotypes: vector toolkit for human gene therapy. Mol Ther. 2006;14(3):316–27.
248. Buj-Bello A, Fougerousse F, Schwab Y, Messaddeq N, Spehner D, Pierson CR, et al. AAV-mediated intramuscular delivery of myotubularin corrects the myotubular myopathy phenotype in targeted murine muscle and suggests a function in plasma membrane homeostasis. Hum Mol Genet. 2008;17(14):2132–43.
249. Childers MK, Joubert R, Poulard K, Moal C, Grange RW, Doering JA, et al. Gene therapy prolongs survival and restores function in murine and canine models of myotubular myopathy. Sci Transl Med. 2014;6(220), 220ra10.
250. Lawlor MW, Armstrong D, Viola MG, Widrick JJ, Meng H, Grange RW, et al. Enzyme replacement therapy rescues weakness and improves muscle pathology in mice with X-linked myotubular myopathy. Hum Mol Genet. 2013;22(8):1525–38.
251. Lawlor MW, Read BP, Edelstein R, Yang N, Pierson CR, Stein MJ, et al. Inhibition of activin receptor type IIB increases strength and lifespan in myotubularin-deficient mice. Am J Pathol. 2011;178(2):784–93.
252. Cowling BS, Chevremont T, Prokic I, Kretz C, Ferry A, Coirault C, et al. Reducing dynamin 2 expression rescues X-linked centronuclear myopathy. J Clin Invest. 2014;124(3):1350–63.
253. Marino M, Stoilova T, Giorgi C, Bachi A, Cattaneo A, Auricchio A, et al. SEPN1, an endoplasmic reticulum-localized selenoprotein linked to skeletal muscle pathology, counteracts hyperoxidation by means of redox-regulating SERCA2 pump activity. Hum Mol Genet. 2015;24(7):1843–55.
254. Arbogast S, Beuvin M, Fraysse B, Zhou H, Muntoni F, Ferreiro A. Oxidative stress in *SEPN1*-related myopathy: from pathophysiology to treatment. Ann Neurol. 2009;65(6):677–86.

Chapter 4
Stem Cell Transplantation for Degenerative Muscle Diseases

Berkcan Akpinar*, Elizabeth C. Stahl*, Aiping Lu, and Johnny Huard

List of Abbreviations

ADM	Aorta-derived mesoangioblast
DGC	Dystrophin-glycoprotein complex
DKO	Double-knockout (dystrophin/utrophin)
DMD	Duchenne muscular dystrophy
ECM	Extracellular matrix
ESC	Embryonic stem cell
FACS	Fluorescence-activated cell sorting
FAP	Fibro-adipogenic progenitor cell
FGF	Fibroblast growth factor
GDF	Growth differentiation factor
GRMD	Golden retriever muscular dystrophy
HGF	Hepatocyte growth factor
HLA	Human leukocyte antigen
HO	Heterotopic ossification
IA	Intra-arterial
IGF	Insulin growth factor
ILK	Integrin-linked kinase
IMCL	Intramyocellular lipid accumulation
iPSC	Induced pluripotent stem cell
MAPK	Mitogen-activating protein kinase

---

*Author contributed equally with all other contributors.

B. Akpinar • E.C. Stahl • A. Lu • J. Huard, Ph.D. (✉)
Department of Orthopaedic Surgery, University of Pittsburgh School of Medicine,
Suite 206, Bridgeside Point II, 450 Technology Drive, Pittsburgh, PA 15213, USA
e-mail: jhuard@pitt.edu

© Springer Science+Business Media New York 2016
M.K. Childers (ed.), *Regenerative Medicine for Degenerative Muscle Diseases*,
Stem Cell Biology and Regenerative Medicine,
DOI 10.1007/978-1-4939-3228-3_4

MDSC	Muscle-derived stem cell
MEC	Myoendothelial cell
MMP	Matrix metalloproteinase
MPC	Muscle progenitor cell
MSC	Mesenchymal stem cell
NF-kB	Nuclear factor kappa B
NO	Nitric oxide
ORAI-1	Calcium release-activated calcium channel protein 1
PAX7	Paired box 7
PDPC	Pericyte-derived progenitor cell
ROS	Reactive oxygen species
SCGA	Alpha-sarcoglycan
SUI	Stress urinary incontinence
TGF	Transforming growth factor
TIMP	Tissue inhibitor of metalloproteinase
VEGF	Vascular endothelial growth factor

4.1   Introduction to Duchenne Muscular Dystrophy

Degenerative muscle disease refers to a group of conditions presenting with acute, subacute, or chronic muscle degeneration ultimately resulting in a loss of musculoskeletal functionality. Neuromuscular, metabolic, inflammatory, infectious, traumatic, or musculoskeletal (myopathy) conditions can all lead to degenerative muscle disease. Effective muscle regeneration depends largely on the severity of degeneration and the ability of the muscle progenitor cell (MPC) pool to regenerate lost muscle mass via activation, proliferation, and myogenic differentiation. The MPC pool is a heterogeneous group of cells with varying levels of myogenic capacity. Consisting mainly of satellite cells, MPCs include, but are not limited to, mesenchymal stem cells (MSCs), muscle-derived stem cells (MDSCs), side population stem cells, and myogenic cells within the muscle vasculature [1–6]. As will be discussed, these cell populations have been shown to contribute to muscle formation during both development and regeneration.

In many degenerative skeletal muscle conditions, the MPC pool is severely affected. In a process known as MPC depletion, MPCs decrease in quantity and lose their proliferation and differentiation capacities which allows muscle deterioration to proceed unchecked. For these reasons, stem cell transplantation therapy has long been considered a potential treatment option to ameliorate ongoing muscle degeneration.

Duchenne muscular dystrophy (DMD) is one of the most common degenerative diseases of skeletal muscle, affecting approximately 1 in 3600 males at birth, and exhibits MPC dysfunction and depletion. As a group, the muscular dystrophies are caused by genetic mutations in genes coding for proteins associated with the cell membrane or the extracellular matrix. DMD, which is the most prevalent and most

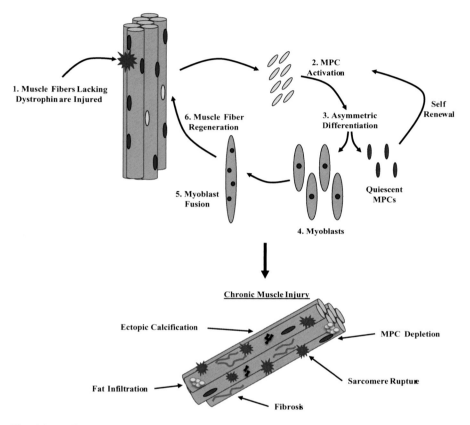

Fig. 4.1 MPC depletion occurs in late-stage DMD

severe muscular dystrophy, results from mutations in the X-linked gene encoding for the dystrophin protein [7]. Dystrophin is a critical component of the dystrophin-glycoprotein complex (DGC), which transmits force during muscle contraction. Nonsense mutations in the dystrophin gene cause the skeletal muscle in patients with DMD to be significantly weaker and lead to a state of constant degeneration and regeneration [8]. In order to offset muscle loss, quiescent MPCs activate and proliferate to regenerate muscle. Through asymmetric differentiation, some of the proliferating MPCs give rise to both MPCs and myoblasts. These myoblasts fuse with one another to form new myofibers, while the newly formed MPCs return to quiescence. Through chronic activation, the MPC pool becomes depleted and muscle is taken over by fibrotic tissue, fat infiltration, and ectopic calcification as illustrated in Fig. 4.1.

The typical clinical manifestations of DMD occur as patients reach 3–5 years of age, presenting with progressive motor skill deficiencies and gait abnormalities related to muscle weakness. Loss of ambulation occurs between 7 and 11 years of age, confining patients to wheelchair assistance for the remainder of their lives [9]. In addition to skeletal muscle pathology, patients with DMD also present with a collection of extramuscular defects such as osteopenia, fragility fractures, mental deficits, and scoliosis [9, 10]. Limited research has attributed osteopenia to a lack of muscular force on the skeleton resulting in decreased bone mineral density; however, these extramuscular aspects of DMD have not been studied extensively [11–13]. Cardiorespiratory failure is the typical cause of death in the second or third decade of life.

As the underlying cause of DMD is genetic, gene therapy studies to restore dystrophin expression or exon-skipping techniques to alter the open reading frame have been examined at the preclinical and clinical stages [14, 15]. Gene therapy for DMD has been hindered by several factors, including the large size of the dystrophin gene and the currently available viral vectors which can cause oncogenic mutations and trigger an immune response that leads to poor expression of dystrophin in the myofibers (muscle fiber units). In addition, delayed restoration of dystrophin expression may induce an autoimmune response requiring immunosuppression, which has many negative side effects.

Constant degeneration and regeneration of the muscle fibers via MPC proliferation and differentiation allow the DMD patient to maintain effective musculoskeletal function in the absence of dystrophin in the first stage of the disease. However, the repeated cycles of satellite cell activation may lead to telomere shortening or changes in myogenesis and/or cell senescence leading to MPC depletion [16, 17]. In a genetically identical mouse model of DMD (the *mdx* mouse), the satellite cell pool was shown to decrease threefold compared with age-matched, wild-type animals [18]. Similarly, proliferation and differentiation defects were observed in muscle-derived stem cells in the dystrophic animals [19]. Hence, the reintroduction of autologous myogenic stem cells could repair tissue damage and improve muscle function and longevity without the disadvantages of current gene therapy approaches.

The double-knockout mouse (dKO) is another experimental model of DMD that more closely recapitulates the clinical phenotype of DMD compared with the *mdx* model, which has a relatively mild phenotype [20, 21]; however, in contrast to patients with DMD, the dKO model is deficient in both dystrophin and utrophin. Many of the histopathological changes present in the damaged muscle of patients with DMD, such as fatty infiltration, calcium deposition, and fibrosis, are also observed in the dKO model starting as early as 5 days of age. As the muscle pathology worsens over the course of 6–8 weeks, there is a significant decline in the dKO stem cell pool as well as the ability of MPCs to proliferate and differentiate in vitro [19]. In tandem, the regenerative capacity of the dKO skeletal muscle also declines as lymphocytic infiltration and necrosis can be seen in the dystrophic skeletal muscle [19].

In order to study the effects of increased telomere shortening on skeletal muscle pathology in dystrophic mice, the *mdx*/mTR mouse was engineered by knocking out telomerase function in the MPCs of an *mdx* mouse. As is observed in clinical DMD, the *mdx*/mTR model exhibits elevated levels of creatine kinase, progressive loss of skeletal muscle volume, as well as an increased susceptibility to muscle fatigue before 8 weeks of age [22]. With regard to calcium deposition and fibrosis, skeletal muscle disease progression in the *mdx*/mTR mouse also displays clinical manifestations resembling DMD, which is in contrast to the relatively mild phenotype of the *mdx* model. Extramuscularly, the *mdx*/mTR exhibits skeletal morphometric abnormalities such as the development of premature kyphosis. In vivo BrdU proliferation assays and in vitro cell culture studies have also shown that MPCs in the *mdx*/mTR mice display progressive defects in replication as the mice age, suggesting a depletion of the replicative capacity of the progenitor cells [22]. Additionally, isolated stem cells from *mdx*/mTR exhibit a reduced capacity for muscle regeneration after injury [22]. Reduction in muscle stem cell telomere length as well as decreased telomere stability in the *mdx*/mTR mice strongly supports the theory of stem cell depletion and is responsible, at least in part, for the dystrophic phenotype [22].

In the following section, we will discuss experimental evidence supporting the use of stem cell transplantation to treat DMD. We will conclude with an in-depth examination of the molecular dysregulation of the dystrophic muscle environment and the challenges and opportunities for stem cell therapy that arise as a result.

4.2   Stem Cell Transplantation for the Treatment of DMD

Stem cell transplantation is a specific form of cell therapy where isolated progenitor cells, derived from autologous (the individual) or allogeneic (same species) sources, are administered to repair tissues or treat diseases. Bone marrow transplantation is a classic example of stem cell transplantation and is commonly used clinically to treat a variety of hematopoietic diseases. In degenerative muscle disease, the goal of stem cell transplantation is to restore musculoskeletal function and integrity through muscle regeneration. For long-term success, stem cell transplants must establish a quiescent pool of progenitor cells for future maintenance of muscle regenerative capacity. In order to achieve both regeneration and a quiescent niche, stem cells must survive implantation, proliferate, undergo myogenesis, and maintain a quiescent cell population in the long-term.

For many years, satellite cells were considered to be the only muscle progenitor cells and were extensively studied for use in stem cell transplantation applications. Satellite cells were first identified by Alexander Mauro in 1961 as mononuclear cells located between the basal lamina and the sarcoplasmic membrane [23]. Satellite cells are present in a quiescent state in healthy adult muscle and represent 2.5–6 % of nuclei of a given myofiber [24]. In the mouse, the transcription factor paired box 7 (Pax7) is required for satellite cell specification and survival and is a commonly used satellite cell marker [25, 26]. In addition, Myf5, MyoD, and Pax3 can be used to isolate or track satellite cells in skeletal muscle [27, 28]. Satellite cells become activated in response to several cues including HGF, FGF, IGF, and NO following myofiber injury [29–32]. Activated satellite cells, also referred to as myoblasts, proliferate and fuse to regenerate myofibers. The Pax7$^+$MyoD$^-$ subset of cells return to quiescence to maintain the satellite cell pool in a process known as asymmetric differentiation [33].

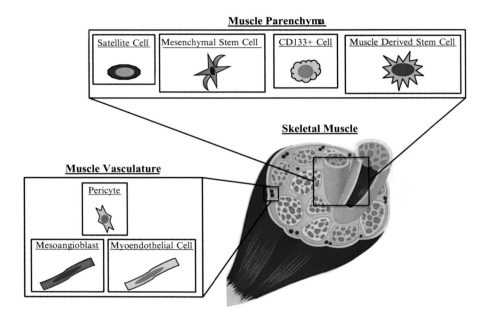

Fig. 4.2 Muscle progenitor cells derived from muscle parenchyma or associated vasculature represent potential sources of stem cells for cell therapy for the treatment of degenerative muscle diseases

More recently, multi-lineage stem cells capable of myogenic differentiation have been identified, as shown in Fig. 4.2. Among these are muscle-derived stem cells (MDSCs), CD133+ progenitor cells, and mesenchymal stem cells. These multi-lineage stem cells have shown extensive myogenic differentiation in controlled experiments and are hypothesized to play a role in muscle turnover during normal physiological conditions [34–36]. In addition to resident MPCs, there exist progenitor cells closely associated with muscle vasculature that also demonstrate myogenic ability, including mesoangioblasts, myoendothelial cells, and pericytes. Mesoangioblasts are located in the dorsal aorta of mammalian species [37]. Myoendothelial cells are associated with blood vessels and express surface markers of both myogenic and endothelial cell types [38, 39]. Pericytes are associated with capillaries, where they modulate vessel permeability [2, 40]. Mesoangioblasts, myoendothelial cells, and pericytes have all been documented to promote muscle regeneration in vivo [41].

Finally, advances in the stem cell field over the past several years have harnessed the power of pluripotent stem cells for cell therapy. Pluripotent stem cells give rise to all three germ layers (endoderm, mesoderm, ectoderm) when coaxed to differentiate under appropriate culture conditions. Embryonic stem cells (ESCs) are pluripotent cells derived from the inner-cell mass of a blastocyst. Protocols to induce the differentiation of ESCs into skeletal myoblasts and other MPCs have been developed in an effort to treat muscle disease [42, 43]. While the generation of new

ESC cell lines is heavily restricted, existing lines have been studied extensively in muscle regeneration. The heavy restrictions placed on ESC use prompted the ingenious induction of pluripotent stem cells from terminally differentiated somatic cells. Induced pluripotent stem cells (iPSCs) were developed in 2006 by reprogramming skin fibroblasts by activation of the Yamanaka genes: Oct3/4, Sox2, c-Myc, and Klf4 [44]. iPSCs offer the versatility of pluripotent differentiation without the ethical disadvantages associated with ESC isolation. These iPSCs have also been used to treat muscle disease by differentiation into various MPC lineages [45, 46].

In the following sections, we describe the critical experiments whereby satellite cells, multi-lineage progenitor cells, or pluripotent stem cells were transplanted to ameliorate symptoms of DMD. While the list of MPCs is not exhaustive, it includes the most well-studied adult stem/progenitor cell sources at the clinical and preclinical stages. Comparing cell sources can be challenging since each research group used different animal models and/or isolation and administration protocols; however, there are very clear advantages and disadvantages of the various cell sources that will be highlighted throughout the discussion [47]. While clinical trials that utilize stem cell transplantation have seen only marginal success in the past, new technologies are on the horizon which offer hope for the treatment of degenerative muscle diseases through the use of cell therapy.

4.2.1   Satellite Cells/Myoblasts

Given that satellite cells represent the major regenerative cell population in adult skeletal muscle, they have been the most extensively studied for use in stem cell transplantation both preclinically and clinically [41, 48]. Transplantation of satellite cells/myoblasts into the skeletal muscle of *mdx* mice restores dystrophin and forms new myofibers while producing a reserve pool of undifferentiated satellite cells within the host muscle [18, 49–51].

Initial attempts at cellular therapy for the treatment of DMD focused on the administration of dystrophin-positive satellite cells or myoblasts from allogeneic sources combined with immunosuppression. Several clinical trials were performed in the early 1990s but were generally met with disappointing results including low levels of engraftment and/or insignificant improvements in muscle function [52–55]. Further in vitro examination of myoblasts found that passaging the cells on tissue culture plastic significantly reduced their myogenic capabilities [56]. In order to obtain a sufficient number of myoblasts for transplantation, they require extensive expansion which may influence the efficacy of the transplanted cells. Recently, Blau's group found that culturing satellite cells on elastic hydrogels achieved a fourfold increase in self-renewal capacity and engraftment up to 1 week, which could improve the feasibility of myoblast expansion prior to transplantation [57].

Although satellite cells are the classic myogenic stem cell, the transplantation of other MPC populations may also lead to improved clinical outcomes for numerous reasons including increased myogenic potential, ability to maintain a quiescent cell pool upon administration, increased tissue engraftment, and paracrine signaling effects. The following is a discussion of promising results from other MPC types currently under investigation for DMD therapy.

4.2.2   Muscle-Derived Stem Cells

Our laboratory has isolated and described a population of muscle-derived stem cells (MDSC) isolated from murine skeletal muscle using a technique that takes advantage of the specific adherence properties of the different cell types, also known as the preplate technique. While fibroblasts and myoblasts rapidly adhere to collagen-coated flasks, the MDSCs are considered slowly adhering cells [47]. MDSCs are known as a primitive muscle progenitor cell population that gives rise to the satellite cell population; therefore, MDSCs play a role in muscle homeostasis and repair. Although the exact origin of mouse MDSCs remains to be determined, these cells express endothelial factors, such as von Willebrand factor, that can spontaneously form blood vessels after transplantation into skeletal muscle. This phenomenon is likely a result of their VEGF expression, suggesting that the actions of MDSCs are closely related to those of endothelial cells [58]. In addition, the cells display stem cell characteristics such as long-term proliferation, high self-renewal, and multi-lineage differentiation capacities [5].

Unlike satellite cells, MDSCs possess a noncommitment cell marker profile resembling adult stem cells; yet the cells possess an inherent myogenic potential marked by their expression of desmin and MyoD [5]. Furthermore, MDSC transplantation offers many more benefits than myoblast transplantation with regard to newly formed myofiber quantity and engraftment in vivo, likely due to an increased resistance to oxidative stress that results in superior cell survival [34, 59].

Several studies have proposed that MDSCs work via a "bystander effect" whereby they secrete paracrine factors that can modulate the microenvironment and promote muscle regeneration [60, 61]. A recent study using an accelerated aging mouse model (*progeria*) found that wild-type MDSC transplantation into progeroid mice improved their lifespan compared with PBS-injected controls. Further studies showed that age-related stem cell defects in MDSCs could be rescued by coculturing the cells with young wild-type MDSCs, which supports the existence of rejuvenating paracrine factors [62, 63]. In the context of natural aging, GDF-11 has been shown to be one of these rejuvenating paracrine factors that exert anti-aging effects in skeletal and cardiac muscle [63, 64],

While MDSCs have not yet moved to clinical trials for degenerative muscle disease, human MDSC-like cells have been studied in phase I and II clinical trials for the treatment of stress urinary incontinence (SUI). After 1 year, improvement of SUI was seen in five out of eight subjects, where one subject achieved total continence [65]. These improvements occurred a median of 10 months post-treatment, with no serious adverse effects reported. The success of MDSCs in clinical trials for SUI will aid in expediting the use of MDSCs for the treatment of other disorders including degenerative muscle diseases.

4.2.3 Myoendothelial Cells and Pericytes

Interestingly, the cells that exist along the walls of the blood vessels within human skeletal muscle, including endothelial cells, myoendothelial cells (MECs), and pericytes demonstrate high myogenic potentials, much like murine MDSCs [38–40, 61]. Myoendothelial cells represent less than 0.5 % of cells in degenerative adult skeletal muscle and express both endothelial (CD34, von Willebrand factor, VE-cadherin) and myogenic (CD56, Pax7) markers [38, 39, 66]. Transplantation of myoendothelial cells was recently shown to be superior to myoblast transplantation in DMD using an *mdx*/SCID model (mice that are both dystrophin deficient and immunocompromised) [6]. Human myoendothelial cell transplantation was also found to reduce muscle atrophy associated with sciatic nerve injury [67]. These findings suggest that MECs may be closely related to murine MDSCs [68].

Studies have demonstrated that pericytes, the contractile cells providing structural support to blood vessels, may contribute to muscle formation during development [2]. Isolated from human skeletal muscle, these pericyte-derived progenitor cells (PDPCs) display a cell marker profile consistent with pericytes (annexin V, vimentin, smooth muscle actin, desmin, PDFGFR-beta, alkaline phosphatase). However, PDPCs inherently lack myogenic cell markers (MyoD, myogenin, Myf5), providing evidence for a non-muscle origin [2]. PDPCs can undergo myogenic differentiation in vitro when cocultured with murine myogenic cells or in muscle-differentiation medium [2]. Although PDPCs display a capacity to rejuvenate the muscle progenitor pool in vivo (similar to MDSCs), limitations regarding poor migration, engraftment, and differentiation capabilities still persist [2, 40].

A recent study directly compared the regenerative capability of endothelial cells, myoendothelial cells, pericytes, and myoblasts isolated from human cryopreserved skeletal muscle by fluorescence-activated cell sorting (FACS). The MECs exhibited the highest regenerative capacity after transplantation into injured mouse skeletal muscle, followed by pericytes, which were both superior to myoblast transplantation [69]. Therefore, pericytes may be most efficient at promoting angiogenesis which may also improve the muscle remodeling process. Therefore, human myoendothelial cells and pericytes offer great hope for improving the success of the next generation of myogenic stem cell transplantation protocols.

4.2.4 Mesoangioblasts

Mesoangioblasts are another population of human vessel-derived progenitor cells that express endothelial markers in the embryonic stage and pericyte markers in postnatal tissue [70]. In addition, mesoangioblasts can cross vessel walls, making them an intriguing candidate for systemic transplantation delivery. Wild-type or genetically-corrected mesoangioblasts delivered intra-arterially (IA) into dystrophic α-sarcoglycan (SCGA) mutant mice significantly improved the functional outcome of the mice [71]. Furthermore, canine mesoangioblasts transduced to express micro-dystrophin were delivered IA into the golden retriever muscular dystrophy (GRMD) canine model, which resulted in widespread expression of dystrophin and significantly improved muscle morphology and function [72]. These studies have led to an ongoing phase I clinical trial in Europe with mesoangioblast allotransplantation in patients with DMD [73].

Because DMD is a disease that affects both skeletal muscle and cardiac muscle, the versatility of mesoangioblasts is of great benefit when developing potential cell-based therapies. Aorta-derived mesoangioblasts (ADMs) expressing cardiomyocyte markers in vitro possess the ability to delay dilated cardiomyopathy formation in dKO mice upon transplantation as determined by heart function, interventricular septal hypertrophy, and ventricular size [74]. Furthermore, ADM transplantation is capable of increasing cardiac angiogenesis alongside the restoration of dystrophin expression in dKO heart tissue, thus supporting the pursuit of mesoangioblast-based cell therapy for the treatment of DMD [74].

In addition to their roles in skeletal and cardiac muscle regeneration, mesoangioblasts also have an intrinsic anti-inflammatory function. Even after upregulation of human leukocyte antigen (HLA) surface proteins, human mesoangioblasts fail to stimulate T-cell proliferation in vitro which implies a lack of immunogenicity [75]. Furthermore, human mesoangioblasts possess the ability to hinder T-cell proliferation and inflammatory cytokine secretion in vitro. Interestingly, mesoangioblast transplantation was recently combined with tissue engineering strategies to treat acute and chronic muscle degeneration [76]. Consequently, mesoangioblasts are an exciting potential candidate for therapeutic stem cell transplantation.

4.2.5   CD133+ Cells

In humans, a subpopulation of blood cells that express CD133 and exhibit myogenic properties when cocultured with myoblasts has been reported [77]. Recently, lentivirus-mediated gene-skipping studies have restored a truncated-yet-functional dystrophin protein in dystrophic CD133+ stem cells. These CD133+ cells were delivered via intramuscular or IA injection into *mdx*/SCID mice, which resulted in significant dystrophin restoration as well as improvement in muscle function and morphology [35]. In addition, autologous CD133+ cells derived from bone marrow induced angiogenesis and myocardial tissue survival following infarction [78]. Clinical trials utilizing these cells have also been initiated for the treatment of DMD [79].

The phase I clinical trial demonstrated that CD133+ cell transplantation was safe, but ineffective in patients with DMD. Autologous muscle biopsies were obtained from 5 patients with DMD enrolled in the trial and CD133+ cells were isolated from each biopsy. Three parallel injections of CD133+ stem cells were performed in the left abductor digiti minimi muscle and muscle function was assessed over 7 months. Interestingly, four of the five patients showed increased numbers of capillaries per myofiber, which may be related to paracrine signaling from the CD133+ stem cells [79].

4.2.6   Mesenchymal Stem Cells

Mesenchymal stem cells (MSCs)are adherent, non-hematopoietic stem cells that express CD90, CD105, and CD73, are negative for CD14, CD34, and CD45 surface markers, and can be isolated from bone marrow or other sources such as adipose tissue [80]. Functionally, MSCs are defined by their ability to differentiate into adipocytes, chondrocytes, myocytes, and osteocytes when appropriate differentiation protocols are followed [81]; however, MSCs have also been reported to differentiate into neurons, hepatocytes, and muscle cells [82–85].

MSCs have two distinct advantages when considering degenerative muscle disease. MSCs have been reported to selectively localize to injured tissues and differentiate into appropriate progeny without the need for ex vivo reprogramming, as observed in myocardium and bone defects [86, 87]. Furthermore, MSCs have well-characterized anti-inflammatory properties and have been shown to modulate chronic inflammation in several diseases including autoimmune arthritis, type I diabetes, and multiple sclerosis [88–90]. The anti-inflammatory mechanism of MSCs includes the suppression of macrophage activation, inhibition of cytotoxic natural killer cell and T-cell function, and regulation of T-cell production [91–93]. The paracrine effects of MSCs include the inhibition of apoptosis, stimulation of endogenous cell proliferation, and activation of resident tissue stem cells [94]. For these reasons, MSCs have been explored as a potential cell source for the treatment of skeletal muscle disease.

MSCs have been examined for their myogenic regeneration capacity in *mdx* mice. By inducing Pax3 gene expression, the MSCs engrafted and formed myofibers in the dystrophic muscle; however, no functional recovery was observed [95]. This discrepancy may have resulted from use of the *mdx* model, which exhibits a less-severe phenotype compared to human DMD as described above.

Recently, studies have explored the use of bone marrow-derived mesenchymal stem cells enriched for CD271 as a potential cell type for the treatment of dystrophic muscles in GRMD dogs [36]. Canine MSCs transduced with MyoD demonstrated increased myogenic differentiation capacity in vitro as well as in vivo after allogeneic transplantation [36]. Furthermore, MSC transplantation displayed long-term engraftment and favorable cell migration properties upon administration. In GRMD dystrophic skeletal muscle, transduced MSCs also restored dystrophin expression while also exhibiting the above transplant-related properties, such as the absence of imunogenicity [36]. Despite the lack of intrinsic myogenic potential, MSCs are a potential cell source for DMD treatment in the future [36].

4.2.7   Embryonic Stem Cells

The above-mentioned multipotent progenitor cells have a variety of benefits in the context of DMD therapy; however, drawbacks related to the use of adult myogenic stem cells may include their limited self-renewal capacity, senescent tendencies, and limited pluripotency [96]. Genetically engineered Pax7-inducible embryonic stem cells (ESCs) have been shown to display robust myogenic differentiation in vitro as well as the ability to regenerate skeletal muscle in vivo [96]. Two separate groups have been successful at deriving myoblasts and satellite cells from human embryonic stem cells [42, 97]. Upon transplantation, the ESC-derived satellite cells regenerated myofibers in the *mdx* model [97]. Not only do ESCs contribute significantly to new myofiber formation in vivo, but ESCs also contribute to the pool of quiescent host stem cells [96]. These results suggest that ESCs may be a promising potential cell source for the treatment of DMD; however, several drawbacks of ESC therapy-based studies include the ethical dilemma surrounding their use, immunogenicity, and the potential for tumor formation once ESCs are transplanted into adult tissue.

In 2013, a human embryonic stem cell line derived from DMD was approved for NIH funding in the United States [98]. Sixteen newly approved lines also include cells that carry genes for other hereditary disorders, such as spinal muscular atrophy, myotonic dystrophy, and neurofibromatosis. These ESCs may offer an alternative strategy to murine modeling of disease and may lead to exciting findings in high-throughput drug screening that will likely affect how degenerative muscle diseases are treated in the future.

4.2.8   Induced Pluripotent Stem Cells

iPSCs offer the versatility of embryonic stem cells without the ethical and medicolegal issues associated with ESC isolation. However, iPSC technology is still in its infancy due to the limited efficiency of cell reprogramming, the potential immunogenicity of cell transplants, and potential for oncogenic transformation. Nonetheless, several protocols have been established that are relevant to degenerative muscle disease [99].

Through the introduction of specific reprogramming genes into dKO fibroblasts, dKO iPSCs can be generated quite efficiently. These dKO iPSCs transduced with micro-utrophin can also be engineered to conditionally express Pax3 with doxycycline stimulation in order to stimulate myogenic potential [100]. Similar to Pax7/3-induced ESCs, gene-corrected micro-utrophin dKO iPSCs possess high myogenic potential (both in vitro and in vivo) as well as a high capacity for regeneration [100]. Systemic transplantation of iPSCs demonstrates skeletal muscle engraftment results comparable to those of ESC transplantation in that the iPSCs can reach many muscles groups throughout the body that are distant from the site of injection. Moreover, the role of iPSCs may not be restricted to to the treatment of skeletal muscle in DMD. Recent evidence suggests that iPSC-mediated systemic delivery of micro-dystrophin in the dKO model may also correct pathologies related to fat/lipid metabolism [101]. Additionally, a protocol to produce mesoangioblasts from human iPSCs was recently described which could provide a new source of mesoangioblasts for subsequent transplantation studies [102].

4.2.9   Acellular/Cellular Combination Strategies

The collective efforts of researchers involved with the development of cell-based therapies for the treatment of DMD are truly admirable; however, it is possible that combining these efforts with acellular therapies may prove even more useful in the future. Recent studies that explored the molecular mechanisms of aging in MDSCs have found that pharmacological blocking of the p38α/β mitogen-activating protein kinase (MAPK) pathway prevented age-related reduction in MPC proliferation [103]. Furthermore, blocking this pathway restored the engraftment capacity of previously defective aged MPCs with respect to graft survival and regenerative capacity within injured host skeletal muscle [103]. Of note, blocking p38α/β also allowed for the long-term rejuvenation of the MPC pool in vivo, which is a desirable characteristic of any cell-based DMD therapy (Table 4.1) [103].

4.3   Interaction between Progenitor Cells and the Surrounding Microenvironment

In order to design effective cellular therapies for the treatment of degenerative muscle diseases, it is necessary to have a thorough understanding of the cellular and molecular environment that exists within the degenerating muscle tissue. In DMD, loss of the functional dystrophin-glycoprotein complex leads to sarcomere rupture and the release of cytoplasmic proteins, ions, and nuclear material into the extracellular space. Abnormal skeletal muscle physiology also results in calcium release-activated calcium channel protein 1 (Orai1) dysregulation, which has been shown to accelerate myofiber breakdown in DMD [104–106]. In response to tissue breakdown, inflammatory cell infiltrates, cytokines, and activated complement create an inflammatory environment to clear the cellular debris [80, 107]. MPCs are then activated primarily by IGF-1, a cytokine produced by invading macrophages and endothelial cells [108, 109]. The MPCs proliferate and fuse with existing fibers (or themselves) to create new myofibers. Furthermore, there is an upregulation of integrin-like kinase (ILK) and cytoskeleton-associated proteins such as vimentin in order to compensate for the lack of dystrophin within the myofibers [105]. These myofibers also downregulate and upregulate the calcium transporters SERCA1 and NCX3, respectively, which has been theorized as a compensatory mechanism to restore calcium homeostasis in patients with DMD [110]. Despite the many mechanisms set in place to curb the evolution of muscle pathology, the absence of dystrophin still leads to eventual disruption of the newly formed fibers.

Increased metabolic/oxidative stress may also contribute to the deterioration of MPC function in DMD [104]. Recently, defects in autophagy in the *mdx* mouse model mediated by mTOR and Src-kinase have been shown to contribute to this oxidative stress [104]. Furthermore, the *mdx* mouse model exhibits generalized mitochondrial dysfunction as evidenced by the reduction of protein expression

Table 4.1 Comparison of muscle progenitor cells (MPCs)

MPC	Anatomical location	Advantages	Disadvantages	DMD clinical trials
Satellite cells/ myoblasts	Sarcoplasmic membrane	Unipotent muscle progenitor Establish quiescent pool	Limited expansion Poor engraftment	Phase II
MDSCs	Unknown	Multipotent muscle progenitor Increased expansion and survival Angiogenesis	Rare population	N/A
MECs	Vasculature	More effective muscle regeneration than myoblasts	Rare population	N/A
Pericytes	Vasculature	More effective muscle regeneration than myoblasts	Lack myogenic markers Limited migration and engraftment Senescence	N/A
Mesoangioblasts	Vasculature	Systemic delivery Anti-inflammatory	Rare population	Phase I
CD133+	Peripheral blood and bone marrow	Easily isolated Promote angiogenesis	Limited restoration of dystrophin	Phase I
MSCs	Mesenchymal tissue and bone marrow	Migrate to injured tissue Anti-inflammatory	Require myogenic gene activation	N/A
ESCs	Blastocyst	Pluripotent	Require myogenic gene activation Ethical implications Teratoma formation	N/A
iPSCs	Reprogrammed terminally differentiated cells	Pluripotent Derived from readily available cell source Withstand oxidative stress	Require myogenic gene activation Oncogenic transformation Low efficiency of reprogramming	N/A

and activity within their mitochondria [105, 111]. Upregulation of reactive oxide-producing enzymes, such as NADPH oxidase, have also been shown to contribute to the muscle degeneration observed in DMD patients [112]. These studies suggest that, in DMD, there exists an ongoing process of metabolic dysregulation which leads to a reduced capacity for aerobic energy production and an excessive production of harmful oxidizing molecules and free radicals. These deficits in metabolic and mitochondrial function in the MPC microenvironment may also contribute to MPC depletion. For these reasons, robust cells capable of withstanding a higher threshold of oxidative stress could represent suitable targets for cell therapy [113].

The onset and progression DMD are associated with the upregulation of inflammatory genes, including cytokines such as TNF-α, IL-1, and IL-4 [114–116]. The nuclear factor kappa B (NF-κB) pathway is strongly implicated among these inflammatory genes. Inhibition of the NF-κB pathway in the *mdx* model results in a therapeutic benefit by decreasing macrophage infiltration and promoting myogenesis [117, 118]; however, the complete depletion of inflammatory cells negatively impacts muscle regeneration [119]. In fact, corticosteroids are commonly prescribed for the treatment of patients with DMD due to their immunosuppressive effects that inhibit long-term muscle degeneration through the reduction of muscle inflammation [120, 121]. Long-term use of corticosteroids, however, also causes negative side effects such as increased appetite, weight gain, loss of bone mass, pathologic fractures, and cataracts [122]. Therefore, an optimal strategy for stem cell therapy in the context of this harsh, pro-inflammatory environment must involve the use of cells that are able to modulate the immune response to prevent chronic inflammation.

Chronic inflammation not only results in monocytic and lymphocytic infiltration of musculoskeletal tissues, but it also severely alters the tissue architecture and microenvironment, both of which hinder healthy muscle regeneration. For example, chronic inflammation has key roles in the induction of fibrosis formation and the dysregulation of adult stem cell compartments, thus leading to ectopic calcification and fat deposition. Chronic inflammation leads to the activation of TGF-β via the renin-angiotensin system, which further promotes the deposition of extracellular matrix (ECM) by myofibroblasts and ultimately replaces functional muscle with scar tissue [123]. In addition, the expression levels of matrix metalloproteinases (MMPs) and their inhibitors (i.e., TIMPs) are skewed in that TIMPs are more highly expressed, leading to an accumulation of connective tissue. The expression of MMPs in *mdx* mice has a potential therapeutic benefit on dystrophic muscle by decreasing the synthesis of new ECM and limiting fibrosis [124]. In addition, patients with DMD often present with heterotopic ossification (HO) and intramyocellular lipid accumulation (IMCL), suggesting that there exists a dysregulation in the pathways for stem cell differentiation [125].

HO is the formation of ectopic pathological calcification within muscle tissue, whereas IMCL arises as a result of dysregulated fatty acid metabolism [126]. Although seemingly unrelated, the aforementioned "side effects" and reduced myogenic potential of underperforming muscle progenitor cells may be partially mediated by the activation of RhoA, a GTPase regulator of the actin cytoskeleton.

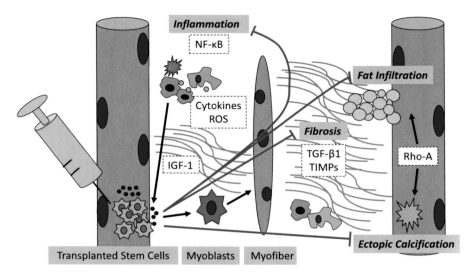

Fig. 4.3 Schematic representation of late-stage DMD muscle with molecular factors regulating inflammation, fibrosis, and fat and bone formation. Transplanted stem cells with paracrine signaling capabilities can modulate the microenvironment and self-facilitate improvements in survival, engraftment, and tissue regeneration capacities

Recent studies have shown that HO and IMCL levels in 8-week-old dKO murine muscle are associated directly with the degree of local myofiber necrosis and inflammation [125]. RhoA is upregulated in HO-containing skeletal muscle with additional upregulation of the inflammatory factors TNF-α and osteogenesis-related BMPs. Strikingly, inactivation of RhoA results in a significant decrease in HO which implies a causal relationship [125]. Additionally, RhoA inactivation lowers IMCL in dystrophic skeletal muscle and increases the myogenic potential of the stem cell pool [125]. Thus, manipulation of RhoA levels may provide an alternate avenue for the treatment of these manifestations observed in DMD patients.

The skeletal muscle environment continuously worsens with the progression of DMD and is characterized by inflammation, fibrosis, dysregulation of calcium homeostasis, metabolic dysfunction, increased oxidative stress, and ectopic bone and fat formation (schematically represented in Fig. 4.3). Thus, exhaustion of the stem cell pool is only one of the challenges we face as we continue to develop new strategies for the treatment of DMD. Transplanted stem cells must not only form new myofibers, but they must also survive in a hostile inflammatory microenvironment and engraft within tissues that contain mixed signaling cues. While stem cells are classically thought to directly contribute to new cell formation, an emerging role for the importance of paracrine signaling has surfaced in the literature as of late. As new information emerges regarding the systemic barriers to muscle/MPC health, such as metabolic and calcium-related dysfunction, paracrine-mediated cell therapies are of significant appeal. An ideal role for these stem cells is to secrete paracrine factors that modulate the microenvironment by regulating the expression of molecules such as NF-kB, RhoA, TGF-β, Src-kinase, mTOR, ILK,

Orai1, and MMPs in order to reestablish tissue homeostasis. Several groups have documented the paracrine effects of MPCs. MSCs and mesoangioblasts have displayed anti-inflammatory effects while also modulating the microenvironment and promoting cell engraftment. It has also been described that MDSCs can secrete rejuvenating factors that influence endogenous progenitor cells. Consideration of the microenvironment and the numerous advantages of paracrine signaling will allow for the success of the next generation of cell therapies for DMD.

4.4 Summary

In degenerative muscle diseases such as DMD, constant activation of MPCs ultimately leads to stem cell depletion and a dramatic decline in muscle repair capacity and function. Stem cell transplantation offers hope in the battle against degenerative muscle diseases through replenishment of the depleted pool of MPCs. Over the past 25 years, many clinical and preclinical studies have attempted to treat muscular dystrophy using satellite cells/myoblasts, MDSCs, myoendothelial cells, pericytes, mesoangioblasts, CD133+ cells, mesenchymal stem cells, ESCs, and iPSCs. Consideration of the interaction between transplanted stem cells and the hostile microenvironment characteristic of degenerative muscle tissue is crucial to the success of any future therapeutic strategies for DMD. This interaction can be modulated using pharmacological agents or intrinsic paracrine signaling pathways that regulate stem cell function.

The MPCs outlined in this chapter have intrinsic advantages and disadvantages with regard to cell-based therapeutics for the treatment of degenerative muscle diseases. In general, the ideal stem cell therapy would include the following characteristics: (1) isolation from an easily accessible source, such as blood, muscle, or bone marrow, (2) expansion in vitro without compromise of myogenic ability or induction of early senescence, (3) efficient transduction capacity for genetic corrections, (4) global muscle engraftment upon administration, and (5) long-term in vivo survival in the absence of immunosuppression. Although there are no curative, long-term therapies for DMD at the time of this writing, there has been an influx of new cell sources and genetic engineering techniques in addition to an improved understanding of the characteristic pro-inflammatory microenvironment in recent years. With this gain in knowledge and technology, the future of stem cell transplantation for the treatment of degenerative muscle diseases appears promising.

References

1. Dellavalle A, Maroli G, Covarello D, Azzoni E, Innocenzi A, Perani L, et al. Pericytes resident in postnatal skeletal muscle differentiate into muscle fibres and generate satellite cells. Nat Commun. 2011;2:499.

2. Dellavalle A, Sampaolesi M, Tonlorenzi R, Tagliafico E, Sacchetti B, Perani L, et al. Pericytes of human skeletal muscle are myogenic precursors distinct from satellite cells. Nat Cell Biol. 2007;9(3):255–67.

3. Mitchell KJ, Pannerec A, Cadot B, Parlakian A, Besson V, Gomes ER, et al. Identification and characterization of a non-satellite cell muscle resident progenitor during postnatal development. Nat Cell Biol. 2010;12(3):257–66.

4. Asakura A, Seale P, Girgis-Gabardo A, Rudnicki MA. Myogenic specification of side population cells in skeletal muscle. J Cell Biol. 2002;159(1):123–34.

5. Qu-Petersen Z, Deasy B, Jankowski R, Ikezawa M, Cummins J, Pruchnic R, et al. Identification of a novel population of muscle stem cells in mice: potential for muscle regeneration. J Cell Biol. 2002;157(5):851–64.

6. Chirieleison SM, Feduska JM, Schugar RC, Askew Y, Deasy BM. Human muscle-derived cell populations isolated by differential adhesion rates: phenotype and contribution to skeletal muscle regeneration in Mdx/SCID mice. Tissue Eng A. 2012;18(3–4):232–41.

7. Hoffman EP, Brown Jr RH, Kunkel LM. Dystrophin: the protein product of the duchenne muscular dystrophy locus. Cell. 1987;51(6):919–28.

8. Zubrzyckagaarn EF, Bulman DE, Karpati G, Burghes AHM, Belfall B, Klamut HJ, et al. The Duchenne muscular-dystrophy gene-product is localized in sarcolemma of human skeletal-muscle. Nature. 1988;333(6172):466–9.

9. Parker AE, Robb SA, Chambers J, Davidson AC, Evans K, O'Dowd J, et al. Analysis of an adult Duchenne muscular dystrophy population. QJM. 2005;98(10):729–36.

10. Larson CM, Henderson RC. Bone mineral density and fractures in boys with Duchenne muscular dystrophy. J Pediatr Orthop. 2000;20(1):71–4.

11. McKay H, Smith E. Winning the battle against childhood physical inactivity: the key to bone strength? J Bone Miner Res. 2008;23(7):980–5.

12. Ward K, Alsop C, Caulton J, Rubin C, Adams J, Mughal Z. Low magnitude mechanical loading is osteogenic in children with disabling conditions. J Bone Miner Res. 2004;19(3): 360–9.

13. Boot AM, de Ridder MA, Pols HA, Krenning EP, de Muinck Keizer-Schrama SM. Bone mineral density in children and adolescents: relation to puberty, calcium intake, and physical activity. J Clin Endocrinol Metab. 1997;82(1):57–62.

14. Romero NB, Braun S, Benveniste O, Leturcq F, Hogrel JY, Morris GE, et al. Phase I study of dystrophin plasmid-based gene therapy in Duchenne/Becker muscular dystrophy. Hum Gene Ther. 2004;15(11):1065–76.

15. van Deutekom JC, Janson AA, Ginjaar IB, Frankhuizen WS, Aartsma-Rus A, Bremmer-Bout M, et al. Local dystrophin restoration with antisense oligonucleotide PRO051. N Engl J Med. 2007;357(26):2677–86.

16. Decary S, Ben Hamida C, Mouly V, Barbet JP, Hentati F, Butler-Browne GS. Shorter telomeres in dystrophic muscle consistent with extensive regeneration in young children. Neuromuscul Disord. 2000;10(2):113–20.

17. Lund TC, Grange RW, Lowe DA. Telomere shortening in diaphragm and tibialis anterior muscles of aged mdx mice. Muscle Nerve. 2007;36(3):387–90.

18. Cerletti M, Jurga S, Witczak CA, Hirshman MF, Shadrach JL, Goodyear LJ, et al. Highly efficient, functional engraftment of skeletal muscle stem cells in dystrophic muscles. Cell. 2008;134(1):37–47.

19. Lu A, Poddar M, Tang Y, Proto JD, Sohn J, Mu X, et al. Rapid depletion of muscle progenitor cells in dystrophic mdx/utrophin−/− mice. Hum Mol Genet. 2014;23(18):4786–800.

20. Grady RM, Teng H, Nichol MC, Cunningham JC, Wilkinson RS, Sanes JR. Skeletal and cardiac myopathies in mice lacking utrophin and dystrophin: a model for Duchenne muscular dystrophy. Cell. 1997;90(4):729–38.

21. Deconinck AE, Rafael JA, Skinner JA, Brown SC, Potter AC, Metzinger L, et al. Utrophin-dystrophin-deficient mice as a model for Duchenne muscular dystrophy. Cell. 1997;90(4): 717–27.

22. Sacco A, Mourkioti F, Tran R, Choi J, Llewellyn M, Kraft P, et al. Short telomeres and stem cell exhaustion model Duchenne muscular dystrophy in mdx/mTR mice. Cell. 2010;143(7): 1059–71.
23. Mauro A. Satellite cell of skeletal muscle fibers. J Biophys Biochem Cytol. 1961;9:493–5.
24. Whalen RG, Harris JB, Butler-Browne GS, Sesodia S. Expression of myosin isoforms during notexin-induced regeneration of rat soleus muscles. Dev Biol. 1990;141(1):24–40.
25. Zammit PS, Relaix F, Nagata Y, Ruiz AP, Collins CA, Partridge TA, et al. Pax7 and myogenic progression in skeletal muscle satellite cells. J Cell Sci. 2006;119(Pt 9):1824–32.
26. Kuang S, Charge SB, Seale P, Huh M, Rudnicki MA. Distinct roles for Pax7 and Pax3 in adult regenerative myogenesis. J Cell Biol. 2006;172(1):103–13.
27. Tajbakhsh S, Bober E, Babinet C, Pournin S, Arnold H, Buckingham M. Gene targeting the myf-5 locus with nlacZ reveals expression of this myogenic factor in mature skeletal muscle fibres as well as early embryonic muscle. Dev Dyn. 1996;206(3):291–300.
28. Day K, Shefer G, Richardson JB, Enikolopov G, Yablonka-Reuveni Z. Nestin-GFP reporter expression defines the quiescent state of skeletal muscle satellite cells. Dev Biol. 2007;304(1): 246–59.
29. Tatsumi R, Anderson JE, Nevoret CJ, Halevy O, Allen RE. HGF/SF is present in normal adult skeletal muscle and is capable of activating satellite cells. Dev Biol. 1998;194(1):114–28.
30. Floss T, Arnold HH, Braun T. A role for FGF-6 in skeletal muscle regeneration. Genes Dev. 1997;11(16):2040–51.
31. Musaro A. Growth factor enhancement of muscle regeneration: a central role of IGF-1. Arch Ital Biol. 2005;143(3–4):243–8.
32. Wozniak AC, Anderson JE. Nitric oxide-dependence of satellite stem cell activation and quiescence on normal skeletal muscle fibers. Dev Dyn. 2007;236(1):240–50.
33. Zammit PS, Golding JP, Nagata Y, Hudon V, Partridge TA, Beauchamp JR. Muscle satellite cells adopt divergent fates: a mechanism for self-renewal? J Cell Biol. 2004;166(3):347–57.
34. Urish KL, Vella JB, Okada M, Deasy BM, Tobita K, Keller BB, et al. Antioxidant levels represent a major determinant in the regenerative capacity of muscle stem cells. Mol Biol Cell. 2009;20(1):509–20.
35. Benchaouir R, Meregalli M, Farini A, D'Antona G, Belicchi M, Goyenvalle A, et al. Restoration of human dystrophin following transplantation of exon-skipping-engineered DMD patient stem cells into dystrophic mice. Cell Stem Cell. 2007;1(6):646–57.
36. Nitahara-Kasahara Y, Hayashita-Kinoh H, Ohshima-Hosoyama S, Okada H, Wada-Maeda M, Nakamura A, et al. Long-term engraftment of multipotent mesenchymal stromal cells that differentiate to form myogenic cells in dogs with Duchenne muscular dystrophy. Mol Ther. 2012;20(1):168–77.
37. Cossu G, Bianco P. Mesoangioblasts--vascular progenitors for extravascular mesodermal tissues. Curr Opin Genet Dev. 2003;13(5):537–42.
38. Tavian M, Zheng B, Oberlin E, Crisan M, Sun B, Huard J, et al. The vascular wall as a source of stem cells. Ann N Y Acad Sci. 2005;1044:41–50.
39. Zheng B, Cao B, Crisan M, Sun B, Li G, Logar A, et al. Prospective identification of myogenic endothelial cells in human skeletal muscle. Nat Biotechnol. 2007;25(9):1025–34.
40. Crisan M, Yap S, Casteilla L, Chen CW, Corselli M, Park TS, et al. A perivascular origin for mesenchymal stem cells in multiple human organs. Cell Stem Cell. 2008;3(3):301–13.
41. Peault B, Rudnicki M, Torrente Y, Cossu G, Tremblay JP, Partridge T, et al. Stem and progenitor cells in skeletal muscle development, maintenance, and therapy. Mol Ther. 2007;15(5):867–77.
42. Barberi T, Bradbury M, Dincer Z, Panagiotakos G, Socci ND, Studer L. Derivation of engraftable skeletal myoblasts from human embryonic stem cells. Nat Med. 2007;13(5): 642–8.
43. Rohwedel J, Maltsev V, Bober E, Arnold HH, Hescheler J, Wobus AM. Muscle cell differentiation of embryonic stem cells reflects myogenesis in vivo: developmentally regulated expression of myogenic determination genes and functional expression of ionic currents. Dev Biol. 1994;164(1):87–101.

44. Takahashi K, Yamanaka S. Induction of pluripotent stem cells from mouse embryonic and adult fibroblast cultures by defined factors. Cell. 2006;126(4):663–76.
45. Darabi R, Arpke RW, Irion S, Dimos JT, Grskovic M, Kyba M, et al. Human ES- and iPS-derived myogenic progenitors restore DYSTROPHIN and improve contractility upon transplantation in dystrophic mice. Cell Stem Cell. 2012;10(5):610–9.
46. Filareto A, Parker S, Darabi R, Borges L, Iacovino M, Schaaf T, et al. An ex vivo gene therapy approach to treat muscular dystrophy using inducible pluripotent stem cells. Nat Commun. 2013;4:1549.
47. Gharaibeh B, Lu A, Tebbets J, Zheng B, Feduska J, Crisan M, et al. Isolation of a slowly adhering cell fraction containing stem cells from murine skeletal muscle by the preplate technique. Nat Protoc. 2008;3(9):1501–9.
48. Wagers AJ, Conboy IM. Cellular and molecular signatures of muscle regeneration: current concepts and controversies in adult myogenesis. Cell. 2005;122(5):659–67.
49. Collins CA, Olsen I, Zammit PS, Heslop L, Petrie A, Partridge TA, et al. Stem cell function, self-renewal, and behavioral heterogeneity of cells from the adult muscle satellite cell niche. Cell. 2005;122(2):289–301.
50. Sacco A, Doyonnas R, Kraft P, Vitorovic S, Blau HM. Self-renewal and expansion of single transplanted muscle stem cells. Nature. 2008;456(7221):502–6.
51. Partridge TA, Morgan JE, Coulton GR, Hoffman EP, Kunkel LM. Conversion of mdx myofibres from dystrophin-negative to -positive by injection of normal myoblasts. Nature. 1989;337(6203):176–9.
52. Huard J, Bouchard JP, Roy R, Malouin F, Dansereau G, Labrecque C, et al. Human myoblast transplantation: preliminary results of 4 cases. Muscle Nerve. 1992;15(5):550–60.
53. Mendell JR, Kissel JT, Amato AA, King W, Signore L, Prior TW, et al. Myoblast transfer in the treatment of Duchenne's muscular dystrophy. N Engl J Med. 1995;333(13):832–8.
54. Vilquin JT, Wagner E, Kinoshita I, Roy R, Tremblay JP. Successful histocompatible myoblast transplantation in dystrophin-deficient mdx mouse despite the production of antibodies against dystrophin. J Cell Biol. 1995;131(4):975–88.
55. Fan Y, Maley M, Beilharz M, Grounds M. Rapid death of injected myoblasts in myoblast transfer therapy. Muscle Nerve. 1996;19(7):853–60.
56. Montarras D, Morgan J, Collins C, Relaix F, Zaffran S, Cumano A, et al. Direct isolation of satellite cells for skeletal muscle regeneration. Science. 2005;309(5743):2064–7.
57. Gilbert PM, Havenstrite KL, Magnusson KE, Sacco A, Leonardi NA, Kraft P, et al. Substrate elasticity regulates skeletal muscle stem cell self-renewal in culture. Science. 2010;329(5995):1078–81.
58. Deasy BM, Feduska JM, Payne TR, Li Y, Ambrosio F, Huard J. Effect of VEGF on the regenerative capacity of muscle stem cells in dystrophic skeletal muscle. Mol Ther. 2009; 17(10):1788–98.
59. Vella JB, Thompson SD, Bucsek MJ, Song M, Huard J. Murine and human myogenic cells identified by elevated aldehyde dehydrogenase activity: implications for muscle regeneration and repair. PLoS ONE. 2011;6(12), e29226.
60. Oshima H, Payne TR, Urish KL, Sakai T, Ling Y, Gharaibeh B, et al. Differential myocardial infarct repair with muscle stem cells compared to myoblasts. Mol Ther. 2005;12(6): 1130–41.
61. Okada M, Payne TR, Zheng B, Oshima H, Momoi N, Tobita K, et al. Myogenic endothelial cells purified from human skeletal muscle improve cardiac function after transplantation into infarcted myocardium. J Am Coll Cardiol. 2008;52(23):1869–80.
62. Lavasani M, Robinson AR, Lu A, Song M, Feduska JM, Ahani B, et al. Muscle-derived stem/progenitor cell dysfunction limits healthspan and lifespan in a murine progeria model. Nat Commun. 2012;3:608.
63. Sinha M, Jang YC, Oh J, Khong D, Wu EY, Manohar R, et al. Restoring systemic GDF11 levels reverses age-related dysfunction in mouse skeletal muscle. Science. 2014;344(6184): 649–52.

64. Loffredo FS, Steinhauser ML, Jay SM, Gannon J, Pancoast JR, Yalamanchi P, et al. Growth differentiation factor 11 is a circulating factor that reverses age-related cardiac hypertrophy. Cell. 2013;153(4):828–39.

65. Carr LK, Steele D, Steele S, Wagner D, Pruchnic R, Jankowski R, et al. 1-year follow-up of autologous muscle-derived stem cell injection pilot study to treat stress urinary incontinence. Int Urogynecol J Pelvic Floor Dysfunct. 2008;19(6):881–3.

66. Crisan M, Deasy B, Gavina M, Zheng B, Huard J, Lazzari L, et al. Purification and long-term culture of multipotent progenitor cells affiliated with the walls of human blood vessels: myo-endothelial cells and pericytes. Methods Cell Biol. 2008;86:295–309.

67. Lavasani M, Thompson SD, Pollett JB, Usas A, Lu A, Stolz DB, et al. Human muscle-derived stem/progenitor cells promote functional murine peripheral nerve regeneration. J Clin Invest. 2014;124(4):1745–56.

68. Gharaibeh B, Lavasani M, Cummins JH, Huard J. Terminal differentiation is not a major determinant for the success of stem cell therapy - cross-talk between muscle-derived stem cells and host cells. Stem Cell Res Ther. 2011;2(4):31.

69. Zheng B, Chen CW, Li G, Thompson SD, Poddar M, Peault B, et al. Isolation of myogenic stem cells from cultures of cryopreserved human skeletal muscle. Cell Transplant. 2012;21(6):1087–93.

70. Minasi MG, Riminucci M, De Angelis L, Borello U, Berarducci B, Innocenzi A, et al. The meso-angioblast: a multipotent, self-renewing cell that originates from the dorsal aorta and differentiates into most mesodermal tissues. Development. 2002;129(11):2773–83.

71. Sampaolesi M, Torrente Y, Innocenzi A, Tonlorenzi R, D'Antona G, Pellegrino MA, et al. Cell therapy of alpha-sarcoglycan null dystrophic mice through intra-arterial delivery of mesoangioblasts. Science. 2003;301(5632):487–92.

72. Sampaolesi M, Blot S, D'Antona G, Granger N, Tonlorenzi R, Innocenzi A, et al. Mesoangioblast stem cells ameliorate muscle function in dystrophic dogs. Nature. 2006; 444(7119):574–9.

73. EU Clinical Trials Register 2011-000176-33 2011. Available from: https://www.clinicaltrial-sregister.eu/ctr-search/trial/2011-000176-33/IT

74. Chun JL, O'Brien R, Song MH, Wondrasch BF, Berry SE. Injection of vessel-derived stem cells prevents dilated cardiomyopathy and promotes angiogenesis and endogenous cardiac stem cell proliferation in mdx/utrn(−/−) but not aged mdx mouse models for duchenne muscular dystrophy. Stem Cell Transl Med. 2013;2(2):68.

75. English K, Tonlorenzi R, Cossu G, Wood KJ. Mesoangioblasts suppress T cell proliferation through IDO and PGE-2-dependent pathways. Stem Cells Dev. 2013;22(3):512–23.

76. Fuoco C, Salvatori ML, Biondo A, Shapira-Schweitzer K, Santoleri S, Antonini S, et al. Injectable polyethylene glycol-fibrinogen hydrogel adjuvant improves survival and differentiation of transplanted mesoangioblasts in acute and chronic skeletal-muscle degeneration. Skelet Muscle. 2012;2(1):24.

77. Torrente Y, Belicchi M, Sampaolesi M, Pisati F, Meregalli M, D'Antona G, et al. Human circulating AC133(+) stem cells restore dystrophin expression and ameliorate function in dystrophic skeletal muscle. J Clin Invest. 2004;114(2):182–95.

78. Stamm C, Westphal B, Kleine HD, Petzsch M, Kittner C, Klinge H, et al. Autologous bone-marrow stem-cell transplantation for myocardial regeneration. Lancet. 2003;361(9351): 45–6.

79. Torrente Y, Belicchi M, Marchesi C, D'Antona G, Cogiamanian F, Pisati F, et al. Autologous transplantation of muscle-derived CD133+ stem cells in Duchenne muscle patients. Cell Transplant. 2007;16(6):563–77.

80. Ichim TE, Alexandrescu DT, Solano F, Lara F, Campion RDN, Paris E, et al. Mesenchymal stem cells as anti-inflammatories: implications for treatment of Duchenne muscular dystrophy. Cell Immunol. 2010;260(2):75–82.

81. Prockop DJ. Marrow stromal cells as stem cells for nonhematopoietic tissues. Science. 1997;276(5309):71–4.

82. Lepski G, Jannes CE, Maciaczyk J, Papazoglou A, Mehlhorn AT, Kaiser S, et al. Limited Ca2+ and PKA-pathway dependent neurogenic differentiation of human adult mesenchymal stem cells as compared to fetal neuronal stem cells. Exp Cell Res. 2010;316(2): 216–31.
83. Trzaska KA, King CC, Li KY, Kuzhikandathil EV, Nowycky MC, Ye JH, et al. Brain-derived neurotrophic factor facilitates maturation of mesenchymal stem cell-derived dopamine progenitors to functional neurons. J Neurochem. 2009;110(3):1058–69.
84. Chivu M, Dima SO, Stancu CI, Dobrea C, Uscatescu V, Necula LG, et al. In vitro hepatic differentiation of human bone marrow mesenchymal stem cells under differential exposure to liver-specific factors. Transl Res. 2009;154(3):122–32.
85. Liu Y, Yan X, Sun Z, Chen B, Han Q, Li J, et al. Flk-1+ adipose-derived mesenchymal stem cells differentiate into skeletal muscle satellite cells and ameliorate muscular dystrophy in mdx mice. Stem Cells Dev. 2007;16(5):695–706.
86. Bruder SP, Fink DJ, Caplan AI. Mesenchymal stem cells in bone development, bone repair, and skeletal regeneration therapy. J Cell Biochem. 1994;56(3):283–94.
87. Berry MF, Engler AJ, Woo YJ, Pirolli TJ, Bish LT, Jayasankar V, et al. Mesenchymal stem cell injection after myocardial infarction improves myocardial compliance. Am J Physiol Heart Circ Physiol. 2006;290(6):H2196–203.
88. Gonzalez MA, Gonzalez-Rey E, Rico L, Buscher D, Delgado M. Treatment of experimental arthritis by inducing immune tolerance with human adipose-derived mesenchymal stem cells. Arthritis Rheum. 2009;60(4):1006–19.
89. Fiorina P, Jurewicz M, Augello A, Vergani A, Dada S, La Rosa S, et al. Immunomodulatory function of bone marrow-derived mesenchymal stem cells in experimental autoimmune type 1 diabetes. J Immunol. 2009;183(2):993–1004.
90. Constantin G, Marconi S, Rossi B, Angiari S, Calderan L, Anghileri E, et al. Adipose-derived mesenchymal stem cells ameliorate chronic experimental autoimmune encephalomyelitis. Stem Cells. 2009;27(10):2624–35.
91. Spaggiari GM, Abdelrazik H, Becchetti F, Moretta L. MSCs inhibit monocyte-derived DC maturation and function by selectively interfering with the generation of immature DCs: central role of MSC-derived prostaglandin E2. Blood. 2009;113(26):6576–83.
92. Selmani Z, Naji A, Zidi I, Favier B, Gaiffe E, Obert L, et al. Human leukocyte antigen-G5 secretion by human mesenchymal stem cells is required to suppress T lymphocyte and natural killer function and to induce CD4+CD25highFOXP3+ regulatory T cells. Stem Cells. 2008;26(1):212–22.
93. Casiraghi F, Azzollini N, Cassis P, Imberti B, Morigi M, Cugini D, et al. Pretransplant infusion of mesenchymal stem cells prolongs the survival of a semiallogeneic heart transplant through the generation of regulatory T cells. J Immunol. 2008;181(6):3933–46.
94. Biswadeep C, Krishna P. Key aspects of mesenchymal stem cells (MSCs) in tissue engineering for in vitro skeletal muscle regeneration. Biotechnol Mol Biol Rev. 2012;7(1):5–15.
95. Gang EJ, Darabi R, Bosnakovski D, Xu Z, Kamm KE, Kyba M, et al. Engraftment of mesenchymal stem cells into dystrophin-deficient mice is not accompanied by functional recovery. Exp Cell Res. 2009;315(15):2624–36.
96. Darabi R, Santos FN, Filareto A, Pan W, Koene R, Rudnicki MA, et al. Assessment of the myogenic stem cell compartment following transplantation of Pax3/Pax7-induced embryonic stem cell-derived progenitors. Stem Cells. 2011;29(5):777–90.
97. Chang H, Yoshimoto M, Umeda K, Iwasa T, Mizuno Y, Fukada S, et al. Generation of transplantable, functional satellite-like cells from mouse embryonic stem cells. FASEB J. 2009;23(6):1907–19.
98. King's College London. Disease-specific human embryonic stem cell lines placed on NIH registry 2013. Available from: http://www.sciencedaily.com/releases/2013/09/130924122815. htm
99. Zhao T, Zhang ZN, Rong Z, Xu Y. Immunogenicity of induced pluripotent stem cells. Nature. 2011;474(7350):212–5.

100. Filareto A, Parker S, Darabi R, Borges L, Iacovino M, Schaaf T, et al. An ex vivo gene therapy approach to treat muscular dystrophy using inducible pluripotent stem cells. Nat Commun. 2013;4.
101. Beck AJ, Vitale JM, Zhao Q, Schneider JS, Chang C, Altaf A, et al. Differential requirement for utrophin in the induced pluripotent stem cell correction of muscle versus fat in muscular dystrophy mice. PLoS ONE. 2011;6(5), e20065.
102. Tedesco FS, Gerli MF, Perani L, Benedetti S, Ungaro F, Cassano M, et al. Transplantation of genetically corrected human iPSC-derived progenitors in mice with limb-girdle muscular dystrophy. Sci Transl Med. 2012;4(140), 140ra89.
103. Cosgrove BD, Gilbert PM, Porpiglia E, Mourkioti F, Lee SP, Corbel SY, et al. Rejuvenation of the muscle stem cell population restores strength to injured aged muscles. Nat Med. 2014;20(3):255–64.
104. Pal R, Palmieri M, Loehr JA, Li S, Abo-Zahrah R, Monroe TO, et al. Src-dependent impairment of autophagy by oxidative stress in a mouse model of Duchenne muscular dystrophy. Nat Commun. 2014;5:4425.
105. Rayavarapu S, Coley W, Cakir E, Jahnke V, Takeda S, Aoki Y, et al. Identification of disease specific pathways using in vivo SILAC proteomics in dystrophin deficient mdx mouse. Mol Cell Proteomics. 2013;12(5):1061–73.
106. Mariol MC, Segalat L. Muscular degeneration in the absence of dystrophin is a calcium-dependent process. Curr Biol. 2001;11(21):1691–4.
107. Spuler S, Engel AG. Unexpected sarcolemmal complement membrane attack complex deposits on nonnecrotic muscle fibers in muscular dystrophies. Neurology. 1998;50(1): 41–6.
108. Christov C, Chretien F, Abou-Khalil R, Bassez G, Vallet G, Authier FJ, et al. Muscle satellite cells and endothelial cells: close neighbors and privileged partners. Mol Biol Cell. 2007;18(4):1397–409.
109. Malerba A, Pasut A, Frigo M, De Coppi P, Baroni MD, Vitiello L. Macrophage-secreted factors enhance the in vitro expansion of DMD muscle precursor cells while preserving their myogenic potential. Neurol Res. 2010;32(1):55–62.
110. Zhao X, Moloughney JG, Zhang S, Komazaki S, Weisleder N. Orai1 mediates exacerbated Ca(2+) entry in dystrophic skeletal muscle. PLoS ONE. 2012;7(11), e49862.
111. Jahnke VE, Van Der Meulen JH, Johnston HK, Ghimbovschi S, Partridge T, Hoffman EP, et al. Metabolic remodeling agents show beneficial effects in the dystrophin-deficient mdx mouse model. Skelet Muscle. 2012;2(1):16.
112. Whitehead NP, Yeung EW, Froehner SC, Allen DG. Skeletal muscle NADPH oxidase is increased and triggers stretch-induced damage in the mdx mouse. PLoS ONE. 2010;5(12), e15354.
113. Prigione A, Fauler B, Lurz R, Lehrach H, Adjaye J. The senescence-related mitochondrial/oxidative stress pathway is repressed in human induced pluripotent stem cells. Stem Cells. 2010;28(4):721–33.
114. Evans NP, Misyak SA, Robertson JL, Bassaganya-Riera J, Grange RW. Dysregulated intracellular signaling and inflammatory gene expression during initial disease onset in Duchenne muscular dystrophy. Am J Phys Med Rehabil. 2009;88(6):502–22.
115. Marotta M, Ruiz-Roig C, Sarria Y, Peiro JL, Nunez F, Ceron J, et al. Muscle genome-wide expression profiling during disease evolution in mdx mice. Physiol Genomics. 2009;37(2): 119–32.
116. Chen YW, Zhao P, Borup R, Hoffman EP. Expression profiling in the muscular dystrophies: identification of novel aspects of molecular pathophysiology. J Cell Biol. 2000;151(6): 1321–36.
117. Acharyya S, Villalta SA, Bakkar N, Bupha-Intr T, Janssen PM, Carathers M, et al. Interplay of IKK/NF-kappaB signaling in macrophages and myofibers promotes muscle degeneration in Duchenne muscular dystrophy. J Clin Invest. 2007;117(4):889–901.

118. Lu A, Proto JD, Guo L, Tang Y, Lavasani M, Tilstra JS, et al. NF-kappaB negatively impacts the myogenic potential of muscle-derived stem cells. Mol Ther. 2012;20(3):661–8.
119. Arnold L, Henry A, Poron F, Baba-Amer Y, van Rooijen N, Plonquet A, et al. Inflammatory monocytes recruited after skeletal muscle injury switch into antiinflammatory macrophages to support myogenesis. J Exp Med. 2007;204(5):1057–69.
120. Manzur AY, Kuntzer T, Pike M, Swan A. Glucocorticoid corticosteroids for Duchenne muscular dystrophy. Cochrane Database Syst Rev. 2008;1, CD003725.
121. Connolly AM, Schierbecker J, Renna R, Florence J. High dose weekly oral prednisone improves strength in boys with Duchenne muscular dystrophy. Neuromuscul Disord. 2002;12(10):917–25.
122. Ricotti V, Ridout DA, Scott E, Quinlivan R, Robb SA, Manzur AY, et al. Long-term benefits and adverse effects of intermittent versus daily glucocorticoids in boys with Duchenne muscular dystrophy. J Neurol Neurosurg Psychiatry. 2013;84(6):698–705.
123. Acuna MJ, Pessina P, Olguin H, Cabrera D, Vio CP, Bader M, et al. Restoration of muscle strength in dystrophic muscle by angiotensin-1-7 through inhibition of TGF-beta signalling. Hum Mol Genet. 2014;23(5):1237–49.
124. Gargioli C, Coletta M, De Grandis F, Cannata SM, Cossu G. PlGF-MMP-9-expressing cells restore microcirculation and efficacy of cell therapy in aged dystrophic muscle. Nat Med. 2008;14(9):973–8.
125. Mu X, Usas A, Tang Y, Lu A, Wang B, Weiss K, et al. RhoA mediates defective stem cell function and heterotopic ossification in dystrophic muscle of mice. FASEB J. 2013;27(9): 3619–31.
126. Weiss R, Dufour S, Groszmann A, Petersen K, Dziura J, Taksali SE, et al. Low adiponectin levels in adolescent obesity: a marker of increased intramyocellular lipid accumulation. J Clin Endocrinol Metabol. 2003;88(5):2014–8.

Chapter 5
Spinal Cord Cellular Therapeutics Delivery: Device Design Considerations

Khalid Medani, Jonathan Riley, Jason Lamanna, and Nicholas Boulis

5.1 Introduction

Three primary delivery approaches may be attempted when delivering a cellular graft to the spinal cord. Intravascular, intrathecal, and intraparenchymal delivery methodologies have each been explored in clinical trials. A list of recent domestic and international trials that utilize each of these techniques is provided in the attached Table 5.1. These vary widely in procedural complexity and invasiveness, potential for associated neurologic morbidity, and anatomic specificity of graft delivery. Indirect delivery approaches (e.g., intrathecal, intravascular) are less invasive than current methodologic approaches for intraparenchymal delivery, respectively, requiring either vascular access or a lumbar puncture. Indirect approaches require a demonstrated capability of CNS homing to the area of treatment within the spinal cord. Preclinical studies have supported a variable capability to achieve homing to the CNS with both intravascular or intrathecal delivery. Some studies have supported engraftment at the site of interest. Others have supported limited parenchymal penetration (e.g., clustering on the pial surface or dependent clustering in the thecal sac) with intrathecal delivery [39, 40] and a lack of homing with intravascular delivery. Preclinical studies that have attempted a direct comparison between the delivery approaches have favored the substantially higher engraftment efficiency

K. Medani, M.D. • J. Riley, M.D. M.S. • N. Boulis, M.D. (✉)
Department of Neurosurgery, Emory University,
1365-B Clifton Road NE, Suite B6200, Atlanta, GA 30322, USA
e-mail: nboulis@emory.edu

J. Lamanna, Ph.D.
Department of Neurosurgery, Emory University,
1365-B Clifton Road NE, Suite B6200, Atlanta, GA 30322, USA

Department of Biomedical Engineering, Georgia Institute of Technology, Emory University,
Atlanta, GA, USA

© Springer Science+Business Media New York 2016   109
M.K. Childers (ed.), *Regenerative Medicine for Degenerative Muscle Diseases*,
Stem Cell Biology and Regenerative Medicine,
DOI 10.1007/978-1-4939-3228-3_5

Table 5.1 Recent completed domestic and international spinal cord cell delivery clinical trials

Delivery method	Year	Location	Indication (# patients)	Cell line	Observed adverse events (# patients)
	2012	Atlanta, GA, USA [1, 2]	ALS (12)	Fetal spinal cord-derived stem cells	Transient radicular-type pain and/or sensory abnormalities (several); repaired CSF leak (1); wound dehiscence (1)
	2012	Multicenter, USA and Israel [3–5]	SCI (26)	Autologous macrophages	Pulmonary embolism (2), osteomyelitis (1); transient anemia (8), urinary tract infection (UTI) (7), fever (7) (was attributed to UTI); surgery for late spinal instability (1); post-op subsegmental atelectasis (1); resolved bacterial meningitis (1); pseudomeningocele (1)
	2012	Murcia, Spain [6]	ALS (11)	Autologous BM MNCs	Transient wound pain (7), intercostal pain (5), hypesthesia (7), paresthesia (4), dysesthesia (2), headache (2) and/or intracranial hypotension (3); persistent hypoesthesia (2)
	2012	Novara, Italy [7–10]	ALS (10)	Autologous BM MSCs	Transient pain (7), light-touch impairment in one leg (4) or sacral region (1), and/or tingling sensation in one leg (6)
	2012	Multicenter, USA [11]	SCI (?)	Human Embryonic stem cells	N/A
	2012	Tianjin, China [12]	SCI (6)	Autologous activated Schwann cells (AASCs)	None observed
	2012	Beijing, China [13]	SCI (108)	Autologous olfactory ensheathing cells	None observed
	2012	Beijing, China [14]	SCI (11)	Fetal olfactory ensheathing cells	None observed

Intraparenchymal	2011	Ankara, Turkey [15]	SCI (4)	Autologous BM MNCs	None observed
	2010	Lisbon, Portugal [16, 17]	SCI (20)	Autologous olfactory mucosal cells	Sensory deficit and transient motor deficit secondary to resolved aseptic meningitis (1); minor resolved subcutaneous CSF collection (3); transient irritable bowel syndrome (1)
	2009	Ankara, Turkey [18]	ALS (13)	Autologous BM MNCs	None observed
	2008	Brisbane, Australia [19, 20]	SCI (6)	Autologous olfactory ensheathing cells	None observed after 1 year of follow-up
	2007	Tehran, Iran [21, 22]	SCI (33)	Autologous Schwann cells	Transient fever, nausea, vomiting, and headache (few), superficial wound dehiscence (1),transient paresthesia (3), transient late-onset (after 4 months) increased muscle spasm (1)
	2007	Incheon, Korea [23]	SCI (35)	Autologous BM cells	Fever (22); transient neurological deterioration (1), spasticity (1), rigidity (3), headache (3), numbness or tingling sensation (6), facial flushing or rash (5); neuropathic pain (7); abdominal discomfort (7); constipation (3); general ache (3)
	2007	Beijing, China [24]	ALS(327)	Olfactory ensheathing cells (OECs)	Headache, short-term fever, seizure attack, central nerve system infection, pneumonia, respiratory failure, urinary tract infection, heart failure, and possible pulmonary embolism (16), death(4) from the 16 patients
	2013	Beijing, China [25]	SCI(22)	Umbilical cord MSCs	Transient headache (1), transient low back pain (1)

(continued)

Table 5.1 (continued)

Delivery method	Year	Location	Indication (# patients)	Cell line	Observed adverse events (# patients)
Intrathecal	2012	Mumbai, India [26]	Multiple: muscular dystrophy (38), SCI (4), cerebral palsy (20), others (9)	Autologous BM MNCs	Transient headache (12), nausea (7), backache (7)
	2012	Moscow, Russia [27]	SCI (20)	Autologous hematopoietic stem cells	None mentioned
	2012	Kerman, Iran [28]	SCI (11)	Autologous BM MSCs	None observed
	2012	Osaka, Japan [29, 30]	SCI (5)	Autologous BM stromal cells	None observed
	2010	Jerusalem, Israel [31]	MS/ALS (34)	Autologous MSCs	Fever (21); transient headache (15), rigidity (2), and leg pain (3); aseptic meningitis attributed to intrathecal injection (1), dyspnea (1), confusion (1), neck pain (1), difficulty walking/ standing (4)
	2009	Chennai, India [32]	SCI (297)	Autologous BM MNCs	Fever (95); transient headache (67); tingling sensation (68), spasm (1), and neuropathic pain (17)
	2009	Bangalore, India [33]	SCI (30)	Autologous BM MSCs	None observed after 1 year of follow-up
	2008	Gujarat, India [34]	SCI (163), cerebral palsy (6), MND (4), encephalopathy (5)	Adipose tissue MSCs (81), embryonic hematopoietic stem cells (99) and autologous BM MSCs (180)	Headache (96), fever (4), meningism (2)
	2011	Seoul, Republic of Korea [35]	SCI (8)	Autologous adipose tissue-derived MSCs	None observed after 3 months of follow-up
	2010	Jerusalem, Israel [31]	MS/ALS (34)	Autologous MSCs	Fever (21); transient headache (15), rigidity (2), and leg pain (3); aseptic meningitis attributed to intrathecal injection (1), dyspnea (1), confusion (1), neck pain (1), difficulty walking/ standing (4)
Intravascular	2011	Beijing, China [36]	MS (36)	Autologous peripheral blood stem cells	Adverse events were not measured or discussed
	2009	Sao Paulo, Brazil [37]	SCI (39)	Autologous peripheral blood stem cells	Pneumothorax associated with stem cell collection (1), local allergic reaction to contrast agent (3)
	2006	Buenos Aires, Argentina [38]	SCI (2)	Autologous BM MSCs	None observed

observed with intraparenchymal administration [41, 42]. Additional considerations relevant to both indirect delivery approaches include the possibility of disseminated tumorigenesis and delivery-associated vascular complications. Early reports of tumorigenesis with endovascular approaches have subsequently been attributed to graft contamination [43, 44]. Intravascular delivery may also be complicated by delivery-associated vascular congestion [45] though this has only been reported to date in small animal studies.

Intraparenchymal delivery is associated with an additional set of considerations that are predominantly technical in nature. These include: design of the injection device, stabilization technique that accounts for device or patient movement, choice of targeting approach, and employed dosing parameters. Our group has developed an intraparenchymal delivery platform and approach that have been explored in pre-clinical studies, in a recently completed phase I clinical trial [46, 47] and now in an ongoing phase II trial for delivery of a cellular therapeutic to the ALS spinal cord. Here we describe design considerations relevant to an intraparenchymal microinjection approach and explore how these considerations have been clinically managed.

5.2   Intraparenchymal Microinjection Design Considerations

5.2.1   Stabilization Approaches

Varied procedural methodologies have been used to deliver different biologic payloads (e.g., cellular grafts, viral vectors) to targeted intraspinal sites of interest in both preclinical and clinical studies. Broadly, intraparenchymal microinjection approaches may be divided into non-stabilized (e.g., freehand) and stabilized (e.g. table mounted, patient mounted) methodologies. Freehand delivery, a non-stabilized approach, has been the predominant approach explored in early preclinical and clinical investigations [3–5, 13, 48–50]. Published clinical reports utilizing a freehand delivery approach are summarized in Table 5.2. This technique employs a cannula that is manually manipulated and held at the injection target area by the surgeon. Graft delivery may take place through either manual injection by the surgeon or through the use of a programmable infusion pump. The primary advantage of this approach is optimized mobility when manipulating the injection cannula. Disadvantages include concerns for shear-related parenchymal injury at the injection site related to patient or needle movement, an inability to accurately target the intended injection site, and poor reproducibility of targeting between injections. With this technique, damage might occur to the spinal cord with the inadvertent movement of the needle or of the patient during the injection procedure. Concomitant use of a manual surgeon-controlled injector (e.g., syringe), as has been employed clinically, further complicates payload delivery. The uncontrolled infusion rate may lead to concern for mass effect with resultant local cord injury or payload reflux through the catheter tract. Broad variability in outcome measure reporting between these trials precludes rigorous evaluation or comparison [51].

Table 5.2 Completed freehand spinal cord cell injection trials

Delivery method	Year	Location	Indication (Pt #)	Cell line	Intraspinal target	Methodology/coordinates	Cannula type	Immunosuppression
	2012	Multicenter, USA and Israel [3–5]	SCI (26)	Autologous macrophages	(1) White matter lateral to the lateral corticospinal tract bilaterally (2) Dorsal column bilaterally	Preoperative MRI and surgical judgment. Intraoperative spinal sonography to identify contusion boundaries (11)	30-gauge needle	None (autologous)
	2012	Tianjin, China [12]	SCI (6)	Autologous activated Schwann cells (AASCs)	Distributed in multiple locations	Not indicated	Not mentioned	None (autologous)
	2012	Beijing, China [13]	SCI (108)	Autologous olfactory ensheathing cells	Not mentioned	Not indicated	Not mentioned	None (autologous)

Intraparenchymal (freehand delivery)							
	Beijing, China [14]	2012	SCI (11)	Fetal olfactory ensheathing cells	Adjacent to rostral and caudal ends of the lesions	Not indicated	Not mentioned	Not mentioned
	Ankara, Turkey [15]	2011	SCI (4)	Autologous BM MNCs	Intralesionally, 5 mm depth from dorsal surface	Preoperative MRI, SEP, and MEP	26-gauge needle	None (autologous)
	Lisbon, Portugal [16, 17]	2010	SCI (20)	Autologous olfactory mucosal cells	Intralesionally	Preoperative MRI	Not mentioned	None (autologous)
	Ankara, Turkey [18]	2009	ALS (13)	Autologous BM MNCs	Multiple locations intraspinally, intra-brain stem, subarachnoid space, and intravascularly	Preoperative MRI	21-gauge needle	None (autologous)
	Tehran, Iran [21, 22]	2007	SCI (33)	Autologous Schwann cells	Intra-syrinx through the posterior median sulcus midline, at three different points 5 mm apart	Preoperative MRI, intraoperative navigation	30.5-gauge needle	None (autologous)
	Incheon, Korea [23]	2007	SCI (35)	Autologous BM cells	Six separate positions around the lesion, 5 mm depth from the dorsal surface and 5 mm lateral from the midline	High-power microscope to locate contusion site	21-gauge needle	None (autologous)
	Beijing, China [24]	2007	ALS(327)	Olfactory ensheathing cells (OECs)	?	N/A	?	None (autologous)

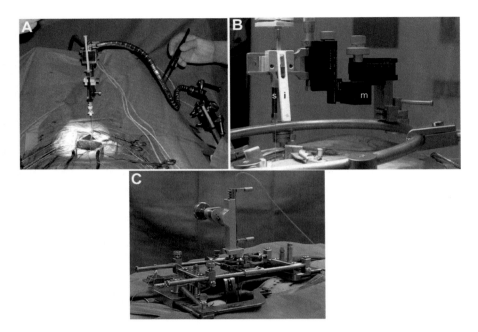

Fig. 5.1 Stabilized intraspinal microinjection platforms. (**a**) A table-mounted microinjection system is employed. A Yasargil retractor system, mounted to the patient bed, was used to hold an injection needle and syringe. This particular system utilized a surgeon-actuated syringe. This image has been reproduced from Blanquer et al. [52]. (**b**) An alternative table-mounted design utilizes the *circular* Synthes Synframe. The *circular table*-mounted frame accommodates an attached micromanipulator, syringe, and injection cannula [19]. (**c**) A patient-stabilized microinjection platform utilized by our group is demonstrated. This platform is stabilized by the use of an integrated retractor system and the use of percutaneous posts at either end. A "floating" microinjection needle operated by a programmable microinjection pump is utilized (*Permission to republish images from Blanquer et al. and Feron et al. has not yet been obtained and will be processed during manuscript review)

Stabilized direct microinjection approaches include the use of both table-mounted and patient-stabilized microinjection platforms. Examples of table- and patient-stabilized platforms are provided in the attached Fig. 5.1a–c. Clinical studies utilizing each stabilized approach are, respectively, summarized in Tables 5.3 and 5.4. Table-mounted microinjection platforms provide advantages not seen with freehand injection. Rigid fixation of an injection platform and microinjection cannula substantially reduces concern for needle tip movement during the injection process. Attachment of the microinjection needle to a platform-mounted micromanipulator stage allows for multidimensional control of the cannula trajectory, improving both the achievable accuracy and precision. Patient-mounted microinjection platforms include the advantages of stabilization shared with table-mounted devices. However, instead of rigid fixation to the operating room table or alternate structure, patient-stabilized microinjection platforms are rigidly affixed to the patient. Stabilization to the patient can be accomplished through attachment to a self-retaining retractor, bony prominence, or both.

Table 5.3 Pending and completed table-stabilized spinal cord cell injection trials

Delivery method	Year	Location	Indication (# patients)	Cell line	Intraspinal target	Methodology/ coordinates	Injection cannula type	Immunosuppression
	2012	Novara, Italy [7–10]	ALS (10)	Autologous BM MSCs	The most central part of the spinal cord towards AHCs in thoracic spine, in three rows, 3 mm apart, 2–5 injection sites per level	None, operating microscope used for injection	18-gauge needle	None (autologous)
	2012	Multicenter, USA [11]	SCI(N/A)	Human embryonic stem cells	N/A	N/A	N/A	N/A

(continued)

Table 5.3 (continued)

Delivery method	Year	Location	Indication (# patients)	Cell line	Intraspinal target	Methodology/ coordinates	Injection cannula type	Immunosuppression
Intraparenchymal (table mounted)	2012	Murcia, Spain [6, 53]	ALS (11)	Autologous BM MNCs	The most avascular area in the posterior funiculus 6 cm depth, 1 cm apart from each injection site (ipsilateral or contralateral)	Graded microwheel	22-gauge needle	None (autologous)
	2008	Brisbane, Australia [19, 20]	SCI (6)	Autologous olfactory ensheathing cells	Adjacent to the lesion or intralesionally, in three rows 5 mm apart and five columns 1 mm apart at four depths 1 mm apart (was more accurate in non-damaged portion of the cord)	Micromanipulator with 4 μm, movement in three planes	28-gauge needle and micromanipulator	None (autologous)

Table 5.4 Pending and completed patient-stabilized spinal cord cell injection trials

Delivery method	Year	Location	Indication (# patients)	Cell line	Injection cannula type	Immunosuppression
Intraparenchymal (patient stabilized)	2012	Atlanta, GA, USA [1, 2]	ALS (12)	Fetal spinal cord-derived stem cells	Floating cannula and microinject or pump, 30-gauge needle	Methylprednisolone, basiliximab, tacrolimus, mycophenolate mofetil

5.2.2 Injection Cannula Design

Delivery of cellular payload in both large animal preclinical studies and in clinical application has utilized some variation of an injection cannula or needle. In this section, factors to be considered when choosing a cannula design, examples of current generation cannula alternatives, and possible future design alternatives are discussed. An optimized cannula design should minimize infusate reflux, reduce local injection-related tissue trauma, maintain cell viability, and result in a homogenous distribution of delivered cells. Modifiable factors that contribute to graft reflux include: targeting accuracy, injection parameters (e.g., volume, rate, graft concentration), and the injection cannula construction. Likely due to the anisotropy encountered in white matter as opposed to the spinal gray matter, large animal data supports preferential graft dispersion rostrocaudally when in white matter while maintaining a more focal distribution in the gray matter (data not published). Further, elevations in volume and rate of delivery result in an increased predilection towards payload reflux. Innovations to improve the tissue seal around the cannula tip have been demonstrated to retard infusate reflux over a range of volumes and rates. These include an increased injection needle/cannula insertion speed [54], the use of a stepped cannula design [55, 56], and coating the outer cannula surface with a hydrogel that expands on contact with tissue [57].

When considering graft delivery for both local (e.g., spinal cord injury) and diffuse (e.g., SMA, ALS, MS) afflictions, care must be taken to minimize local tissue trauma. This is a priority both because of the functional importance of all spinal cord tissue and the possible need to complete serial unilateral or bilateral injections to achieve adequate cell delivery to the sites of interest. In vitro data supports an elevated needle insertion speed as a modifiable factor to reduce local tissue trauma [54]. Cannula design represents another modifiable parameter to minimize trauma during the injection process. As compared to traditional rigid injection cannulas, our group has developed a "floating cannula"-based design that allows the cannula and silastic tubing to move with cardioballistic and cardiorespiratory-associated cord movements. The injection cannula enters the spinal cord in rigid conformation. The needle is introduced to a predetermined depth, physically limited by a flange. Once firmly seated, the outer cannula is retracted allowing the injection needle to move freely during the injection process. In our experience, preclinical large animal studies support a faster recovery to neurologic baseline and toleration of multiple bilateral injections with this process as opposed to with a rigid cannula. An alternate approach initially designed for cranial application includes the use of a "steerable" injection cannula. Steerable injection cannulas have been described that allow an injection cannula to be inserted at up to 25° from the axis of the guide cannula [58, 59]. More recently, Silvestrini et al. [54] describe the use of dual opposed side ports and rotation of a guide cannula to allow radial branched deployment (RBD) of the inner injection cannula at up to 90° from the primary guide cannula axis. While the described RBD cannula is 20 gauge, future reductions in cannula caliber may allow broadened application to intraspinal parenchymal delivery.

The delivery of cellular suspensions is associated with a unique set of constraints that can affect both cellular viability during the injection process, homogeneity of the delivered suspension, and long-term cell survival at the engraftment site. These issues have recently been reviewed [60]. The comparative differences in size when considering delivery of a cellular versus viral or peptide-based therapeutics result in cellular payloads behaving like a suspension in their carrier fluid. By contrast, viral- and peptide-based payloads act as a solution. As a consequence, cell suspensions are prone to sedimentation during the delivery process. This introduces the possibility of wide variations in the cell concentration delivered, with very high concentrations delivered during initial injections and progressive decreases during later injections. The unintentional delivery of high cellular concentration into an engraftment site can result in decreased post-transplant cellular viability due to a lack of necessary nutrients in the comparatively avascular engraftment site [61]. Gentle agitation has been explored as a method to reduce cell suspension sedimentation [62]. Graft viability may also be affected by the extensional forces experienced during transition zones (e.g., wide syringe to narrow cannula) and along the lengths of narrow injection cannulas. To minimize exposure to these extensional forces, effort should be given to empiric optimization of infusion parameters, minimization of cell interaction with shifts in internal cannula diameter, and use of minimum necessary lengths of cannula tubing. Further, recent data supports the use of a hydrogel-based cell carrier as a means to reduce cell exposure to extensional forces during the injection process and to improve cell injection viability [63]. Fibrin-based matrices have also been published as a basis for improving cellular engraftment [64].

5.2.3   Targeting Methodologies

Tables 5.2, 5.3, and 5.4 provide the methodologies utilized for targeting intraspinal sites of interest, when described in the literature. Regardless of injection stabilization method employed, published literature generally indicates the use of anatomic surface landmarks to guide introduction of an injection cannula to a prespecified depth. In some instances, investigators have indicated histopathological findings from human specimens or data from preclinical injections in large animals to support the choice of chosen injection coordinates. Our preclinical experience initially attempted the use of microelectrode recording to target the ventral horn by mapping the borders of the gray and white matter [65]. While achievable, this technique required making one or more passes with a recording microelectrode, in addition to the pass required for payload delivery. Subsequent shift to the use of anatomic surface landmarks and an injection platform that allows correction for anatomic variables to ensure orthogonal injection cannula entry to the spinal cord has improved the speed, accuracy, and precision of payload delivery to the ventral horn in preclinical delivery studies to swine. In our clinical experience, somatosensory evoked potentials (SSEPs) have been utilized as a surrogate for possible spinal cord injury associated with the injection process. We have chosen to include a 50 % decrement in SSEP signal that persists over a 30 min period as an indication to terminate the injection process. In a series of 120 spinal cord injection penetrations in a recently completed phase I trial that utilized a patient-stabilized device and a "floating" microinjection cannula, prolonged SSEP decrement during the injection process was not observed [46, 47]. A paucity of histopathological data exists from the trials listed in Tables 5.2, 5.3, and 5.4 to evaluate the accuracy of anatomic landmark-based targeting. A combination of histologic graft site identification and development of methods to achieve in vivo graft tracking will help to optimize utilized targeting approaches.

5.3   Future Considerations, Cell Labeling, and In Vivo Graft Tracking

The in vivo fate of transplanted intraspinal cell grafts is poorly understood. Delivery method, transplanted cell dose, in vitro cell viability, graft location, host immune response, and many other factors interact to modify in vivo graft survival. Much of the data analyzing graft survival comes from postmortem histopathology in small animal xenograft models. The applicability of these reports to human cell transplantation is debatable. To date, published clinical reports lack histopathological evidence of in vivo graft survival. Furthermore, many of these reports lack confirmation of successful graft delivery and initial post-transplant graft location. The difficulties in histological identification of transplanted allografts are the result of methodological limitations in identifying the origins of chimeric tissue, that is, tissue of the same species but from different hosts. Confirmation of successful graft delivery and determination of initial graft location require a method of identifying and tracking cells in vivo. These limitations highlight the utility of a method for labeling cells for identification and tracking.

Several methods have been proposed for tracking transplanted cells in vivo. Cells can be genetically modified with viral vectors engineered to express reporter genes. The expression of reporter genes can be used to place markers on the cell surface for identification of the transplanted cell graft with injected probes for imaging with positron emission tomography (PET) [66, 67] or magnetic resonance imaging (MRI) [68, 69]. A case report has shown the utility of genetic modification of transplanted cells with a PET reporter gene in the brain of a patient with glioblastoma multiforme [70]. However, the scalability of this approach to large animal models and eventually to the clinic requires further investigation. Alternatively, the vector can be designed to allow the cell to accumulate additional iron to produce contrast for identification with MRI [71, 72]. Initial results in small animal models have demonstrated the ability to identify the transplanted graft in vivo. The main advantage of this approach is the specific identification of transplanted cells with a reduced risk of label transfer to host cells. Limitations of this approach include the low signal to noise ratio produced and the increased risk of oncological transformation and tumorigenesis from random integration of the viral vector in to the cell.

Cells may be labeled ex vivo, prior to transplantation, with physical particles that produce contrast for in vivo tracking with MRI or PET. This is comparatively straightforward as the cells can be forced to internalize the particles. However, the intracellular concentration of particles can be reduced from cell division and externalization. The externalized particles can be internalized by host cells and create a false-positive signal that can be incorrectly interpreted as a surrogate of cell graft survival [73]. Superparamagnetic iron oxide (SPIO) nanoparticles have been used to track transplanted cell grafts in vivo with MRI in the central nervous system in many small animal models [74, 75]. Furthermore, SPIO nanoparticles have been used to histologically identify a cell graft 1 year after transplantation in a rodent model of stroke [76]. SPIO nanoparticles have been successfully used in clinical trials to track cell grafts vivo in the brain [77] and spinal cord [31, 78]. However, further investigation must be taken to confirm the cells internalize the particles, assess both in vitro and in vivo cytotoxicity, and determine the feasibility of this approach in large animal studies.

The continued development of in vivo cell graft tracking methodologies and improvement of intraoperative imaging techniques raise the prospect of a minimally invasive, image-guided approach to the spinal cord. A minimally invasive approach could alleviate the need for dural opening, laminectomy, and incision. Clinical experience with percutaneous cordotomy [79–81] provides foundation and precedent for an MRI- or CT-guided approach. Furthermore, percutaneous intraspinal transplantation has been performed under fluoroscopic guidance in a canine model [82]. The spatial resolution of MRI is unparalleled, and MRI is regularly used to guide the placement of DBS electrodes with millimetric accuracy [83–85]. The development of a minimally invasive, MRI-based injection device and targeting approach for intraspinal transplant has the potential to both reduce the risk for surgical complications associated with access to the spine (e.g., infection, postoperative kyphosis) and to greatly improve targeting accuracy and precision. However, potential limitations specific to a minimally invasive approach include pial vessel hemorrhage and hematoma formation, CSF leak, and inaccurate targeting due to cord displacement. While these concerns must be addressed in preclinical large animal studies, the approach remains promising.

5.4 Conclusion

Multiple international trials have been completed with the intent to deliver a putative cell-based therapeutic to the spinal cord. The published literature and our personal experience have informed an iterative approach towards an intraparenchymal microinjection approach. This has resulted in a design that has successfully completed evaluation in a phase I clinical trial and is currently being employed in a phase II clinical trial. Each of the design elements discussed, however, holds continuing challenges for future innovation and improvement. Efforts will continue towards: creating a lower profile microinjection platform and less invasive surgical approach, improving the targeting methodology and achieved accuracy, and optimizing the utilized dosing parameters. In future applications, intraparenchymal microinjection strategies may be expected to incorporate image-guided delivery approaches and mechanisms to track the delivered cellular grafts for viability and distribution. Additionally, current experiences with allograft delivery will help to clarify the needs for immunosuppression on a graft specific basis. Finally, continued technological improvement and an increased experiential understanding will help to better elucidate the roles for both direct and indirect cell delivery approaches for treatment of intrinsic spinal cord pathologies.

References

1. Riley J et al. Intraspinal stem cell transplantation in amyotrophic lateral sclerosis: a phase I safety trial, technical note, and lumbar safety outcomes. Neurosurgery. 2012;71(2):405–16. discussion 416.
2. Glass JD et al. Lumbar intraspinal injection of neural stem cells in patients with amyotrophic lateral sclerosis: results of a phase I trial in 12 patients. Stem Cells. 2012;30(6):1144–51.
3. Knoller N et al. Clinical experience using incubated autologous macrophages as a treatment for complete spinal cord injury: phase I study results. J Neurosurg Spine. 2005;3(3):173–81.
4. Jones LA et al. A phase 2 autologous cellular therapy trial in patients with acute, complete spinal cord injury: pragmatics, recruitment, and demographics. Spinal Cord. 2010;48(11): 798–807.
5. Lammertse DP et al. Autologous incubated macrophage therapy in acute, complete spinal cord injury: results of the phase 2 randomized controlled multicenter trial. Spinal Cord. 2012;50(9): 661–71.
6. Blanquer M et al. Neurotrophic bone marrow cellular nests prevent spinal motoneuron degeneration in amyotrophic lateral sclerosis patients: a pilot safety study. Stem Cells. 2012;30(6): 1277–85.
7. Mazzini L et al. Mesenchymal stromal cell transplantation in amyotrophic lateral sclerosis: a long-term safety study. Cytotherapy. 2012;14(1):56–60.
8. Mazzini L et al. Mesenchymal stem cell transplantation in amyotrophic lateral sclerosis: a phase I clinical trial. Exp Neurol. 2010;223(1):229–37.
9. Mazzini L et al. Mesenchymal stem cells for ALS patients. Amyotroph Lateral Scler. 2009; 10(2):123–4.
10. Mazzini L et al. Stem cell therapy in amyotrophic lateral sclerosis: a methodological approach in humans. Amyotroph Lateral Scler Other Motor Neuron Disord. 2003;4(3):158–61.

11. Chapman AR, Scala CC. Evaluating the first-in-human clinical trial of a human embryonic stem cell-based therapy. Kennedy Inst Ethics J. 2012;22(3):243–61.
12. Zhou XH et al. Transplantation of autologous activated Schwann cells in the treatment of spinal cord injury: six cases, more than five years of follow-up. Cell Transplant. 2012;21 Suppl 1:S39–47.
13. Huang H et al. Long-term outcome of olfactory ensheathing cell therapy for patients with complete chronic spinal cord injury. Cell Transplant. 2012;21 Suppl 1:S23–31.
14. Wu J et al. Clinical observation of fetal olfactory ensheathing glia transplantation (OEGT) in patients with complete chronic spinal cord injury. Cell Transplant. 2012;21 Suppl 1:S33–7.
15. Attar A et al. An attempt to treat patients who have injured spinal cords with intralesional implantation of concentrated autologous bone marrow cells. Cytotherapy. 2011;13(1):54–60.
16. Lima C et al. Olfactory mucosal autografts and rehabilitation for chronic traumatic spinal cord injury. Neurorehabil Neural Repair. 2010;24(1):10–22.
17. Lima C et al. Olfactory mucosa autografts in human spinal cord injury: a pilot clinical study. J Spinal Cord Med. 2006;29(3):191–203. discussion 204–6.
18. Deda H et al. Treatment of amyotrophic lateral sclerosis patients by autologous bone marrow-derived hematopoietic stem cell transplantation: a 1-year follow-up. Cytotherapy. 2009;11(1):18–25.
19. Feron F et al. Autologous olfactory ensheathing cell transplantation in human spinal cord injury. Brain. 2005;128(Pt 12):2951–60.
20. Mackay-Sim A et al. Autologous olfactory ensheathing cell transplantation in human paraplegia: a 3-year clinical trial. Brain. 2008;131(Pt 9):2376–86.
21. Saberi H et al. Treatment of chronic thoracic spinal cord injury patients with autologous Schwann cell transplantation: an interim report on safety considerations and possible outcomes. Neurosci Lett. 2008;443(1):46–50.
22. Saberi H et al. Safety of intramedullary Schwann cell transplantation for postrehabilitation spinal cord injuries: 2-year follow-up of 33 cases. J Neurosurg Spine. 2011;15(5):515–25.
23. Yoon SH et al. Complete spinal cord injury treatment using autologous bone marrow cell transplantation and bone marrow stimulation with granulocyte macrophage-colony stimulating factor: phase I/II clinical trial. Stem Cells. 2007;25(8):2066–73.
24. Chen L et al. Short-term outcome of olfactory ensheathing cells transplantation for treatment of amyotrophic lateral sclerosis. Zhongguo Xiu Fu Chong Jian Wai Ke Za Zhi. 2007;21(9):961–6.
25. Liu J et al. Clinical analysis of the treatment of spinal cord injury with umbilical cord mesenchymal stem cells. Cytotherapy. 2013;15(2):185–91.
26. Sharma A et al. Administration of autologous bone marrow-derived mononuclear cells in children with incurable neurological disorders and injury is safe and improves their quality of life. Cell Transplant. 2012;21 Suppl 1:S79–90.
27. Frolov AA, Bryukhovetskiy AS. Effects of hematopoietic autologous stem cell transplantation to the chronically injured human spinal cord evaluated by motor and somatosensory evoked potentials methods. Cell Transplant. 2012;21 Suppl 1:S49–55.
28. Karamouzian S et al. Clinical safety and primary efficacy of bone marrow mesenchymal cell transplantation in subacute spinal cord injured patients. Clin Neurol Neurosurg. 2012;114(7):935–9.
29. Saito F et al. Administration of cultured autologous bone marrow stromal cells into cerebrospinal fluid in spinal injury patients: a pilot study. Restor Neurol Neurosci. 2012;30(2):127–36.
30. Saito F et al. Spinal cord injury treatment with intrathecal autologous bone marrow stromal cell transplantation: the first clinical trial case report. J Trauma. 2008;64(1):53–9.
31. Karussis D et al. Safety and immunological effects of mesenchymal stem cell transplantation in patients with multiple sclerosis and amyotrophic lateral sclerosis. Arch Neurol. 2010;67(10):1187–94.
32. Kumar AA et al. Autologous bone marrow derived mononuclear cell therapy for spinal cord injury: a phase I/II clinical safety and primary efficacy data. Exp Clin Transplant. 2009;7(4):241–8.

33. Pal R et al. Ex vivo-expanded autologous bone marrow-derived mesenchymal stromal cells in human spinal cord injury/paraplegia: a pilot clinical study. Cytotherapy. 2009;11(7):897–911.
34. Mehta T et al. Subarachnoid placement of stem cells in neurological disorders. Transplant Proc. 2008;40(4):1145–7.
35. Ra JC et al. Safety of intravenous infusion of human adipose tissue-derived mesenchymal stem cells in animals and humans. Stem Cells Dev. 2011;20(8):1297–308.
36. Xu J et al. Clinical outcome of autologous peripheral blood stem cell transplantation in optico-spinal and conventional forms of secondary progressive multiple sclerosis in a Chinese population. Ann Hematol. 2011;90(3):343–8.
37. Cristante AF et al. Stem cells in the treatment of chronic spinal cord injury: evaluation of somatosensitive evoked potentials in 39 patients. Spinal Cord. 2009;47(10):733–8.
38. Moviglia GA et al. Combined protocol of cell therapy for chronic spinal cord injury. Report on the electrical and functional recovery of two patients. Cytotherapy. 2006;8(3):202–9.
39. Habisch HJ et al. Intrathecal application of neuroectodermally converted stem cells into a mouse model of ALS: limited intraparenchymal migration and survival narrows therapeutic effects. J Neural Transm. 2007;114(11):1395–406.
40. Mothe AJ et al. Intrathecal transplantation of stem cells by lumbar puncture for thoracic spinal cord injury in the rat. Spinal Cord. 2011;49(9):967–73.
41. Neuhuber B et al. Stem cell delivery by lumbar puncture as a therapeutic alternative to direct injection into injured spinal cord. J Neurosurg Spine. 2008;9(4):390–9.
42. Takahashi Y et al. Comparative study of methods for administering neural stem/progenitor cells to treat spinal cord injury in mice. Cell Transplant. 2011;20(5):727–39.
43. Garcia S et al. Pitfalls in spontaneous in vitro transformation of human mesenchymal stem cells. Exp Cell Res. 2010;316(9):1648–50.
44. Torsvik A et al. Spontaneous malignant transformation of human mesenchymal stem cells reflects cross-contamination: putting the research field on track - letter. Cancer Res. 2010; 70(15):6393–6.
45. Furlani D et al. Is the intravascular administration of mesenchymal stem cells safe? Mesenchymal stem cells and intravital microscopy. Microvasc Res. 2009;77(3):370–6.
46. Riley J et al. Intraspinal stem cell transplantation in ALS: a phase I safety trial, technical note & lumbar safety outcomes. Neurosurgery. 2012;71(2):405–16.
47. Riley J et al. Intraspinal stem cell transplantation in ALS: a phase I trial, cervical microinjection and final surgical safety outcomes. Neurosurgery. 2013;74(1):77–87.
48. Safety Study of GRNOPC1 in Spinal Cord Injury. [cited 9 Jan 2011]. Available from: http://www.clinicaltrials.gov/ct2/show/NCT01217008?term=Geron&rank=9
49. Huang H et al. Safety of fetal olfactory ensheathing cell transplantation in patients with chronic spinal cord injury. A 38-month follow-up with MRI. Zhongguo Xiu Fu Chong Jian Wai Ke Za Zhi. 2006;20(4):439–43.
50. Huang H et al. Influence of patients' age on functional recovery after transplantation of olfactory ensheathing cells into injured spinal cord injury. Chin Med J (Engl). 2003; 116(10):1488–91.
51. Dobkin BH, Curt A, Guest J. Cellular transplants in China: observational study from the largest human experiment in chronic spinal cord injury. Neurorehabil Neural Repair. 2006; 20(1):5–13.
52. Blanquer M et al. Bone marrow stem cell transplantation in amyotrophic lateral sclerosis: technical aspects and preliminary results from a clinical trial. Methods Find Exp Clin Pharmacol. 2010;32(Suppl A):31–7.
53. Blanquer M et al. A surgical technique of spinal cord cell transplantation in amyotrophic lateral sclerosis. J Neurosci Methods. 2010;191(2):255–7.
54. Casanova F, Carney PR, Sarntinoranont M. Influence of needle insertion speed on backflow for convection-enhanced delivery. J Biomech Eng. 2012;134(4):041006.
55. Krauze MT et al. Reflux-free cannula for convection-enhanced high-speed delivery of therapeutic agents. J Neurosurg. 2005;103(5):923–9.

56. Yin D, Forsayeth J, Bankiewicz KS. Optimized cannula design and placement for convection-enhanced delivery in rat striatum. J Neurosci Methods. 2010;187(1):46–51.
57. Vazquez LC et al. Polymer-coated cannulas for the reduction of backflow during intraparenchymal infusions. J Mater Sci Mater Med. 2012;23(8):2037–46.
58. Cunningham MG et al. Preclinical evaluation of a novel intracerebral microinjection instrument permitting electrophysiologically guided delivery of therapeutics. Neurosurgery. 2004;54(6):1497–507. discussion 1507.
59. Bjarkam CR et al. Safety and function of a new clinical intracerebral microinjection instrument for stem cells and therapeutics examined in the Gottingen minipig. Stereotact Funct Neurosurg. 2010;88(1):56–63.
60. Potts MB, Silvestrini MT, Lim DA. Devices for cell transplantation into the central nervous system: design considerations and emerging technologies. Surg Neurol Int. 2013;4 Suppl 1:S22–30.
61. Skuk D et al. Ischemic central necrosis in pockets of transplanted myoblasts in nonhuman primates: implications for cell-transplantation strategies. Transplantation. 2007; 84(10):1307–15.
62. Parsa S et al. Effects of surfactant and gentle agitation on inkjet dispensing of living cells. Biofabrication. 2010;2(2):025003.
63. Aguado BA et al. Improving viability of stem cells during syringe needle flow through the design of hydrogel cell carriers. Tissue Eng Part A. 2012;18(7–8):806–15.
64. Lu P et al. Long-distance growth and connectivity of neural stem cells after severe spinal cord injury. Cell. 2012;150(6):1264–73.
65. Riley J et al. Targeted spinal cord therapeutics delivery: stabilized platform and microelectrode recording guidance validation. Stereotact Funct Neurosurg. 2008;86(2):67–74.
66. Kang JH et al. Development of a sodium/iodide symporter (NIS)-transgenic mouse for imaging of cardiomyocyte-specific reporter gene expression. J Nucl Med. 2005;46(3):479–83.
67. MacLaren DC et al. Repetitive, non-invasive imaging of the dopamine D2 receptor as a reporter gene in living animals. Gene Ther. 1999;6(5):785–91.
68. Arena F et al. beta-Gal gene expression MRI reporter in melanoma tumor cells. Design, synthesis, and in vitro and in vivo testing of a Gd(III) containing probe forming a high relaxivity, melanin-like structure upon beta-Gal enzymatic activation. Bioconjug Chem. 2011; 22(12):2625–35.
69. Bengtsson NE et al. lacZ as a genetic reporter for real-time MRI. Magn Reson Med. 2010;63(3):745–53.
70. Yaghoubi SS et al. Noninvasive detection of therapeutic cytolytic T cells with 18F–FHBG PET in a patient with glioma. Nat Clin Pract Oncol. 2008;6(1):53–8.
71. Zurkiya O, Chan AWS, Hu X. MagA is sufficient for producing magnetic nanoparticles in mammalian cells, making it an MRI reporter. Magn Reson Med. 2008;59(6):1225–31.
72. Genove G et al. A new transgene reporter for in vivo magnetic resonance imaging. Nat Med. 2005;11(4):450–4.
73. Li Z et al. Comparison of reporter gene and iron particle labeling for tracking fate of human embryonic stem cells and differentiated endothelial cells in living subjects. Stem Cells. 2008;26(4):864–73.
74. Gonzalez-Lara LE et al. The use of cellular magnetic resonance imaging to track the fate of iron-labeled multipotent stromal cells after direct transplantation in a mouse model of spinal cord injury. Mol Imaging Biol. 2011;13(4):702–11.
75. Guzman R et al. Long-term monitoring of transplanted human neural stem cells in developmental and pathological contexts with MRI. Proc Natl Acad Sci. 2007;104(24):10211–6.
76. Obenaus A et al. Long-term magnetic resonance imaging of stem cells in neonatal ischemic injury. Ann Neurol. 2011;69(2):282–91.
77. Zhu J, Zhou L, XingWu F. Tracking neural stem cells in patients with brain trauma. N Engl J Med. 2006;355(22):2376–8.

78. Callera F, de Melo CM. Magnetic resonance tracking of magnetically labeled autologous bone marrow CD34+ cells transplanted into the spinal cord via lumbar puncture technique in patients with chronic spinal cord injury: CD34+ cells' migration into the injured site. Stem Cells Dev. 2007;16(3):461–6.
79. Kanpolat Y. Percutaneous destructive pain procedures on the upper spinal cord and brain stem in cancer pain: CT-guided techniques, indications and results. Adv Tech Stand Neurosurg. 2007;32:147–73.
80. Kanpolat Y et al. CT-guided percutaneous selective cordotomy. Acta Neurochir (Wien). 1993;123(1–2):92–6.
81. McGirt MJ et al. MRI-guided frameless stereotactic percutaneous cordotomy. Stereotact Funct Neurosurg. 2002;78(2):53–63.
82. Lee JH et al. Percutaneous transplantation of human umbilical cord blood-derived multipotent stem cells in a canine model of spinal cord injury. J Neurosurg Spine. 2009;11(6):749–57.
83. Larson PS et al. An optimized system for interventional magnetic resonance imaging-guided stereotactic surgery: preliminary evaluation of targeting accuracy. Neurosurgery. 2012;70(1 Suppl Operative):95–103. discussion 103.
84. Martin AJ et al. Placement of deep brain stimulator electrodes using real-time high-field interventional magnetic resonance imaging. Magn Reson Med. 2005;54(5):1107–14.
85. Starr PA et al. Subthalamic nucleus deep brain stimulator placement using high-field interventional magnetic resonance imaging and a skull-mounted aiming device: technique and application accuracy. J Neurosurg. 2010;112(3):479–90.

Chapter 6
Patient-Derived Induced Pluripotent Stem Cells Provide a Regenerative Medicine Platform for Duchenne Muscular Dystrophy Heart Failure

Xuan Guan, David Mack, and Martin K. Childers

6.1 Introduction

Mutations in the dystrophin gene cause dystrophinopathy, a hereditary disorder with variable allelic clinical presentations including Duchenne muscular dystrophy (DMD), Becker muscular dystrophy (BMD), and X-linked dilated cardiomyopathy (XLDC). Though the disease is well known for the skeletal muscle involvement, most patients develop cardiomyopathy and eventually succumb to congestive heart failure. With ventilatory support preventing respiratory-related mortality, the greater cardiac workload associated with longer life expectancy is believed to increase the incidence of heart failure. Currently there are no effective therapies to contain the decline of heart function in these patients. Thus, it is urgent to devise new strategies to prevent, halt, or reverse the cardiomyopathy. To address this unmet medical need, we modeled dystrophin-deficient cardiomyopathy using cellular reprogramming technology, which involves converting adult somatic cells into pluripotent stem

X. Guan
Institute for Stem Cell and Regenerative Medicine, School of Medicine,
University of Washington, Seattle, WA, USA

D. Mack
Department of Rehabilitation Medicine, University of Washington, Seattle, WA, USA

Institute for Stem Cell and Regenerative Medicine, School of Medicine,
University of Washington, Seattle, WA, USA

M.K. Childers (✉)
Department of Rehabilitation Medicine, University of Washington,
Campus Box 358056, Seattle, WA, USA

Institute for Stem Cell and Regenerative Medicine, 850 Republican Street, S421,
Seattle, WA, 98109 USA
e-mail: mkc8@uw.edu

© Springer Science+Business Media New York 2016
M.K. Childers (ed.), *Regenerative Medicine for Degenerative Muscle Diseases*,
Stem Cell Biology and Regenerative Medicine,
DOI 10.1007/978-1-4939-3228-3_6

cells, termed induced pluripotent stem cells (iPSCs). Upon further induction, iPSCs can give rise to vast number of heart cells that can be used to study disease etiology and screen therapeutic compounds.

6.2 Dystrophin-Deficient Cardiomyopathy

6.2.1 Background

Dystrophinopathy refers to a group of genetic disorders, encompassing DMD, its milder variant BMD, and XLDC [1]. Though varied in clinical presentation, these diseases share common gene defects in dystrophin resulting in varying levels of dystrophin deficiency. Cardiac symptoms are invariably associated with all dystrophinopathy patients. Rare mutations cause localized dystrophin protein defects restricted to the heart, making the heart the only affected organ in XLDC. More commonly, mutations cause devastating skeletal muscle weakness for muscular dystrophy patients that overshadows any underlying cardiac abnormality. 98 % of DMD patients develop cardiac abnormalities, while congestive heart failure (CHF) and sudden cardiac death account for 10–20 % of the mortality in DMD patients. In contrast to their relatively mild skeletal muscle involvement, many BMD patients develop evident cardiac symptoms likely due to the cardiac workload imposed by the longer life span and vigorous physical activities. Cardiac complications are estimated to account for up to 50 % of the mortality in patients with DMD [1]. The mortality associated with cardiac failure is expected to rise even further, due to improved respiratory management that decreases fatal respiratory failure and extends the patients' life span.

Other reports have highlighted the linkage of dystrophin with several forms of acquired cardiomyopathy [2–4], suggesting that dystrophin protein remodeling may represent a common pathway underlying contractile dysfunction in failing hearts [5]. Thus, restoring normal function of the dystrophin-associated glycoprotein complex (DGC) could serve as a potential therapeutic target for heart failure patients.

6.2.1.1 Dystrophin Gene and the Mutations

The gene encoding dystrophin locates to the X chromosome, spanning 79 exons and covering 2.4 Mbp [6]. While the shortest isoform, DP71, is ubiquitously expressed in multiple tissues, the full-length transcript variant Dp427m is mainly expressed in muscle tissue, including the heart [7]. By forming a dystrophin-associated glycoprotein complex (DGC) together with the sarcolemma, dystrophin mainly functions as the hub to connect the intracellular actin filament with the extracellular matrix (ECM), providing mechanical support to reinforce the sarcolemma. On the other hand, dystrophin also serves as a scaffold protein to organize molecules in proper position for function, such as membrane receptors and signaling proteins like neuronal nitric oxide synthase (nNOS) [8]. Hence, dystrophin plays a critical role in both mechanical membrane support and in proper function of certain cell signaling pathways.

Numerous genetic mutations have been identified across the whole length of the dystrophin gene, but mainly enriched within two "hotspots." The most common region, 3' end hotspot lies within exon 45–55 with genomic breakpoints at intron 44, while the 5' end hotspot covers exons 2–19 with breakpoints in intron 2 and 7 [7]. Exon deletions and duplications are the most common forms of mutations. The severity of symptoms heavily depends on the maintenance of the open reading frame, rather than the size of mutated genomic regions. The frame shift hypothesis suggests that mutations maintaining the original open reading frame lead to the production of a truncated but partially functional protein, which usually leads to a milder clinical presentation in BMD. On the other hand, mutations shifting the reading frame completely cease protein production [7] with prominent disease manifestations.

6.2.1.2   Clinical Symptoms and Management

Cardiomyopathy in dystrophic patients is largely underdiagnosed and poorly managed [9], partly due to the fact that symptoms dynamically progress over time. To monitor disease progression, electrocardiogram, echocardiogram, and magnetic resonance imaging (MRI) can be used to determine the appropriate time and course of intervention. Cardiac manifestations associated with dystrophin deficiency include rhythmic disturbance, organ structural alteration, and hemodynamic abnormalities. A typical disease course includes three distinct but continuous stages [10]. The preclinical stage usually presents with an abnormal electrocardiogram, demonstrating a variety of findings such as sinus tachycardia, premature contractions, and conduction delays [1, 11]. As the disease progresses, imaging finds evidence of cardiac hypertrophy such as increase of ventricular septal thickness and left ventricular free wall/septum ratio in the hypertrophic stage. In the advanced dilated cardiomyopathy stage, echocardiogram usually reveals ventricular dilation coupled with hemodynamic disturbances that eventually progresses to congestive heart failure.

Available therapies are limited and palliative. Conventional anti-heart failure regimens, including ACE inhibitors (ACEI), angiotensin II receptor blockers (ARBs), beta-adrenergic receptor blockers, and aldosterone antagonists, are typically prescribed in an attempt to delay heart function decline [12–14]. Corticosteroids also demonstrated benefit in several reports, although most of these studies are retrospective observations with limited sample size. More recently a cohort of 86 patients was retrospectively analyzed, and the investigators concluded on top of ACEI therapy that the use of steroids was associated with a 76 % decrease of mortality, largely driven by the reduction of heart failure-associated death. However, the corticosteroid-treated group received ACEI treatment 3 years earlier, which may be the alternative explanation for the observed effect [15]. Other therapeutic modalities, such as pacemaker [16–18], ventricular assist device (VAD) [19–22], and cardiac resynchronization therapy (CRT) [23–25], may be beneficial in decreasing fatal arrhythmias and temporarily boost heart function.

With existing regimens, the majority of patients still face inevitable cardiac failure. This cruel reality makes it imperative to pursue new strategies. However, this attempt is largely hindered by the lack of a reliable disease model. Existing animal models such as the *mdx* mouse fail to precisely reproduce human pathophysiology. For example, pharmacotherapy proven effective in *mdx* mice failed to demonstrate the equivalent efficacy in DMD patients and even worsened heart performance [26]. On the other hand, utilization of primary human cardiomyocytes is limited by risky isolation procedures and poor proliferation capacity of cells captured from human biopsy material. Therefore, a disease model system that closely mimics human symptoms and is capable of predicting in vivo efficacy is invaluable.

6.2.2 Pathogenesis

A plethora of evidence suggests that the outer cell membrane of the skeletal or cardiac muscle cell, the sarcolemma, is abnormally susceptible to mechanical stress in the face of dystrophin deficiency. This "vulnerable membrane" is characterized by the decrease of membrane stability when subjected to mechanical stretch during contraction, predisposing muscle cells to rupture. On the other hand, a spectrum of abnormal phenotypes across multiple physiological domains have also been linked to the absence of dystrophin, including but not limited to disturbance of calcium homeostasis, mitochondria dysfunction, and aberrant nNOS-cGMP signaling (Fig. 6.1). The pleiotropic effects of dystrophin deficiency are likely due to the

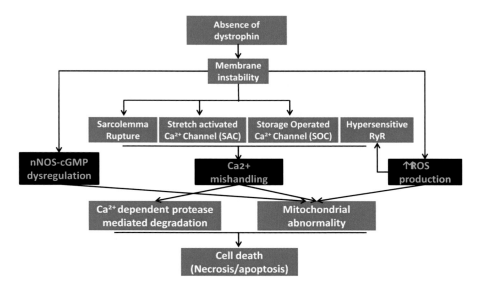

Fig. 6.1 Mechanistic scheme of the dystrophin-deficient cardiomyopathy. The absence of dystrophin causes membrane instability, triggering the abnormal calcium influx through various calcium channels. Improper accumulation of extracellular calcium together with hypersensitive ryanodine receptor leads to calcium mishandling. Coupled with dysregulated nNOS-cGMP pathway and heightened oxidative stress, the disease network triggers cell death through calcium-dependent protein degradation and mitochondria-mediated apoptosis

multifaceted function of dystrophin and to some degree complicate the clear delineation of the pathogenic process. Consequently, it is still elusive how dystrophin deficiency leads to abnormal phenotypes. Earlier work suggests that those afore-mentioned phenotypes are not merely concomitants of dystrophin deficiency, but actively contributing to disease progression. Corrections of these abnormalities, such as amplifying cGMP signaling or correction of calcium mishandling, are accompanied by ameliorated tissue pathology and improved muscle function. Because successful disease modeling in vitro is determined by faithful reproduction of in vivo disease characteristics, confirmation of the presence of the disease features in dystrophin-deficient cardiac cells renders credibility to further exploration.

6.2.2.1   Impairment of Membrane Barrier Function

Dystrophin is a cell membrane anchoring protein. Based on its sarcolemma local-ization, others postulated that the absence of dystrophin causes membrane barrier dysfunction. This idea is supported by the findings of "leaky sarcolemma." For instance, measuring serum levels of intracellular proteins, such as creatine kinase (CK), has been widely employed to assess the degree of muscle damage in muscular dystrophy. Experimentally introduced membrane impermeable substances such as albumin [27–31] and Evans Blue dye were found to accumulate within damaged dystrophin-deficient myofibers. Although these early studies confirmed the pres-ence of an altered membrane barrier function, they failed to precisely define the biophysical nature of the membrane lesion. Normally, skeletal and cardiac muscle endures mechanical strain during contraction. In dystrophin-deficient DMD muscle, discontinuation of the normal sarcolemmal membrane structure in non-necrotic muscle fibers was observed by transmission electron microscopy. This observation was coupled with pathological intracellular changes and hyper-contracture of the surrounding myofibers leading to the speculation that physical breakage of the membrane, or "micro-ruptures," induced by mechanical strain during muscle con-traction is the underlying membrane defect in dystrophin deficiency [32]. This notion is further supported by the observation that a synthetic polymer, poloxamer 188, seals membrane ruptures and reversed the dystrophic phenotype [33, 34]. In contrast to this prevailing view, Allen et al. argued that in dystrophin-deficient mus-cle, slow kinetics of trans-sarcolemmal calcium ingress following injury could not be accounted for by abrupt physical breakage. Instead, the investigators proposed that pathological activation of preexisting membrane channels are responsible for heightened membrane permeability [35]. Various calcium-permeable channels have been scrutinized for this purpose and will be discussed in the following section.

6.2.2.2   Dysregulated Calcium Handling

Calcium is a critical ion with diversified biological functions in both physiological and pathological processes of muscle. As a result, muscle has evolved highly regu-lated machinery to keep calcium flow in check. Thus, malfunction of critical proteins in this system may jeopardize the delicate balance with a detrimental effect. Increase of intracellular calcium concentration triggers muscle contraction and attenuates cellular compliance [36]. Excessive activation of calcium signaling networks could lead to cell death [11], partly through the mitochondrial death pathway [5, 37]. It is also worth noting that skeletal muscle differs from cardiac muscle in calcium-handling processes. Opening of the membrane-bound L-type calcium channel in skeletal myocytes physically interacts and activates the ryanodine receptor (RyR), the gatekeeper of sarcolemma reticulum (SR) calcium storage. While in cardiomyocytes, it is the local elevation of subsarcolemmal calcium concentration, following calcium influx through the dihydropyridine receptor (DHPR), that activates the SR RyR. This difference in calcium handling could potentially contribute to different pathophysiologies and account for organ-specific phenotypes such as heart rhythm disturbances.

1. $[Ca2+]_i$ handling in dystrophin-deficient cardiomyocytes
   Aged *mdx* cardiomyocytes demonstrate elevated resting $[Ca2+]_i$ [38] and attenuated SR calcium storage [39] (Fig. 6.2), not present in young *mdx* mice. When young *mdx* cells were stressed by mechanical stretch, they responded with a profound calcium transient that largely surpassed the WT controls [34, 39, 40] (Fig. 6.2). Both adult and young *mdx* cardiomyocytes demonstrated prolonged calcium reuptake [38, 41, 42]. On the other hand, overexpression of sarco-/endoplasmic reticulum Ca^{2+}-ATPase 2 (SERCA2), the pump responsible for sequestering calcium in SR during repolarization, normalized intracellular calcium load and corrected the abnormal EKG [43].

2. Extracellular calcium entry
   It is generally accepted that abnormal extracellular calcium entry through a disrupted membrane triggers downstream disease networks. In support of this

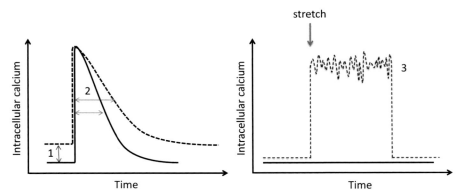

- Elevated basal calcium concentration
- Prolonged calcium decay time
- Abnormal stretch induced calcium oscillation

Fig. 6.2 Schematic representation of the abnormal calcium handling detected on cellular level

idea, removal of extracellular calcium completely abrogates pathological $[Ca^{2+}]_i$ accumulation [40]. However, the biophysical identity of the entry route is still obscure. Though membrane tears resulting from mechanical injury has long been suspected to be the culprit instigator, recently Whitehead et al. argued that the delayed $[Ca2+]i$ uprise following mechanical stretch contradicts membrane micro-rupture to be the leading path of calcium entry [44]. Alternatively, they suggested a group of calcium-permeable cation channels. The so-called stretch-activated channels (SACs) were first described by Franco and Lansman to be abnormally active in *mdx* mice myotubes [45]. Squire et al. later confirmed that the occurrence and opening probability of this channel were greater in *mdx* myofibers [46]. Blocking this channel by gadolinium, streptomycin, and GsMT-4 significantly ameliorated intracellular calcium overload induced by mechanical stretch [38, 47, 48]. Unfortunately, though the functionality of SAC has been confirmed by electrophysiological exams, little is known about its protein identity. Some suggested involvement of a transient receptor potential channel (TRPC). For example, a recent study revealed in *mdx* heart a twofold increase of TRPC vanilloid channels type 2 (TRPV2) expression and a mislocalization from the cytoplasm to the sarcolemma. Knocking down TRPV2 by siRNA or antibody blocking TRPV2 channel abrogated mechanically induced intracellular calcium accumulation [49].

Other membrane-localized Ca^{2+} channels have also been investigated in the context of dystrophin deficiency. The sarcolemma L-type calcium channel, DHPR, is the main calcium channel that triggers excitation-contraction coupling and directly interacts with dystrophin in the t-tubule system [50] [51]. DHPR channel inactivation was delayed in cardiomyocytes from neonatal [50] and adult *mdx* mice [52, 53]. Moreover, on *mdx* cardiomyocytes, the L-type calcium channel mediates an enhanced calcium influx that contributes to prolonged action potential duration [52]. These findings inspired studies to test the effect of calcium inhibitors in DMD patients. However, neither animal experiments nor clinical trials have demonstrated significant benefit [54, 55].

3. Intracellular calcium release through hyperactive RyR
Trans-sarcolemmal calcium influx is augmented in DMD cardiomyocytes. However, this change alone is insufficient to account for the cyclic abnormal calcium oscillations observed in DMD cardiomyocytes, as shown in Fig. 6.2 [36, 40]. The self-sustaining nature of the calcium oscillation suggests the involvement of intracellular calcium storage of the SR. Several groups have reported that the function of RyR, the molecule gating SR calcium release, is altered in DMD cardiomyocytes [36, 56]. A "leaky" RyR loses control over intracellular SR calcium release, contributing to calcium dysregulation. Ullrich et al. reported that the sensitivity of RyR is increased in *mdx* cardiomyocytes, responding to low concentration of extracellular calcium that could not activate wild-type RyR [37]. Prosser et al. further demonstrated that sensitized RyR responds to stimuli by a profound SR calcium liberation, which potentially accounts for DMD cardiac rhythm disturbances [36].

6.2.3 Excessive ROS Production

Cumulative evidence suggests a key role of elevated oxidative stress in the degeneration of dystrophin-deficient skeletal muscle [44, 57, 58]. Exposing *mdx* cardiomyocytes to hypoosmotic or stretch stress triggers higher-than-normal reactive oxygen species (ROS) production [36, 40], likely through upregulating key enzymes in ROS production such as NAPDH oxidase 2 (NOX2) [36], NOX4, and Lysyl oxidase (LOX) [59]. Antagonizing ROS by supplementing the ROS scavenger, N-acetylcysteine (NAC), attenuates calcium-handling abnormalities, preserves heart function, and mitigates inflammation and fibrosis [60].

The role of oxidative stress and its downstream molecular consequences has been well documented in other forms of cardiomyopathy [61]. Though direct evidence is lacking, it is reasonable to argue that a similar pathway may also be activated in dystrophin-deficient cardiomyocytes. Excessive oxidation of functional proteins inflicts cell damage and deteriorates heart function. Those susceptible targets include membrane channels, contractile proteins [62], signaling proteins [63], and EC coupling components [64]. For example, Prosser et al. revealed that the ROS generated via NOX2 oxidize RyR, which is the molecular basis of its hypersensitivity [36].

6.2.4 Mitochondria Dysfunction

Functional imaging studies through positron emission tomography (PET) identified regional abnormalities in the DMD heart, characterized by energy substrate shift from fatty acid to glucose [65–67]. Lately, alterations of metabolic substrate were experimentally confirmed in young *mdx* heart, prior to overt cardiomyopathy, coupled with increased oxygen consumption, glycolysis, and ATP production rate [68]. The same group also discovered that the mitochondria permeability transition pore (mPTP), an important mediator of mitochondrial apoptosis pathway, is more susceptible to open when challenged by stressors [5], suggesting that mitochondria are actively involved in disease progression and may represent an intervention target.

6.2.5 Abnormal nNOS-cGMP Signaling

In skeletal muscle, the neuronal nitric oxide synthase (nNOS) is anchored to the c-terminus of dystrophin to control nitric oxide (NO) production. While active NO modulates target proteins by binding to a thiol residue via s-nitrosylation, NO can also function through activating soluble guanylyl cyclase (sGC) to produce cGMP as the secondary messenger. In DMD patients, the absence of dystrophin causes nNOS mislocalization, which greatly dampens NOS function by up to 80 %. The benefits of nNOS overexpression in skeletal muscle include anti-inflammation, anti-fibrosis, and additional protection of myofibers against mechanical stress [69]. nNOS-mediated NO production also exerts a protective effect in the heart [8]. Differing from its sarcolemmal localization in skeletal myofiber, nNOS mainly associates with SR [70] and mitochondria [71] in the heart, without direct interaction with dystrophin. Surprisingly, Bia et al. discovered in dystrophin/utrophin double knockout mice an 80 % decrease of cardiac nNOS activity. Encouraged by their previous success in skeletal muscle [69], overexpressing nNOS in *mdx* heart demonstrated salutary effect to decrease inflammation and fibrosis [72]. Another group overexpressed the sGC and administration of the PDE5 inhibitor sildenafil with the goal to activate the sGC pathway. Remarkably, both these strategies improved mitochondrial function, attenuated stress-induced mPTP opening, and exerted sarcolemma protection against workload-induced damage [73, 74]. Sildenafil has also been shown capable to reverse cardiomyopathy in aged *mdx* mice, boosting depressed cardiac function to comparable levels to age-matched controls [75]. Taken together, evidence suggests that cGMP-mediated effects account for benefits associated with modulating the nNOS pathway. Commercially available PDE5 inhibitors may hold promise in treating heart disease in DMD.

6.3   Regenerative Cellular Technologies

Advances in genetics, epigenetics, and stem cell biology have empowered scientists with new tools to control cell fate. A variety of human tissues, including cardiomyocytes, neurons, hepatocytes, and endothelial cells can now be produced from stem cells in large quantities. This new technology paves the way for a myriad of downstream applications, such as studying disease mechanisms, screening therapeutic compounds, and enabling cellular replacement therapy.

6.3.1   *The Germination of Nuclear Reprogramming*

Mammalian development is accompanied by diversification of progeny cell identity, realized by a gradual loss of cell plasticity. It was once believed that development is an irreversible process. Unidirectionality was ensured by a gradual loss of genetic material [76]. Consequently, cell fate was considered to be static and inter-lineage switch of cellular identity impossible. In the middle twentieth century, this dogma of a "one-way street" was challenged. John Gurdon discovered that differentiated tadpole muscle and intestinal nuclei could generate mature fertile adults after being transplanted into enucleated Xenopus eggs. This discovery unequivocally proved that the genome is relatively stable across the life span of individual organisms. The development process is not coupled with attrition of genetic material. When provided with the correct cue, an adult nucleus is capable to initiate and maintain normal development in a similar fashion as zygotes [77]. Miller and Ruddle had also shown that differentiated thymus cells could be reset to a primitive stage equivalent to pluripotent embryonic stem cells by fusing with embryonic carcinoma cells [78]. These findings broadened the "one-way street" to "two-way traffic" but were still constrained within a longitudinal developmental path. Later, Helen Blau utilized a similar cell fusion technique, termed heterokaryon, to successfully convert human non-muscle cell nuclei to acquire a skeletal muscle gene expression profile. This was the first demonstration of an inter-lineage switch [79]. Together, these findings revealed the surprising plasticity of nuclei and foreshadowed an era of nuclear reprogramming to manipulate cellular fate by altering gene expression.

6.3.2 Nuclear Reprogramming 2.0

6.3.2.1 Induced Pluripotency

A lesson learned from the somatic nuclear transfer and heterokaryon experiments is that certain molecules within the recipient's cytoplasm redirect a terminally differentiated cell type to acquire a different identity. Consequently, identifying these limited molecules and overexpressing them in donor cells might simplify the process. The cosmologically vast number of molecules contained in the cytoplasm made this a formidable task. After decades of exploration, a seminal breakthrough occurred in 2006. Yamanaka and Takahashi identified ectopic expression of four transcription factors, Oct-4, Sox-2, c-Myc, and Klf4 [80], that converted both mouse [80] and human [81] fibroblasts to pluripotent stem cells. Another report around the same time demonstrated that nanog and Lin28 could replace c-Myc and Klf4 [82]. Converted somatic cells, termed, induced pluripotent stem cells (iPSCs), possess indefinite self-renewal capacity and the potential to generate virtually any cell types within the body. To date, although small differences have been reported between iPSC and hESC [83–87], these two behave largely the same. Not only do iPSC and ES cells share similar morphology and gene expression profiles, more importantly, iPSC cells acquire bona fide pluripotency, forming teratoma tumors after engrafting into immune-compromised animals. Mouse iPSCs could even pass the most stringent tetraploid complementation assay to generate progenies solely composed of cells differentiated from iPSCs [88].

Recent technique evolution has overcome some major hurdles for clinical translation of iPSC technology. The poor reprogramming efficiency and slow kinetics once the bottlenecks of iPSC derivation were overcome by optimizing the reprogramming factor combination and starting cell population [89–90]. Lately, a remarkable 100 % efficiency was achieved by simultaneously knocking down the MDB3 gene [91] together with reprogramming factor delivery. On the other hand, the concern of insertional mutagenesis, elicited by random viral integration, is addressed by "footprint-free" reprogramming technology. Non-integration delivery vectors, such as sendai virus, episome, or mRNA, can achieve efficient reprogramming without any genome perturbation.

6.3.2.2 Inter-Lineage Cell Fate Conversion

Derivation of iPSCs by transient, ectopic expression of transgenes demarcates the new era of nuclear reprogramming. Compared to previous SCNT experiments, the major improvement of this new version of nuclear reprogramming is that cell fate conversion is achieved through manipulating a small set of predefined factors. This breakthrough considerably lowered the technical barrier and encouraged further exploration. Following experiments demonstrated that other cell fates could also be rerouted via ectopic expression of a handful transgenes (Table 6.1). Inter-lineage cell fate conversion bypasses the pluripotent stem cell stage. This process, termed "trans-differentiation," usually occurred with a donor cell developmentally related to the target cells, although trans-germ layer conversions have also been documented.

Table 6.1 Summary of human cells inter-lineage conversions

Donor cells	Target cells	Reprogramming factors
Fibroblasts	Melanocyte	Mitf, Sox10, and Pax3 [92]
Pancreatic exocrine cells	Pancreatic beta cells	Activated MAPK and STAT3 [93]
Fibroblasts	Hematopoietic progenitor with macrophage potential	Sox2 [94]
Fibroblasts	Multipotent neural crest progenitor	Sox10, Wnt activation [95]
Endothelial cells	Hematopoietic progenitors	Fosb, Gfi1, Runx1, and Spi1 with vascular niche monolayers [96]
Fibroblasts	Hepatocytes	FoxA3, Hnf1a, and Hnf4a [97]
Fibroblasts	Hepatocytes	Hnf1a, Hnf4a, Hnf6, Atf5, Prox1, and CEBPA [98]
Fibroblasts	Retinal pigment epithelium like cells	c-Myc, Mitf, Otx2, Rax, and Crx [99]
Fetal lung fibroblasts	Cholinergic neurons	Neurogenin2 supplemented with forskolin and dorsomorphin [100]
Cardiac fibroblasts	Cardiomyocytes	Gata4, Mef2c, and Tbx5 Mesp1 and Myocd [101]
Proximal tubule cells	Nephron progenitors	Six1, Six2, Osr1, Eya1, HoxA11, and Snai2 [102]
Fibroblasts	Cardiomyocytes	GATA4, Hand2, Tbx5, myocardin, miR1, and miR133 [103]
Fibroblasts	Cardiomyocytes	Ets2 and Mesp1 [104]
Fibroblasts	Multipotent neural stem cells	Sox2 [105]
Fibroblasts	Dopaminergic neurons	Mash1, Ngn2, Sox2, Nurr1, and Pitx3 [106]
Fibroblasts	Motor neurons	Brn2, Ascl1, Myt1l, Lhx3, HB9, Isl1, Ngn2 [107]
Fibroblasts	Neurons	Brn2, Myt1l, miR124 [108]
Fibroblasts	Neurons	Ascl1, Myt1ll, NeuroD2, miR9/9*, miR124 [109]

The majority of these trans-differentiation transgenes are transcriptional factors, chromatin modifiers, and miRNAs critical to normal development of certain lineages. Surprisingly only a handful of factors are required to achieve these transformations, which may emphasize that cell identities are maintained through certain "master regulators," such as MyoD for skeletal muscle [110]. Several studies have demonstrated that similar cell fate conversion could be achieved in vivo, such as cardiomyocytes [111], pancreatic beta cells [112, 113], and neurons [114, 115], showcasing the versatility of the reprogramming technology.

6.3.3   Guided Differentiation

An alternative strategy to achieve in vitro cell fate control is to guide pluripotent stem cells to differentiate into desired cell types. Insights gained from embryonic development have enabled scientists to harness the physiological cues to hijack the intrinsic developmental program. The key is to recapitulate those early events governing lineage commitment and germ layers' specification, by temporally modulating critical signaling pathways [116]. A broad spectrum of cell types including cardiomyocytes, hepatocytes, pancreatic beta cells, endothelium, and dopaminergic neurons have been generated in this fashion. The advantages of this method are twofold. First, large quantity, highly purified target cells can be manufactured through an optimized induction protocol. For example, cardiomyocyte production can be over 90 % in purity with 3 output cardiomyocytes for every input stem cell. This scale is compatible with an industrial setting to produce clinically relevant cell numbers, usually in the billions [117]. Secondly, starting from a pluripotent stem cell offers a unique opportunity for genetic engineering to generate homogenous cell line with desired genetic modifications. The following section will focus on cardiac induction. The specification of other germ layers is well reviewed by Murry et al. [118].

It was noticed that beating cardiomyocytes could be generated from embryoid bodies (EB), stem cell aggregates mimicking early developing embryo structure. The efficiency of spontaneous cardiogenesis is low (less than 10 %) with large batch-to-batch variation [118, 119]. Not until insights gained from development were applied have investigators started to consistently generate adequate cardiomyocytes for downstream studies. In the embryo, cardiac progenitors are derived from *brachyury T-* and *Emos*-positive mesoderm cells, which further give rise to *KDR-* and *PDGFRα*-positive cardiac mesoderm cells and gradually turn on the cardiac master regulator *MESP1*. Further specification of cardiac mesoderm generates cardiac progenitors, characterized by the expression of a panel of cardiac-specific transcriptional factors such as *Nkx2.5, GATA4, Tbx5,* and *Isl1*. Temporally and spatially orchestrated Nodal, BMP4, and Wnt3 signaling ensure this organized sequential progression. Artificially supplementing these ligands in culture medium, mimicking the physiological strength and timing, can also differentiate pluripotent stem cells into cardiomyocytes. With some technical variation, this scheme consistently generates cardiomyocytes in around 2 weeks, with an efficiency ranging from 30 to 90 % [120]. It is important to note that the Wnt signaling has a unique biphasic role in cardiac development. Lately the Palecek group had shown that by fine-tuning Wnt pathway alone by two molecules, GSK-3β inhibitor CHIRO-99021 and Wnt inhibitor IWP-4 or IWR-2, cardiomyocytes can be generated from multiple cell lines with an efficiency over 90 % [121].

Various strategies have demonstrated that in vitro cardiac induction follows the same gene expression pattern that would occur during embryonic development. Downregulation of pluripotent genes is accompanied by upregulation of mesoderm markers Brachyury T and MESP1, which peak around day 2. Cardiac progenitor markers KDR, ISL1, and PDGFR-α peak at day 5, followed by a steady increase of mature myocyte markers such as Nkx2.5, myosin heavy chain (MHC), and troponin [122–125]. Differentiated cells manifest sarcomere striation, typical action potentials and associated ion channels [126], potent gap junctions for synchronized contraction, as well as positive humeral regulation response following isoproterenol stimulation [119, 127, 128]. Klug et al. demonstrated that stem cell-derived cardiomyocytes express dystrophin [129]. While this evidence suggests that these are bona fide cardiomyocytes, stem cell-derived cardiomyocytes largely mimic a developmentally immature cardiac phenotype. Transcriptional profiling suggested that the cardiac gene expression pattern was similar to 20-gestational-week fetal cardiomyocytes [130]. Electrophysiological assays indicated that stem cell-derived cardiomyocytes possess fetal-type ion channels [122], leading to a fetal-like negative force frequency relationship, blunted post-rest potentiation [131], higher resting potential, and slow action potential upstroke [130, 132]. Ultrastructure analysis revealed that the characteristic components, such as contraction machinery, sarcomeric reticulum, transverse tubules (t-tubule), and mitochondria, were all present, but their abundance, distribution, and organization failed to reach the level of adult cardiomyocytes [133].

6.3.4 From Urine to Dystrophin-Deficient Cardiomyocytes

Our group demonstrated that urine harbors a unique cell population capable to adhere to plastic and undergo extensive proliferation. These urine-derived cells manifest spindle-shape morphology and express classic MSC surface markers including CD44, CD73, CD90, CD105, and CD146 [134, 135]. Functional assays reveal that urine-derived cells possess progenitor properties and give rise to several somatic cell types [136]. When transduced with lentiviral vector containing classic OKSM reprogramming factors, these cells demonstrate faster reprogramming kinetics compared to mesenchymal stem cells, generating iPSC colonies within 2 weeks [134]. Urine-derived iPSC colonies, from both normal individual and DMD patient urine, can give rise to embryonic tumors containing characteristic tissue structures of all three germ layers, demonstrating bona fide pluripotency.

Pluripotent stem cells

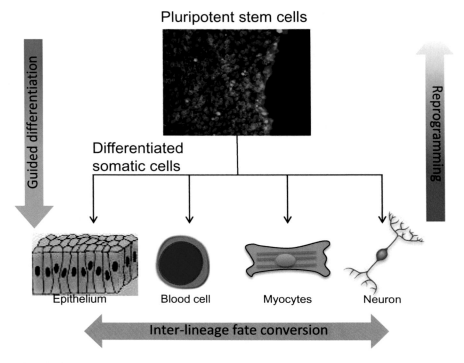

Fig. 6.3 Cell fate controlling. Strategies to ex vivo manipulate cellular identities, including guided differentiation following developmental cues, reprogramming to pluripotency by forced overexpression of reprogramming factors, and inter-lineages by various techniques

Exposing the urine-derived iPSC cells to a combination of growth factors, including Activin A, bone morphogenetic protein 4, and dickkopf-1, in a strict sequence and defined duration [137] forms beating cardiomyocytes with an efficiency from 40 to 90 %. These differentiated cardiomyocytes show typical sarcomere structure and are positive for sarcomeric α-actinin, cardiac myosin heavy chain, as well as membrane-localized connexin43. They also exhibit functional cardiac properties, spontaneous action potentials characteristic of nodal, ventricular, and atrial subtypes. On the other hand, dystrophin expression was absent from DMD cardiomyocytes, recapitulating the essential aspect of the DMD disease [134] (Figs. 6.3 and 6.4).

6.4 Dystrophin-Deficient Cardiomyocytes Generated Via Cellular Reprogramming Are a Novel Biological Reagent for the Study of DMD Cardiomyopathy

The list of iPSC-based disease models is steadily growing. Impressively, different diseased cells manifest phenotypes analogous to symptoms in patients in culture dishes. A partial list includes models for amyotrophic lateral sclerosis, spinal

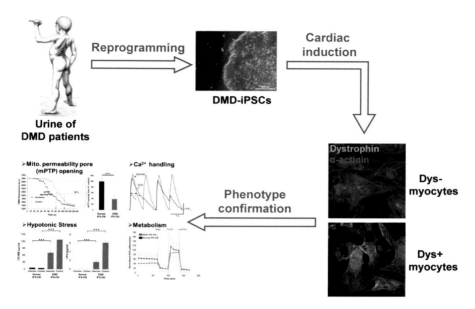

Fig. 6.4 A strategy to model dystrophin-deficient cardiomyopathy via cellular reprogramming

muscular atrophy, familial dysautonomia [138], Rett syndrome [139], schizophrenia [140], Parkinson disease [141–144], Timothy syndrome [145], and long QT syndrome [146]. The explosion of the iPSC application is largely due to the fact that cellular reprogramming is a powerful tool that strongly resonates in the era of personalized medicine, which enables identifying and optimizing solutions for specific mutations. DMD cardiomyopathy is well suited to iPSC-based "disease-in-a-dish" platform. First, there is currently no effective treatment for DMD cardiomyopathy. Secondly, the molecular mechanism of the cardiomyopathy is not well understood. Though the disease root cause is clear, a variety of subdomains, such as membrane stability, calcium handling, and mitochondria abnormalities, have been suggested in the pathogenic process. The complexity of the underlying pathogenic network makes identifying therapeutic targets a daunting task. Thirdly, the most widely used DMD animal model, the *mdx* mouse, does not accurately reproduce the cardiac pathology observed in human patients [148]. Thus it is imperative to study this disease in a human context to render clinical relevance. Furthermore, the risk associated with cardiac biopsy makes acquiring samples from DMD patients nearly impossible. These limiting factors impose significant constraints in developing therapies for human patients. As a result, human cardiomyocytes differentiated from patient iPSCs represent an attractive option to overcome these limitations.

It is technically favorable to model Duchenne cardiomyopathy using the iPSC approach mainly because DMD is a classic Mendelian monogenic disease with complete penetrance in male patients. A large body of literature suggests that the disease phenotype is cell-autonomous, indicating that disease traits can be readily observed independent of the diseased body environment. Compared to polygenic genetic disorder or complex disorders involving strong environmental attributes, the technical complexity associated with modeling dystrophin-deficient cardiomyopathy ex vivo is considerably lower. Moreover, the existence of a highly efficient cardiac induction regimen bolsters this strategy by lowing the technical hurdle [121, 137].

6.4.1 Personalized Diagnosis: Exploring Disease Etiology

Presently, disease diagnosis has evolved from clinical and functional to molecular diagnosis using biomarkers [148, 149] and genome sequencing [150, 151]. However, genotype-phenotype relationships are not always strongly correlated, and sometimes it is difficult to elucidate a causal relationship. Toward this end, iPSC technology represents a unique opportunity to converge information from multiple levels and directly link them to a change in cellular function. For example, in iPSC-differentiated disease target cells, the mutated gene can be sequenced, biochemical reactions can be measured, and abnormal metabolites can be quantified. Readouts from these measurements can be further linked to cellular responses in a quantifiable fashion. In addition, cell culture systems allow intricate experimental interventions at virtually any level, from genetic to environmental, greatly aiding in determining the disease root cause. Lastly, because clinically relevant cells are differentiated stepwise from stem cells in a fashion that recapitulates the embryonic development, disease initiation and early progression, hard to assess by other modalities, can now be closely monitored. It is possible to exploit these models to uncover early disease markers before overt symptoms, increasing the opportunity for early intervention or even prevention. For instance, Kim et al. identified in patient iPSC-derived cardiomyocytes an abnormal peroxisome proliferator-activated receptor gamma (PPAR-γ) activation that underlies the pathogenesis of arrhythmogenic right ventricular dysplasia (ARVC), a previously unrecognized mechanism [152]. iPSC-based platforms also simplifies the diagnosis procedure since tests only need to be conducted on single diseased lineage in isolation, removed from potential secondary effects of residing within a sick animal.

6.4.2 Personalized Therapy: In Search for the Ideal Treatment

Because of its individualized nature, iPSC technology can also serve as an invaluable tool to identify new therapies. First, it can be employed as a platform to predict the potency and toxicity of existing therapies [153, 154]. It is not uncommon that a spectrum of drugs is available for a single disease, yet not every drug is equally effective for all patients. The potency and toxicity, which largely depend on individual's unique genetic background, are hard to predict based on existing modalities. On the other hand, iPSC cells have the potential to be the stage for such forecasts, since the patients' own cells are tested. Such tests are not necessarily limited to chemical compounds. As a cellular assay, other therapeutics, such as gene therapy vectors, can also be assayed as a quality control measure to predict its clinical efficacy. For those disorders without effective treatment, like DMD cardiomyopathy, de novo screening against an extensive compound library can potentially identify new treatments with no risk to the patient [155].

The differentiated cells themselves can also serve as therapeutics. One route to cure a genetic disorder may involve replacing the diseased cells with genetically corrected counterparts. The emergence of novel genetic engineering tools, such as CRISPR and TALEN enzymes, has greatly improved the feasibility of complex genome modification of human stem cells [144, 156–159]. Barrier of immunogenicity could be overcome by autologous transplantation of the progeny differentiated from iPSC. This strategy has been applied to improve skeletal muscle function in dystrophin-deficient *mdx* mice [160]. In terms of the heart, though previously it has been shown that genetic heart disease benefited from embryonic stem cell transplantation [161], the delivery method still imposes the biggest challenge for inherited disorders including DMD.

6.4.3 Dystrophin-Deficient Cardiomyocytes Generated Via Cellular Reprogramming Recapitulate Certain Aspects of the Disease Phenotype

Though studies have shown that disease target cells differentiated from patient's iPS cells faithfully reproduce disease-associated abnormalities, it is still questionable whether the cells created in laboratory, at a developmentally naïve stage, will be able to fully demonstrate a phenotype of late-onset pathology observed in DMD. Therefore the first task is to demonstrate progeny differentiated from iPSC cells that retain certain disease hallmarks, which will then lend credibility to novel insights gained on an iPSC platform. For example, ventricular cells differentiated from a long QT syndrome patient demonstrated the characteristic prolongation of the QT interval as well as an abnormal potassium channel [146, 154, 162].

In our hands, to model DMD cardiomyopathy, cardiomyocytes were assessed both molecularly and phenotypically for well-established DMD features. For example, the absence of dystrophin protein is the root cause of DMD. Confirmed both by immunostaining (Fig. 6.4), cardiomyocytes differentiated from DMD iPS cells were negative for dystrophin expression, recapitulating an essential aspect of the disease. It is well established that the absence of dystrophin causes a spectrum of physiologies deviating from normal control. As discussed earlier, membrane fragility is suspected to be a direct consequence of the absence of dystrophin and initiates downstream pathological cascades. One way to assess the membrane barrier function is measuring the release of intracellular content during a hypotonic stress challenge. In vitro, an increase of intracellular hydrostatic pressure passively stretches the cell's sarcolemma. DMD cells are abnormally susceptible to mechanical stress, and their fragile membrane predisposes cells to rupture, leading to the liberation of intracellular contents. As an example of a hypotonic stress assay, red blood cells are subjected to a series of graded hypotonic solutions, and intracorpuscular hemoglobin release is assayed to diagnose *hereditary spherocytosis* [163]. We employed a similar strategy to mechanically stretch the cardiac sarcolemma by incubating normal and DMD cells in hypotonic solutions ranging from normal tonicity to 1/8 tonicity. The cardiac-specific injury marker CK-MB, a widely used clinical marker, was measured after 30 min. Dystrophin-deficient cardiomyocytes demonstrated an abnormally high release CK-MB profile. Measured as CK-MB concentration, increased levels were demonstrated even at relatively normal tonicity and became marked greater than normal cells as tonicity decreased (Fig. 6.4). This evidence supports the notion that dystrophin deficiency leads to sarcolemma fragility, which predisposes cardiomyocytes to mechanical stretch damage.

Calcium mishandling is another characteristic feature of DMD. Loaded with the calcium indicator Fluo-4, iPSC-derived DMD cardiomyocytes were paced by external field stimulation. Compared to dystrophin replete (normal) control cells, DMD cells demonstrated a prolonged calcium decay time T_{50}, consistent with previous reports in the dystrophin-deficient mouse [38, 41, 42].

Mitochondria dysfunction observed in DMD patients was recently recognized as an important mediator of disease progression. Two prominent features, a lowered opening threshold for mitochondrial permeability transition pore (mPTP) and altered bioenergetics, have been linked to the disease in the *mdx* mouse. To evaluate mPTP pore opening, cardiomyocytes were loaded with mitochondria potential ($\Delta\psi m$) indicator tetramethylrhodamine ethyl ester (TMRE) and then exposed to focused laser to induce oxidative stress-mediated mPTP opening. The opening of mPTP leads to the decrease of $\Delta\psi m$, reflected by lowering of TMRE fluorescence. Dystrophin-deficient cardiomyocytes manifested a 50 % shorter mPTP time compared to normal cells. In other experiments, an extracellular biochemical analyzer (Seahorse) was employed to examine the bioenergetics of cardiomyocytes. Oxygen consumption rate (OCR) was measured as the parameter of metabolism activity. Surprisingly, DMD cells demonstrated augmented basal and maximal oxygen consumption, different from the report on *mdx* skeletal muscle but in agreement with a study using a Langendorff perfused *mdx* adult heart, in which the *mdx* heart exhibited elevated level of glycolysis, carbohydrate utilization, oxygen consumption, and overall ATP production [5].

6.5   Summary

Duchenne muscular dystrophy causes degeneration of both skeletal and cardiac muscle. All DMD patients inevitably develop heart disease, but the mechanism of DMD heart failure is still elusive, partly due to the hard-to-access heart cells from patients. We modeled heart disease in DMD using cellular reprogramming by converting patient urine cells into induced pluripotent stem cells or iPSCs. IPSCs were further transformed into heart cells for study. This regenerative medicine technology provided an unprecedented opportunity to study DMD heart disease from several physiological perspectives including membrane fragility, calcium handling, and energy metabolism. This regenerative medicine platform will facilitate exploration of DMD disease etiology and provide unlimited patient biologic material to screen new therapeutic compounds.

References

1. Hermans MC, Pinto YM, Merkies IS, de Die-Smulders CE, Crijns HJ, Faber CG. Hereditary muscular dystrophies and the heart. Neuromuscul Disord. 2010;20:479–92.
2. Armstrong SC, Latham CA, Shivell CL, Ganote CE. Ischemic loss of sarcolemmal dystrophin and spectrin: correlation with myocardial injury. J Mol Cell Cardiol. 2001;33:1165–79.
3. Kido M, Otani H, Kyoi S, Sumida T, Fujiwara H, Okada T, Imamura H. Ischemic preconditioning-mediated restoration of membrane dystrophin during reperfusion correlates with protection against contraction-induced myocardial injury. Am J Physiol Heart Circ Physiol. 2004;287:H81–90.
4. Lee GH, Badorff C, Knowlton KU. Dissociation of sarcoglycans and the dystrophin carboxyl terminus from the sarcolemma in enteroviral cardiomyopathy. Circ Res. 2000;87:489–95.
5. Burelle Y, Khairallah M, Ascah A, Allen BG, Deschepper CF, Petrof BJ, Des Rosiers C. Alterations in mitochondrial function as a harbinger of cardiomyopathy: lessons from the dystrophic heart. J Mol Cell Cardiol. 2010;48:310–21.
6. Hoffman EP, Brown Jr RH, Kunkel LM. Dystrophin: the protein product of the Duchenne muscular dystrophy locus. Cell. 1987;51:919–28.
7. Muntoni F, Torelli S, Ferlini A. Dystrophin and mutations: one gene, several proteins, multiple phenotypes. Lancet Neurol. 2003;2:731–40.
8. Percival JM, Adamo CM, Beavo JA, Froehner SC. Evaluation of the therapeutic utility of phosphodiesterase 5A inhibition in the mdx mouse model of duchenne muscular dystrophy. Handb Exp Pharmacol. 2011;323–344.
9. Spurney C, Shimizu R, Morgenroth LP, Kolski H, Gordish-Dressman H, Clemens PR, Investigators C. Cooperative International Neuromuscular Research Group Duchenne Natural History Study demonstrates insufficient diagnosis and treatment of cardiomyopathy in Duchenne muscular dystrophy. Muscle Nerve. 2014;50:250–6.
10. Nigro G, Comi LI, Politano L, Bain RJ. The incidence and evolution of cardiomyopathy in Duchenne muscular dystrophy. Int J Cardiol. 1990;26:271–7.
11. Townsend D, Yasuda S, Metzger J. Cardiomyopathy of Duchenne muscular dystrophy: pathogenesis and prospect of membrane sealants as a new therapeutic approach. Expert Rev Cardiovasc Ther. 2007;5:99–109.
12. Ishikawa Y, Bach JR, Minami R. Cardioprotection for Duchenne's muscular dystrophy. Am Heart J. 1999;137:895–902.
13. Matsumura T, Tamura T, Kuru S, Kikuchi Y, Kawai M. Carvedilol can prevent cardiac events in Duchenne muscular dystrophy. Intern Med. 2010;49:1357–63.
14. Rhodes J, Margossian R, Darras BT, Colan SD, Jenkins KJ, Geva T, Powell AJ. Safety and efficacy of carvedilol therapy for patients with dilated cardiomyopathy secondary to muscular dystrophy. Pediatr Cardiol. 2008;29:343–51.
15. Schram G, Fournier A, Leduc H, Dahdah N, Therien J, Vanasse M, Khairy P. All-cause mortality and cardiovascular outcomes with prophylactic steroid therapy in Duchenne muscular dystrophy. J Am Coll Cardiol. 2013;61:948–54.
16. Fayssoil A, Orlikowski D, Nardi O, Annane D. Complete atrioventricular block in Duchenne muscular dystrophy. Europace. 2008;10:1351–2.

17. Fayssoil A, Orlikowski D, Nardi O, Annane D. Pacemaker implantation for sinus node dysfunction in a young patient with Duchenne muscular dystrophy. Congest Heart Fail. 2010;16:127–8.
18. Takano N, Honke K, Hasui M, Ohno I, Takemura H. A case of pacemaker implantation for complete atrioventricular block associated with Duchenne muscular dystrophy. No To Hattatsu. 1997;29:476–80.
19. Amodeo A, Adorisio R. Left ventricular assist device in Duchenne cardiomyopathy: can we change the natural history of cardiac disease? Int J Cardiol. 2012;161, e43.
20. Davies JE, Winokur TS, Aaron MF, Benza RL, Foley BA, Holman WL. Cardiomyopathy in a carrier of Duchenne's muscular dystrophy. J Heart Lung Transplant. 2001;20:781–4.
21. Smith MC, Arabia FA, Tsau PH, Smith RG, Bose RK, Woolley DS, Rhenman BE, Sethi GK, Copeland JG. CardioWest total artificial heart in a moribund adolescent with left ventricular thrombi. Ann Thorac Surg. 2005;80:1490–2.
22. Webb ST, Patil V, Vuylsteke A. Anaesthesia for non-cardiac surgery in patient with Becker's muscular dystrophy supported with a left ventricular assist device. Eur J Anaesthesiol. 2007;24:640–2.
23. Andrikopoulos G, Kourouklis S, Trika C, Tzeis S, Rassias I, Papademetriou C, Katsivas A, Theodorakis G. Cardiac resynchronization therapy in Becker muscular dystrophy. Hellenic J Cardiol. 2013;54:227–9.
24. Fayssoil A, Abasse S. Cardiac resynchronization therapy in Becker muscular dystrophy: for which patients? Hellenic J Cardiol. 2010;51:377–8.
25. Stollberger C, Finsterer J. Left ventricular synchronization by biventricular pacing in Becker muscular dystrophy as assessed by tissue Doppler imaging. Heart Lung. 2005;34:317–20.
26. Leung DG, Herzka DA, Thompson WR, He B, Bibat G, Tennekoon G, Russell SD, Schuleri KH, Lardo AC, Kass DA, Thompson RE, Judge DP, Wagner KR. Sildenafil does not improve cardiomyopathy in Duchenne/Becker muscular dystrophy. Ann Neurol. 2014;76(4):541–9.
27. Childers MK, Okamura CS, Bogan DJ, Bogan JR, Petroski GF, McDonald K, Kornegay JN. Eccentric contraction injury in dystrophic canine muscle. Arch Phys Med Rehabil. 2002;83:1572–8.
28. Childers MK, Okamura CS, Bogan DJ, Bogan JR, Sullivan MJ, Kornegay JN. Myofiber injury and regeneration in a canine homologue of Duchenne muscular dystrophy. Am J Phys Med Rehabil. 2001;80:175–81.
29. Childers MK, Staley JT, Kornegay JN, McDonald KS. Skinned single fibers from normal and dystrophin-deficient dogs incur comparable stretch-induced force deficits. Muscle Nerve. 2005;31:768–71.
30. McNeil PL, Khakee R. Disruptions of muscle fiber plasma membranes. Role in exercise-induced damage. Am J Pathol. 1992;140:1097–109.
31. Petrof BJ, Shrager JB, Stedman HH, Kelly AM, Sweeney HL. Dystrophin protects the sarcolemma from stresses developed during muscle contraction. Proc Natl Acad Sci U S A. 1993;90:3710–4.
32. Mokri B, Engel AG. Duchenne dystrophy: electron microscopic findings pointing to a basic or early abnormality in the plasma membrane of the muscle fiber. Neurology. 1975;25: 1111–20.
33. Townsend D, Turner I, Yasuda S, Martindale J, Davis J, Shillingford M, Kornegay JN, Metzger JM. Chronic administration of membrane sealant prevents severe cardiac injury and ventricular dilatation in dystrophic dogs. J Clin Invest. 2010;120:1140–50.
34. Yasuda S, Townsend D, Michele DE, Favre EG, Day SM, Metzger JM. Dystrophic heart failure blocked by membrane sealant poloxamer. Nature. 2005;436:1025–9.
35. Allen DG, Whitehead NP. Duchenne muscular dystrophy--what causes the increased membrane permeability in skeletal muscle? Int J Biochem Cell Biol. 2011;43:290–4.
36. Prosser BL, Ward CW, Lederer WJ. X-ROS signaling: rapid mechano-chemo transduction in heart. Science. 2011;333:1440–5.
37. Ullrich ND, Fanchaouy M, Gusev K, Shirokova N, Niggli E. Hypersensitivity of excitation-contraction coupling in dystrophic cardiomyocytes. Am J Physiol Heart Circ Physiol. 2009;297:H1992–2003.

38. Williams IA, Allen DG. Intracellular calcium handling in ventricular myocytes from mdx mice. Am J Physiol Heart Circ Physiol. 2007;292:H846–55.
39. Kyrychenko S, Polakova E, Kang C, Pocsai K, Ullrich ND, Niggli E, Shirokova N. Hierarchical accumulation of RyR post-translational modifications drives disease progression in dystrophic cardiomyopathy. Cardiovasc Res. 2013;97:666–75.
40. Jung C, Martins AS, Niggli E, Shirokova N. Dystrophic cardiomyopathy: amplification of cellular damage by Ca2+ signalling and reactive oxygen species-generating pathways. Cardiovasc Res. 2008;77:766–73.
41. Cheng YJ, Lang D, Caruthers SD, Efimov IR, Chen J, Wickline SA. Focal but reversible diastolic sheet dysfunction reflects regional calcium mishandling in dystrophic mdx mouse hearts. Am J Physiol Heart Circ Physiol. 2012;303:H559–68.
42. Koenig X, Dysek S, Kimbacher S, Mike AK, Cervenka R, Lukacs P, Nagl K, Dang XB, Todt H, Bittner RE, Hilber K. Voltage-gated ion channel dysfunction precedes cardiomyopathy development in the dystrophic heart. PLoS ONE. 2011;6, e20300.
43. Shin JH, Bostick B, Yue Y, Hajjar R, Duan D. SERCA2a gene transfer improves electrocardiographic performance in aged mdx mice. J Transl Med. 2011;9:132.
44. Whitehead NP, Yeung EW, Allen DG. Muscle damage in mdx (dystrophic) mice: role of calcium and reactive oxygen species. Clin Exp Pharmacol Physiol. 2006;33:657–62.
45. Franco Jr A, Lansman JB. Calcium entry through stretch-inactivated ion channels in mdx myotubes. Nature. 1990;344:670–3.
46. Squire S, Raymackers JM, Vandebrouck C, Potter A, Tinsley J, Fisher R, Gillis JM, Davies KE. Prevention of pathology in mdx mice by expression of utrophin: analysis using an inducible transgenic expression system. Hum Mol Genet. 2002;11:3333–44.
47. Fanchaouy M, Polakova E, Jung C, Ogrodnik J, Shirokova N, Niggli E. Pathways of abnormal stress-induced Ca2+ influx into dystrophic mdx cardiomyocytes. Cell Calcium. 2009;46:114–21.
48. Teichmann MD, Wegner FV, Fink RH, Chamberlain JS, Launikonis BS, Martinac B, Friedrich O. Inhibitory control over Ca(2+) sparks via mechanosensitive channels is disrupted in dystrophin deficient muscle but restored by mini-dystrophin expression. PLoS ONE. 2008;3, e3644.
49. Lorin C, Vögeli I, Niggli E. Dystrophic cardiomyopathy - role of TRPV2 channels in stretch-induced cell damage. Cardiovasc Res. 2015;106(1):153–62.
50. Sadeghi A, Doyle AD, Johnson BD. Regulation of the cardiac L-type Ca2+ channel by the actin-binding proteins alpha-actinin and dystrophin. Am J Physiol Cell Physiol. 2002;282: C1502–11.
51. Friedrich O, von Wegner F, Chamberlain JS, Fink RH, Rohrbach P. L-type Ca2+ channel function is linked to dystrophin expression in mammalian muscle. PLoS ONE. 2008;3, e1762.
52. Koenig X, Rubi L, Obermair GJ, Cervenka R, Dang XB, Lukacs P, Kummer S, Bittner RE, Kubista H, Todt H, Hilber K. Enhanced currents through L-type calcium channels in cardiomyocytes disturb the electrophysiology of the dystrophic heart. Am J Physiol Heart Circ Physiol. 2014;306:H564–73.
53. Woolf PJ, Lu S, Cornford-Nairn R, Watson M, Xiao XH, Holroyd SM, Brown L, Hoey AJ. Alterations in dihydropyridine receptors in dystrophin-deficient cardiac muscle. Am J Physiol Heart Circ Physiol. 2006;290:H2439–45.
54. Cohn RD, Durbeej M, Moore SA, Coral-Vazquez R, Prouty S, Campbell KP. Prevention of cardiomyopathy in mouse models lacking the smooth muscle sarcoglycan-sarcospan complex. J Clin Invest. 2001;107:R1–7.
55. Phillips MF, Quinlivan R. Calcium antagonists for Duchenne muscular dystrophy. Cochrane Database Syst Rev. 2008, CD004571.
56. Fauconnier J, Thireau J, Reiken S, Cassan C, Richard S, Matecki S, Marks AR, Lacampagne A. Leaky RyR2 trigger ventricular arrhythmias in Duchenne muscular dystrophy. Proc Natl Acad Sci U S A. 2010;107:1559–64.
57. Menazza S, Blaauw B, Tiepolo T, Toniolo L, Braghetta P, Spolaore B, Reggiani C, Di Lisa F, Bonaldo P, Canton M. Oxidative stress by monoamine oxidases is causally involved in myofiber damage in muscular dystrophy. Hum Mol Genet. 2010;19:4207–15.

58. Rando TA. Oxidative stress and the pathogenesis of muscular dystrophies. Am J Phys Med Rehabil. 2002;81:S175–86.
59. Spurney CF, Knoblach S, Pistilli EE, Nagaraju K, Martin GR, Hoffman EP. Dystrophin-deficient cardiomyopathy in mouse: expression of Nox4 and Lox are associated with fibrosis and altered functional parameters in the heart. Neuromuscul Disord. 2008;18:371–81.
60. Williams IA, Allen DG. The role of reactive oxygen species in the hearts of dystrophin-deficient mdx mice. Am J Physiol Heart Circ Physiol. 2007;293:H1969–77.
61. Nediani C, Raimondi L, Borchi E, Cerbai E. Nitric oxide/reactive oxygen species generation and nitroso/redox imbalance in heart failure: from molecular mechanisms to therapeutic implications. Antioxid Redox Signal. 2011;14:289–331.
62. Gao WD, Liu Y, Marban E. Selective effects of oxygen free radicals on excitation-contraction coupling in ventricular muscle. Implications for the mechanism of stunned myocardium. Circulation. 1996;94:2597–604.
63. Aikawa R, Komuro I, Yamazaki T, Zou Y, Kudoh S, Tanaka M, Shiojima I, Hiroi Y, Yazaki Y. Oxidative stress activates extracellular signal-regulated kinases through Src and Ras in cultured cardiac myocytes of neonatal rats. J Clin Invest. 1997;100:1813–21.
64. Zima AV, Blatter LA. Redox regulation of cardiac calcium channels and transporters. Cardiovasc Res. 2006;71:310–21.
65. Momose M, Iguchi N, Imamura K, Usui H, Ueda T, Miyamoto K, Inaba S. Depressed myocardial fatty acid metabolism in patients with muscular dystrophy. Neuromuscul Disord. 2001;11:464–9.
66. Perloff JK, Henze E, Schelbert HR. Alterations in regional myocardial metabolism, perfusion, and wall motion in Duchenne muscular dystrophy studied by radionuclide imaging. Circulation. 1984;69:33–42.
67. Quinlivan RM, Lewis P, Marsden P, Dundas R, Robb SA, Baker E, Maisey M. Cardiac function, metabolism and perfusion in Duchenne and Becker muscular dystrophy. Neuromuscul Disord. 1996;6:237–46.
68. Khairallah M, Labarthe F, Bouchard B, Danialou G, Petrof BJ, Des RC. Profiling substrate fluxes in the isolated working mouse heart using 13C-labeled substrates: focusing on the origin and fate of pyruvate and citrate carbons. Am J Physiol Heart Circ Physiol. 2004;286:H1461–70.
69. Wehling M, Spencer MJ, Tidball JG. A nitric oxide synthase transgene ameliorates muscular dystrophy in mdx mice. J Cell Biol. 2001;155:123–31.
70. Xu KY, Huso DL, Dawson TM, Bredt DS, Becker LC. Nitric oxide synthase in cardiac sarcoplasmic reticulum. Proc Natl Acad Sci U S A. 1999;96:657–62.
71. Elfering SL, Sarkela TM, Giulivi C. Biochemistry of mitochondrial nitric-oxide synthase. J Biol Chem. 2002;277:38079–86.
72. Wehling-Henricks M, Jordan MC, Roos KP, Deng B, Tidball JG. Cardiomyopathy in dystrophin-deficient hearts is prevented by expression of a neuronal nitric oxide synthase transgene in the myocardium. Hum Mol Genet. 2005;14:1921–33.
73. Ascah A, Khairallah M, Daussin F, Bourcier-Lucas C, Godin R, Allen BG, Petrof BJ, Des Rosiers C, Burelle Y. Stress-induced opening of the permeability transition pore in the dystrophin-deficient heart is attenuated by acute treatment with sildenafil. Am J Physiol Heart Circ Physiol. 2011;300:H144–53.
74. Khairallah M, Khairallah RJ, Young ME, Allen BG, Gillis MA, Danialou G, Deschepper CF, Petrof BJ, Des RC. Sildenafil and cardiomyocyte-specific cGMP signaling prevent cardiomyopathic changes associated with dystrophin deficiency. Proc Natl Acad Sci U S A. 2008;105:7028–33.
75. Adamo CM, Dai DF, Percival JM, Minami E, Willis MS, Patrucco E, Froehner SC, Beavo JA. Sildenafil reverses cardiac dysfunction in the mdx mouse model of Duchenne muscular dystrophy. Proc Natl Acad Sci U S A. 2010;107:19079–83.
76. Briggs R, King TJ. Nuclear transplantation studies on the early gastrula (Rana pipiens). I. Nuclei of presumptive endoderm. Dev Biol. 1960;2:252–70.

77. Gurdon JB, Elsdale TR, Fischberg M. Sexually mature individuals of Xenopus laevis from the transplantation of single somatic nuclei. Nature. 1958;182:64–5.
78. Miller RA, Ruddle FH. Pluripotent teratocarcinoma-thymus somatic cell hybrids. Cell. 1976;9:45–55.
79. Blau HM, Chiu CP, Webster C. Cytoplasmic activation of human nuclear genes in stable heterocaryons. Cell. 1983;32:1171–80.
80. Takahashi K, Yamanaka S. Induction of pluripotent stem cells from mouse embryonic and adult fibroblast cultures by defined factors. Cell. 2006;126:663–76.
81. Takahashi K, Tanabe K, Ohnuki M, Narita M, Ichisaka T, Tomoda K, Yamanaka S. Induction of pluripotent stem cells from adult human fibroblasts by defined factors. Cell. 2007;131:861–72.
82. Yu J, Vodyanik MA, Smuga-Otto K, Antosiewicz-Bourget J, Frane JL, Tian S, Nie J, Jonsdottir GA, Ruotti V, Stewart R, Slukvin II, Thomson JA. Induced pluripotent stem cell lines derived from human somatic cells. Science. 2007;318:1917–20.
83. Kim K, Doi A, Wen B, Ng K, Zhao R, Cahan P, Kim J, Aryee MJ, Ji H, Ehrlich LI, Yabuuchi A, Takeuchi A, Cunniff KC, Hongguang H, McKinney-Freeman S, Naveiras O, Yoon TJ, Irizarry RA, Jung N, Seita J, Hanna J, Murakami P, Jaenisch R, Weissleder R, Orkin SH, Weissman IL, Feinberg AP, Daley GQ. Epigenetic memory in induced pluripotent stem cells. Nature. 2010;467:285–90.
84. Kim K, Zhao R, Doi A, Ng K, Unternaehrer J, Cahan P, Hongguang H, Loh YH, Aryee MJ, Lensch MW, Li H, Collins JJ, Feinberg AP, Daley GQ. Donor cell type can influence the epigenome and differentiation potential of human induced pluripotent stem cells. Nat Biotechnol. 2011;29(12):1117–9.
85. Marchetto MC, Yeo GW, Kainohana O, Marsala M, Gage FH, Muotri AR. Transcriptional signature and memory retention of human-induced pluripotent stem cells. PLoS ONE. 2009;4, e7076.
86. Ohi Y, Qin H, Hong C, Blouin L, Polo JM, Guo T, Qi Z, Downey SL, Manos PD, Rossi DJ, Yu J, Hebrok M, Hochedlinger K, Costello JF, Song JS, Ramalho-Santos M. Incomplete DNA methylation underlies a transcriptional memory of somatic cells in human iPS cells. Nat Cell Biol. 2011;13:541–9.
87. Polo JM, Liu S, Figueroa ME, Kulalert W, Eminli S, Tan KY, Apostolou E, Stadtfeld M, Li Y, Shioda T, Natesan S, Wagers AJ, Melnick A, Evans T, Hochedlinger K. Cell type of origin influences the molecular and functional properties of mouse induced pluripotent stem cells. Nat Biotechnol. 2010;28:848–55.
88. Zhao XY, Li W, Lv Z, Liu L, Tong M, Hai T, Hao J, Guo CL, Ma QW, Wang L, Zeng F, Zhou Q. iPS cells produce viable mice through tetraploid complementation. Nature. 2009;461: 86–90.
89. Kim JB, Greber B, Arauzo-Bravo MJ, Meyer J, Park KI, Zaehres H, Scholer HR. Direct reprogramming of human neural stem cells by OCT4. Nature. 2009;461:649–53.
90. Liu H, Ye Z, Kim Y, Sharkis S, Jang YY. Generation of endoderm-derived human induced pluripotent stem cells from primary hepatocytes. Hepatology. 2010;51:1810–9.
91. Mukamel Z, Hagai T, Gilad S, Amann-Zalcenstein D, Tanay A, Amit I, Novershtern N, Hanna JH. Deterministic direct reprogramming of somatic cells to pluripotency. Nature. 2013;502:65–70.
92. Yang R, Zheng Y, Li L, Liu S, Burrows M, Wei Z, Nace A, Herlyn M, Cui R, Guo W, Cotsarelis G, Xu X. Direct conversion of mouse and human fibroblasts to functional melanocytes by defined factors. Nat Commun. 2014;5:5807.
93. Lemper M, Leuckx G, Heremans Y, German MS, Heimberg H, Bouwens L, Baeyens L. Reprogramming of human pancreatic exocrine cells to beta-like cells. Cell Death Differ. 2014;22(7):1117–30.
94. Pulecio J, Nivet E, Sancho-Martinez I, Vitaloni M, Guenechea G, Xia Y, Kurian L, Dubova I, Bueren J, Laricchia-Robbio L, Izpisua Belmonte JC. Conversion of human fibroblasts into monocyte-like progenitor cells. Stem Cells. 2014;32:2923–38.

95. Kim YJ, Lim H, Li Z, Oh Y, Kovlyagina I, Choi IY, Dong X, Lee G. Generation of multipotent induced neural crest by direct reprogramming of human postnatal fibroblasts with a single transcription factor. Cell Stem Cell. 2014;15:497–506.
96. Sandler VM, Lis R, Liu Y, Kedem A, James D, Elemento O, Butler JM, Scandura JM, Rafii S. Reprogramming human endothelial cells to haematopoietic cells requires vascular induction. Nature. 2014;511:312–8.
97. Huang P, Zhang L, Gao Y, He Z, Yao D, Wu Z, Cen J, Chen X, Liu C, Hu Y, Lai D, Hu Z, Chen L, Zhang Y, Cheng X, Ma X, Pan G, Wang X, Hui L. Direct reprogramming of human fibroblasts to functional and expandable hepatocytes. Cell Stem Cell. 2014;14:370–84.
98. Du Y, Wang J, Jia J, Song N, Xiang C, Xu J, Hou Z, Su X, Liu B, Jiang T, Zhao D, Sun Y, Shu J, Guo Q, Yin M, Sun D, Lu S, Shi Y, Deng H. Human hepatocytes with drug metabolic function induced from fibroblasts by lineage reprogramming. Cell Stem Cell. 2014;14:394–403.
99. Zhang K, Liu GH, Yi F, Montserrat N, Hishida T, Esteban CR, Izpisua Belmonte JC. Direct conversion of human fibroblasts into retinal pigment epithelium-like cells by defined factors. Protein Cell. 2014;5:48–58.
100. Liu ML, Zang T, Zou Y, Chang JC, Gibson JR, Huber KM, Zhang CL. Small molecules enable neurogenin 2 to efficiently convert human fibroblasts into cholinergic neurons. Nat Commun. 2013;4:2183.
101. Wada R, Muraoka N, Inagawa K, Yamakawa H, Miyamoto K, Sadahiro T, Umei T, Kaneda R, Suzuki T, Kamiya K, Tohyama S, Yuasa S, Kokaji K, Aeba R, Yozu R, Yamagishi H, Kitamura T, Fukuda K, Ieda M. Induction of human cardiomyocyte-like cells from fibroblasts by defined factors. Proc Natl Acad Sci U S A. 2013;110:12667–72.
102. Hendry CE, Vanslambrouck JM, Ineson J, Suhaimi N, Takasato M, Rae F, Little MH. Direct transcriptional reprogramming of adult cells to embryonic nephron progenitors. J Am Soc Nephrol. 2013;24:1424–34.
103. Nam YJ, Song K, Luo X, Daniel E, Lambeth K, West K, Hill JA, DiMaio JM, Baker LA, Bassel-Duby R, Olson EN. Reprogramming of human fibroblasts toward a cardiac fate. Proc Natl Acad Sci U S A. 2013;110:5588–93.
104. Islas JF, Liu Y, Weng KC, Robertson MJ, Zhang S, Prejusa A, Harger J, Tikhomirova D, Chopra M, Iyer D, Mercola M, Oshima RG, Willerson JT, Potaman VN, Schwartz RJ. Transcription factors ETS2 and MESP1 transdifferentiate human dermal fibroblasts into cardiac progenitors. Proc Natl Acad Sci U S A. 2012;109:13016–21.
105. Ring KL, Tong LM, Balestra ME, Javier R, Andrews-Zwilling Y, Li G, Walker D, Zhang WR, Kreitzer AC, Huang Y. Direct reprogramming of mouse and human fibroblasts into multipotent neural stem cells with a single factor. Cell Stem Cell. 2012;11:100–9.
106. Liu X, Li F, Stubblefield EA, Blanchard B, Richards TL, Larson GA, He Y, Huang Q, Tan AC, Zhang D, Benke TA, Sladek JR, Zahniser NR, Li CY. Direct reprogramming of human fibroblasts into dopaminergic neuron-like cells. Cell Res. 2012;22:321–32.
107. Son EY, Ichida JK, Wainger BJ, Toma JS, Rafuse VF, Woolf CJ, Eggan K. Conversion of mouse and human fibroblasts into functional spinal motor neurons. Cell Stem Cell. 2011;9:205–18.
108. Ambasudhan R, Talantova M, Coleman R, Yuan X, Zhu S, Lipton Stuart A, Ding S. Direct reprogramming of adult human fibroblasts to functional neurons under defined conditions. Cell Stem Cell. 2011;9:113–8.
109. Yoo AS, Sun AX, Li L, Shcheglovitov A, Portmann T, Li Y, Lee-Messer C, Dolmetsch RE, Tsien RW, Crabtree GR. MicroRNA-mediated conversion of human fibroblasts to neurons. Nature. 2011;476:228–31.
110. Tapscott SJ, Davis RL, Thayer MJ, Cheng PF, Weintraub H, Lassar AB. MyoD1: a nuclear phosphoprotein requiring a Myc homology region to convert fibroblasts to myoblasts. Science. 1988;242:405–11.
111. Qian L, Huang Y, Spencer CI, Foley A, Vedantham V, Liu L, Conway SJ, Fu J-D, Srivastava D. In vivo reprogramming of murine cardiac fibroblasts into induced cardiomyocytes. Nature. 2012;485:593–8.
112. Li W, Nakanishi M, Zumsteg A, Shear M, Wright C, Melton DA, Zhou Q. In vivo reprogramming of pancreatic acinar cells to three islet endocrine subtypes. Elife. 2014;3, e01846.

113. Zhou Q, Brown J, Kanarek A, Rajagopal J, Melton DA. In vivo reprogramming of adult pancreatic exocrine cells to beta-cells. Nature. 2008;455:627–32.
114. De la Rossa A, Bellone C, Golding B, Vitali I, Moss J, Toni N, Luscher C, Jabaudon D. In vivo reprogramming of circuit connectivity in postmitotic neocortical neurons. Nat Neurosci. 2013;16:193–200.
115. Guo Z, Zhang L, Wu Z, Chen Y, Wang F, Chen G. In vivo direct reprogramming of reactive glial cells into functional neurons after brain injury and in an Alzheimer's disease model. Cell Stem Cell. 2014;14:188–202.
116. Murry CE, Keller G. Differentiation of embryonic stem cells to clinically relevant populations: lessons from embryonic development. Cell. 2008;132:661–80.
117. Laflamme MA, Murry CE. Regenerating the heart. Nat Biotechnol. 2005;23:845–56.
118. Kehat I, Kenyagin-Karsenti D, Snir M, Segev H, Amit M, Gepstein A, Livne E, Binah O, Itskovitz-Eldor J, Gepstein L. Human embryonic stem cells can differentiate into myocytes with structural and functional properties of cardiomyocytes. J Clin Invest. 2001;108:407–14.
119. Zhang J, Wilson GF, Soerens AG, Koonce CH, Yu J, Palecek SP, Thomson JA, Kamp TJ. Functional cardiomyocytes derived from human induced pluripotent stem cells. Circ Res. 2009;104:e30–41.
120. Burridge Paul W, Keller G, Gold Joseph D, Wu Joseph C. Production of de novo cardiomyocytes: human pluripotent stem cell differentiation and direct reprogramming. Cell Stem Cell. 2012;10:16–28.
121. Lian X, Hsiao C, Wilson G, Zhu K, Hazeltine LB, Azarin SM, Raval KK, Zhang J, Kamp TJ, Palecek SP. Robust cardiomyocyte differentiation from human pluripotent stem cells via temporal modulation of canonical Wnt signaling. Proc Natl Acad Sci U S A. 2012;109(27): E1848–57.
122. Beqqali A, Kloots J, Ward-van Oostwaard D, Mummery C, Passier R. Genome-wide transcriptional profiling of human embryonic stem cells differentiating to cardiomyocytes. Stem Cells. 2006;24:1956–67.
123. Burridge PW, Thompson S, Millrod MA, Weinberg S, Yuan X, Peters A, Mahairaki V, Koliatsos VE, Tung L, Zambidis ET. A universal system for highly efficient cardiac differentiation of human induced pluripotent stem cells that eliminates interline variability. PLoS ONE. 2011;6, e18293.
124. Uosaki H, Fukushima H, Takeuchi A, Matsuoka S, Nakatsuji N, Yamanaka S, Yamashita JK. Efficient and scalable purification of cardiomyocytes from human embryonic and induced pluripotent stem cells by VCAM1 surface expression. PLoS ONE. 2011;6, e23657.
125. Yang L, Soonpaa MH, Adler ED, Roepke TK, Kattman SJ, Kennedy M, Henckaerts E, Bonham K, Abbott GW, Linden RM, Field LJ, Keller GM. Human cardiovascular progenitor cells develop from a KDR+ embryonic-stem-cell-derived population. Nature. 2008;453:524–8.
126. Ma J, Guo L, Fiene SJ, Anson BD, Thomson JA, Kamp TJ, Kolaja KL, Swanson BJ, January CT. High purity human-induced pluripotent stem cell-derived cardiomyocytes: electrophysiological properties of action potentials and ionic currents. Am J Physiol Heart Circ Physiol. 2011;301:H2006–17.
127. Liu J, Sun N, Bruce MA, Wu JC, Butte MJ. Atomic force mechanobiology of pluripotent stem cell-derived cardiomyocytes. PLoS ONE. 2012;7, e37559.
128. Pillekamp F, Haustein M, Khalil M, Emmelheinz M, Nazzal R, Adelmann R, Nguemo F, Rubenchyk O, Pfannkuche K, Matzkies M, Reppel M, Bloch W, Brockmeier K, Hescheler J. Contractile properties of early human embryonic stem cell-derived cardiomyocytes: beta-adrenergic stimulation induces positive chronotropy and lusitropy but not inotropy. Stem Cells Dev. 2012;21(12):2111–21.
129. Klug MG, Soonpaa MH, Koh GY, Field LJ. Genetically selected cardiomyocytes from differentiating embronic stem cells form stable intracardiac grafts. J Clin Invest. 1996;98:216–24.
130. Cao F, Wagner RA, Wilson KD, Xie X, Fu JD, Drukker M, Lee A, Li RA, Gambhir SS, Weissman IL, Robbins RC, Wu JC. Transcriptional and functional profiling of human embryonic stem cell-derived cardiomyocytes. PLoS ONE. 2008;3, e3474.

131. Binah O, Dolnikov K, Sadan O, Shilkrut M, Zeevi-Levin N, Amit M, Danon A, Itskovitz-Eldor J. Functional and developmental properties of human embryonic stem cells-derived cardiomyocytes. J Electrocardiol. 2007;40:S192–6.
132. He JQ, Ma Y, Lee Y, Thomson JA, Kamp TJ. Human embryonic stem cells develop into multiple types of cardiac myocytes: action potential characterization. Circ Res. 2003;93:32–9.
133. Satin J, Itzhaki I, Rapoport S, Schroder EA, Izu L, Arbel G, Beyar R, Balke CW, Schiller J, Gepstein L. Calcium handling in human embryonic stem cell-derived cardiomyocytes. Stem Cells. 2008;26:1961–72.
134. Guan X, Mack DL, Moreno CM, Strande JL, Mathieu J, Shi Y, Markert CD, Wang Z, Liu G, Lawlor MW, Moorefield EC, Jones TN, Fugate JA, Furth ME, Murry CE, Ruohola-Baker H, Zhang Y, Santana LF, Childers MK. Dystrophin-deficient cardiomyocytes derived from human urine: new biologic reagents for drug discovery. Stem Cell Res. 2014;12:467–80.
135. Zhang Y, McNeill E, Tian H, Soker S, Andersson KE, Yoo JJ, Atala A. Urine derived cells are a potential source for urological tissue reconstruction. J Urol. 2008;180:2226–33.
136. Bharadwaj S, Liu G, Shi Y, Wu R, Yang B, He T, Fan Y, Lu X, Zhou X, Liu H, Atala A, Rohozinski J, Zhang Y. Multipotential differentiation of human urine-derived stem cells: potential for therapeutic applications in urology. Stem Cells. 2013;31:1840–56.
137. Laflamme MA, Chen KY, Naumova AV, Muskheli V, Fugate JA, Dupras SK, Reinecke H, Xu C, Hassanipour M, Police S, O'Sullivan C, Collins L, Chen Y, Minami E, Gill EA, Ueno S, Yuan C, Gold J, Murry CE. Cardiomyocytes derived from human embryonic stem cells in pro-survival factors enhance function of infarcted rat hearts. Nat Biotechnol. 2007;25:1015–24.
138. Lee G, Papapetrou EP, Kim H, Chambers SM, Tomishima MJ, Fasano CA, Ganat YM, Menon J, Shimizu F, Viale A, Tabar V, Sadelain M, Studer L. Modelling pathogenesis and treatment of familial dysautonomia using patient-specific iPSCs. Nature. 2009;461:402–6.
139. Marchetto MC, Carromeu C, Acab A, Yu D, Yeo GW, Mu Y, Chen G, Gage FH, Muotri AR. A model for neural development and treatment of Rett syndrome using human induced pluripotent stem cells. Cell. 2010;143:527–39.
140. Brennand KJ, Simone A, Jou J, Gelboin-Burkhart C, Tran N, Sangar S, Li Y, Mu Y, Chen G, Yu D, McCarthy S, Sebat J, Gage FH. Modelling schizophrenia using human induced pluripotent stem cells. Nature. 2011;473:221–5.
141. Cooper O, Seo H, Andrabi S, Guardia-Laguarta C, Graziotto J, Sundberg M, McLean JR, Carrillo-Reid L, Xie Z, Osborn T, Hargus G, Deleidi M, Lawson T, Bogetofte H, Perez-Torres E, Clark L, Moskowitz C, Mazzulli J, Chen L, Volpicelli-Daley L, Romero N, Jiang H, Uitti RJ, Huang Z, Opala G, Scarffe LA, Dawson VL, Klein C, Feng J, Ross OA, Trojanowski JQ, Lee VM-Y, Marder K, Surmeier DJ, Wszolek ZK, Przedborski S, Krainc D, Dawson TM, Isacson O. Pharmacological rescue of mitochondrial deficits in iPSC-derived neural cells from patients with familial Parkinson's disease. Sci Transl Med. 2012;4, 141ra190.
142. Kaplitt MG, Feigin A, Tang C, Fitzsimons HL, Mattis P, Lawlor PA, Bland RJ, Young D, Strybing K, Eidelberg D, During MJ. Safety and tolerability of gene therapy with an adeno-associated virus (AAV) borne GAD gene for Parkinson's disease: an open label, phase I trial. Lancet. 2007;369:2097–105.
143. Ryan Scott D, Dolatabadi N, Chan Shing F, Zhang X, Akhtar Mohd W, Parker J, Soldner F, Sunico Carmen R, Nagar S, Talantova M, Lee B, Lopez K, Nutter A, Shan B, Molokanova E, Zhang Y, Han X, Nakamura T, Masliah E, Yates John R, Nakanishi N, Andreyev Aleksander Y, Okamoto S-i, Jaenisch R, Ambasudhan R, Lipton Stuart A. Isogenic human iPSC Parkinson s model shows nitrosative stress-induced dysfunction in MEF2-PGC1± transcription. Cell. 2013;155:1351–64.
144. Soldner F, Laganière J, Cheng Albert W, Hockemeyer D, Gao Q, Alagappan R, Khurana V, Golbe Lawrence I, Myers Richard H, Lindquist S, Zhang L, Guschin D, Fong Lauren K, Vu BJ, Meng X, Urnov Fyodor D, Rebar Edward J, Gregory Philip D, Zhang HS, Jaenisch R. Generation of isogenic pluripotent stem cells differing exclusively at two early onset Parkinson point mutations. Cell. 2011;146:318–31.
145. Yazawa M, Hsueh B, Jia X, Pasca AM, Bernstein JA, Hallmayer J, Dolmetsch RE. Using induced pluripotent stem cells to investigate cardiac phenotypes in Timothy syndrome. Nature. 2011;471:230–4.

146. Malan D, Friedrichs S, Fleischmann BK, Sasse P. Cardiomyocytes obtained from induced pluripotent stem cells with long-QT syndrome 3 recapitulate typical disease-specific features in vitro. Circ Res. 2011;109:841–7.
147. Sacco A, Mourkioti F, Tran R, Choi J, Llewellyn M, Kraft P, Shkreli M, Delp S, Pomerantz JH, Artandi SE, Blau HM. Short telomeres and stem cell exhaustion model Duchenne muscular dystrophy in mdx/mTR mice. Cell. 2010;143:1059–71.
148. Braunwald E. Biomarkers in heart failure. N Engl J Med. 2008;358:2148–59.
149. Coca SG, Yalavarthy R, Concato J, Parikh CR. Biomarkers for the diagnosis and risk stratification of acute kidney injury: a systematic review. Kidney Int. 2008;73:1008–16.
150. Berg JS, Khoury MJ, Evans JP. Deploying whole genome sequencing in clinical practice and public health: meeting the challenge one bin at a time. Genet Med. 2011;13:499–504.
151. Rizzo JM, Buck MJ. Key principles and clinical applications of "next-generation" DNA sequencing. Cancer Prev Res. 2012;5:887–900.
152. Kim C, Wong J, Wen J, Wang S, Wang C, Spiering S, Kan NG, Forcales S, Puri PL, Leone TC, Marine JE, Calkins H, Kelly DP, Judge DP, Chen HS. Studying arrhythmogenic right ventricular dysplasia with patient-specific iPSCs. Nature. 2013;494:105–10.
153. Lan F, Lee AS, Liang P, Sanchez-Freire V, Nguyen PK, Wang L, Han L, Yen M, Wang Y, Sun N, Abilez OJ, Hu S, Ebert AD, Navarrete EG, Simmons CS, Wheeler M, Pruitt B, Lewis R, Yamaguchi Y, Ashley EA, Bers DM, Robbins RC, Longaker MT, Wu JC. Abnormal calcium handling properties underlie familial hypertrophic cardiomyopathy pathology in patient-specific induced pluripotent stem cells. Cell Stem Cell. 2013;12:101–13.
154. Wang Y, Liang P, Lan F, Wu H, Lisowski L, Gu M, Hu S, Kay MA, Urnov FD, Shinnawi R, Gold JD, Gepstein L, Wu JC. Genome editing of isogenic human induced pluripotent stem cells recapitulates long QT phenotype for drug testing. J Am Coll Cardiol. 2014;64:451–9.
155. Yang YM, Gupta SK, Kim KJ, Powers BE, Cerqueira A, Wainger BJ, Ngo HD, Rosowski KA, Schein PA, Ackeifi CA, Arvanites AC, Davidow LS, Woolf CJ, Rubin LL. A small molecule screen in stem-cell-derived motor neurons identifies a kinase inhibitor as a candidate therapeutic for ALS. Cell Stem Cell. 2013;12:713–26.
156. Cong L, Ran FA, Cox D, Lin S, Barretto R, Habib N, Hsu PD, Wu X, Jiang W, Marraffini LA, Zhang F. Multiplex genome engineering using CRISPR/Cas systems. Science. 2013;339: 819–23.
157. Gonzalez F, Zhu Z, Shi ZD, Lelli K, Verma N, Li QV, Huangfu D. An iCRISPR platform for rapid, multiplexable, and inducible genome editing in human pluripotent stem cells. Cell Stem Cell. 2014;15:215–26.
158. Li HL, Fujimoto N, Sasakawa N, Shirai S, Ohkame T, Sakuma T, Tanaka M, Amano N, Watanabe A, Sakurai H, Yamamoto T, Yamanaka S, Hotta A. Precise correction of the dystrophin gene in Duchenne muscular dystrophy patient induced pluripotent stem cells by TALEN and CRISPR-Cas9. Stem Cell Rep. 2014;4(1):143–54.
159. Veres A, Gosis BS, Ding Q, Collins R, Ragavendran A, Brand H, Erdin S, Cowan CA, Talkowski ME, Musunuru K. Low incidence of off-target mutations in individual CRISPR-Cas9 and TALEN targeted human stem cell clones detected by whole-genome sequencing. Cell Stem Cell. 2014;15:27–30.
160. Filareto A, Parker S, Darabi R, Borges L, Iacovino M, Schaaf T, Mayerhofer T, Chamberlain JS, Ervasti JM, McIvor RS, Kyba M, Perlingeiro RC. An ex vivo gene therapy approach to treat muscular dystrophy using inducible pluripotent stem cells. Nat Commun. 2013;4:1549.
161. Yamada S, Nelson TJ, Crespo-Diaz RJ, Perez-Terzic C, Liu XK, Miki T, Seino S, Behfar A, Terzic A. Embryonic stem cell therapy of heart failure in genetic cardiomyopathy. Stem Cells. 2008;26:2644–53.
162. Moretti A, Bellin M, Welling A, Jung CB, Lam JT, Bott-Flugel L, Dorn T, Goedel A, Hohnke C, Hofmann F, Seyfarth M, Sinnecker D, Schomig A, Laugwitz KL. Patient-specific induced pluripotent stem-cell models for long-QT syndrome. N Engl J Med. 2010;363:1397–409.
163. Young LE, Platzer RF, Ervin DM, Izzo MJ. Hereditary spherocytosis. II. Observations on the role of the spleen. Blood. 1951 Nov;6(11):1099–1113.

Chapter 7
Overview of Chemistry, Manufacturing, and Controls (CMC) for Pluripotent Stem Cell-Based Therapies

Amy Lynnette Van Deusen and Michael Earl McGary

7.1   Introduction

The necessary requirements for completing the Chemistry, Manufacturing, and Control (CMC) section of a Food and Drug Administration (FDA) Investigational New Drug (IND) application as it pertains to the production of pluripotent stem cell (PSC)-based cell therapy products will be described in this chapter. The expectations for IND content are located in 21 CFR 312.23 [1] and regulations enacted for Phase 1 investigational products are described in an FDA Guidance for Industry titled "cGMP for Phase 1 Investigational Drugs" [2].

PSC-based products, generated from either embryonic or adult cell sources, are regulated under the general classification of Human Cells, Tissues, or Cellular or Tissue-Based Products (HCT/Ps) under Title 21 of the Code of Federal Regulations Part 1271 (21 CFR 1271) [3]. Additional regulations intended to prevent the introduction, transmission, or spread of communicable disease are contained within section 351 and 361 of the Public Health Service (PHS) Act [4, 5]. Regulatory implementation is primarily determined through risk-based assessments at each clinical phase and varies with the source, manipulation, and intended application of cells used in PSC-based therapies. As the clinical cohort increases in size, the scope and expectations of regulations will expand significantly.

While there are several criteria set out in 21 CFR 1271 Subpart A to determine if a cell therapeutic is exempt from any part of these regulations [6], the necessary ex vivo manipulation of cells and their intended use for transplantation into human patients renders the majority of PSC-based cell therapeutics fully regulated under

A.L. Van Deusen (✉) • M.E. McGary
Regenerative Medicine Strategy Group, LLC,
10401 Venice Boulevard, Suite 354, Los Angeles, CA 90034, USA
e-mail: avandeus@gmail.com; michael.mcgary@icloud.com

© Springer Science+Business Media New York 2016
M.K. Childers (ed.), *Regenerative Medicine for Degenerative Muscle Diseases*,
Stem Cell Biology and Regenerative Medicine,
DOI 10.1007/978-1-4939-3228-3_7

all previously mentioned statutes [7]. Examples of biologically similar HCT/Ps not regulated under 21 CFR 1271 include minimally manipulated bone marrow and blood products. However, the phrase "minimally manipulated" has been controversial, even resulting in a legal battle before the United States District court between the FDA and a cell therapy manufacturer in 2010 [8].

During preclinical and IND phases of development, emphasis is placed on generation of verifiable proof-of-concept studies and prevention of communicable diseases within the laboratory through Good Laboratory Practice (GLP) [9] and Good Tissue Practice (GTP) [10]. As a trial advances to Phase 1, current Good Manufacturing Procedures (cGMP) must be more rigorously implemented. As studies progress into even later stages, there is an increased focus beyond safety onto control of the manufacturing process including assessments of product stability and consistency. This largely occurs through evaluation of all generated documentation, including manufacturing and quality control records.

While this chapter will frequently refer to applicable regulations and guidelines, it is not intended to fully recapitulate any section or subsection of the 21 CFR or any FDA Guidance for Industry issued by the Center for Biologics Evaluation and Research (CBER). Rather, this chapter is intended as a general overview of the many regulatory requirements that must be considered in order to generate a complete CMC section for a successful IND application. As there are a multitude of regulations distributed throughout the CFR and US Pharmacopeial Convention (USP), we have provided a list of relevant statutes in Table 7.1.

Table 7.1 Regulatory Statutes for Development of an IND for PSC-based Cell Therapies

Title of Regulatory Statute	Location
Good Laboratory Practice for Nonclinical Laboratory Studies	21 CFR Part 58
Current Good Manufacturing Practice in Manufacturing, Processing, Packing or Hold of Drug	21 CFR Part 210
Current Good Manufacturing Practice for Finished Pharmaceuticals	21 CFR Part 211
Investigational New Drug Application	21 CFR Part 312
General Biological Products Standards	21 CFR Part 610
Quality System Regulation	21 CFR Part 820
Human Cells, Tissues and Cellular and Tissue-Based Products	21 CFR Part 1271
Sterility Tests	USP <71>
Bacterial Endotoxins Test	USP <85>
Growth Factors and Cytokines Used in Cell Therapy Manufacturing	USP <92>
Transfusion and Infusion Assemblies and Similar Medical Devices	USP <161>
Flow Cytometry	USP <1027>
Cell and Gene Therapy Products	USP <1046>
Gene Therapy Products	USP <1047>
Microbiological Evaluation of Clean Rooms and Other Controlled Environments	USP <1116>
Validation of Alternative Microbiological Methods	USP <1223>

7.2 Pluripotent Stem Cell Platforms

The most essential component of any cell-based therapeutic is the cellular platform from which the therapeutic is generated. Stem cells from adult and embryonic sources have individualized considerations with regard to their potential use in the clinic, making the origin of stem cells integral to the regulatory risk assessment. For example, risks associated with human embryonic stem cells (hESCs) are primarily linked to the health of the donor, quality of the derivation process, and subsequent storage and handling procedures. However, induced pluripotent stem cells (iPSCs) pose additional safety concerns due to the nature of current technologies used to transform adult somatic cells into PSCs. Additionally, some evidence suggests iPSCs may be more tumorigenic than their hESC counterparts [11].

Safety concerns about iPSC-based therapies largely emanate from the methodology used to reprogram cells. The original method (for which Shinya Yamanaka received the Nobel Prize in Medicine in 2012) is not suitable for the production of clinical-grade cells due to incorporation of retrovirally induced expression of c-Myc, a known oncogenic factor [12]. However, since the discovery of induced pluripotency, various alternative reprogramming strategies have been investigated to address this safety concern including the use of episomal plasmids [13], cell-permeable proteins [14], sendai viruses [15], minicircle vectors [16], synthetic mRNAs [17], and non-integrating microRNAs [18, 19]. Recently, a group at the University of California in Los Angeles reported the generation of "putative clinical-grade status" iPSCs derived from adult human dermal fibroblasts using a polycistronic "stem cell cassette" (STEMCCA) method that incorporates an excisable lentiviral reprogramming system. By utilizing cGMP to develop and characterize transgene-free iPSCs, this group has provided proof-of-principle for the generation of clinically relevant iPSCs [20].

Whether cells are hESC, iPSC, or derived from other sources, federal and state governments administer registration and licenses for facilities that host HCT/Ps [21]. Compliance with federal and state regulatory statutes entails proper disposal and handling of both biohazardous and medical waste, as well as adherence to all applicable occupational health and safety requirements [22]. These regulations must be identified and met within the individual state where production of HCT/Ps is to occur, and compliance with these procedures should be acknowledged in the IND application. Specific state restrictions should be investigated prior to obtaining, storing, or distributing these materials.

7.2.1 Autologous Cell Sources

Potential sources of PSCs include a patient's own (autologous) cells or cells obtained from another human donor (allogenic). There are many benefits to using autologous cell sources for PSC-based therapeutics including reduced potential for negative immune responses and the ability to generate patient-specific products. Further, autologous cell-based therapies are exempt from donor eligibility determination [23].

Autologous cells for regenerative medicine research have traditionally been harvested from dermal fibroblasts for reprogramming. However, in recent years, several additional adult cell sources have been successfully reprogrammed into iPSCs. Depending on the intended clinical use and cell manipulation protocols required, iPSCs reprogrammed from blood, adipose tissue, or other sources may provide more versatility or stability in the clinic. While clinical procedures used to obtain these cell types may be slightly more invasive than for dermal keratinocytes or fibroblasts, it is possible that the genetic condition of blood and adipose cells remains more stable over a human lifetime because of reduced exposure to environmental factors such as UV light [24].

There are logistical benefits to using blood cells as a source of reprogrammable cells instead of dermal fibroblasts. First, the ability to utilize established blood bank systems and relative ease of cell harvesting processes for the creation of new banks potentiates immediate and consistent supplies of iPSCs and derived cells. Second, iPSCs generated from blood sources may prove an excellent source for generating mass quantities of de novo hematopoietic stem cells or mature blood cell types [25]. Considering the Red Cross uses only donated sources to supply approximately 40 % of the blood and blood products used in the USA [26], the ability to generate and expand blood from iPSCs could potentially solve problems associated with limited donor pools and issues with a comparably small ratio of product generated per donor.

In 2013, a research group at the National Institute of Health's (NIH) Laboratory of Host Defenses reported using a STEMCCA method to generate iPSCs from small volumes of peripheral blood [27]. Additional groups have also generated iPSCs from blood mononuclear cells (BMCs) [28] and CD34+ cord blood progenitor cells using this method. Though integration-free cells were produced, residual reprogramming factors that impaired subsequent differentiation procedures were reported in the latter instance [29].

More recently, a research group from the University of California in San Francisco reported the creation of iPSCs from adult mobilized CD34+ and peripheral BMCs using a non-integrating Sendai viral vector. After several passages in culture, the viral vector could no longer be detected in cells by real-time quantitative reverse transcription-polymerase chain reaction (RT-PCR), suggesting that iPSCs produced by this method could potentially be used for future clinical applications if reproduced under cGMP [30].

It is important to note that while adipose-derived stromal cells may have been touted by some for their use in many "stem cell" therapies, these cells are multipotent and more similar in nature to mesenchymal stem cells (MSCs) found within bone marrow stromal cells than to hESCs or iPSCs [31]. However, several groups have successfully reprogrammed adipose cells into iPSCs using a traditional retroviral method [32], a STEMCCA method [33], and a nonviral DNA minicircle technique. These cells are readily isolated from adults and, reportedly, are easy to expand and maintain in culture even under xeno-free conditions [34].

Beyond initial donor eligibility requirements, the use of autologous cells does not decrease the regulatory burden for production of cellular therapeutics in which cells are more than "minimally manipulated." All of the previously described methods used to reprogram adult cells constitute significant manipulation and, therefore, render products generated from these cells "drugs" as defined by the FDA. Though, until 2005, the FDA did not regulate autologous adult stem cell therapies under its standard regulations for cell therapies, which state "regulations for this chapter do not apply to autologous human tissue" [35].

The status of regulatory oversight changed in July 2008 when the FDA sent a letter to Regenexx (Broomfield, CO) stating that their "Regenexx™ Procedure" constituted a "drug" under section 201(g) of the Federal Food, Drug and Cosmetic Act and a "biological product" under section 351(i) of the Public Health Service Act [36]. After several attempts by Regenexx to prove in court that their product was only "minimally manipulated," in August 2012, the FDA's jurisdiction over autologous stem cell therapies was solidified with a victory in the US District Court for the District of Columbia and codification [5]. Currently, Regenexx offers a range of procedures that take only a few hours to process at its Colorado facilities. However, its RegenexxC™ Cultured Stem Cell Procedure, which incorporates more extensive ex vivo culturing, is now only available in the Cayman Islands [37].

The increased authority of the FDA over autologous cell therapies has been viewed somewhat controversially [38], but in September 2012, the FDA again asserted its authority by sending a similar letter to Celltex Therapeutics (Houston, TX). This company specializes in isolation of mesenchymal stem cells (MSCs) to "treat" a range of disorders including multiple sclerosis, arthritis, back pain, and Parkinson's disease [39]. With the governor of Texas as a vocal advocate and their first patient, Celltex waged a public battle against the FDA's decision, but has since announced plans to relocate their practice to Mexico [40].

In an effort to raise awareness about unapproved stem cell therapies, the FDA has published informational materials through its Consumer Health Information initiative to warn about the safety risks of unregulated treatments [41] and clearly states on its website:

> Stem cells, like other medical products that are intended to treat, cure, or prevent disease, generally require FDA approval before they can be marketed. At this time, there are no licensed stem cell treatments [42].

Virtually all PSC-based therapies, whether because of reprogramming or differentiation processes, will require more than minimal manipulation for formulation into clinical products. Thus, regardless of the origin and source of cells, there are numerous steps to qualify autologous PSC-based cell therapies.

7.2.2 Allogenic Cell Sources

The use of hESCs, adult-derived PSCs, or cultured PSC lines not harvested directly from the intended patient constitutes an allogenic cell therapy product. For all allogenic HCT/Ps, there are several criteria that must be met to ensure suitability of donors. Procedures for all steps of eligibility determination and testing must be established in compliance with 21 CFR 1271, Subpart C, and departure from these procedures in any manner must be documented with justification [43]. All human cells and/or tissues collected must be accompanied by appropriate informed consent documentation in order to be used for clinical research or therapeutics generation. The general requirements and established language for informed consent documentation procedures can be found in 21 CFR 50.20 [44] as well as an additional Guidance for Industry published by CBER in August 2007.

The two main components of donor eligibility determination for allogenic cell therapies include a comprehensive review of the donor's medical history and screening for pathogenic agents. Relevant risk factors observed in medical histories include illicit drug use, high-risk sexual activities, previous diagnosis with a communicable disease, blood transfusions, and transplantation recipients [45]. There are extensive recommendations regarding review of medical records in 21 CFR 1271.75 [46] as well as additional recommendations regarding physical assessments of living and cadaveric donors outlined in 21 CFR 1271.3 [7].

To prevent source-related contamination, donor cells must be screened for the presence of adventitious agents that may potentially infect downstream materials and patients. General recommendations regarding donor testing are covered in 21 CFR 1271.80 [47] and an additional list of required screening for cell therapy products including cytomegalovirus (CMV), human immunodeficiency virus (HIV-1 and -2), Epstein-Barr Virus (EBV), Parvovirus B19, hepatitis B virus (HBV), hepatitis C virus (HCV), and human transmissible spongiform encephalopathy/Creutzfeldt-Jakob disease [48]. Additional pathogenic screening is required for donors of certain cell types based on specific potential downstream risks to patients. For example, donation of leukocyte-rich cells and tissues requires a negative screening result for human T-lymphoropic virus, and donors of reproductive cells or tissue must be screened for Chlamydia and gonorrhea prior to determination of eligibility.

There are also nonspecific provisions for diseases that pose a risk of transmission when used in cell therapeutics that include potentially life-threatening diseases for which appropriate screening measures are available. Relevancy is determined by risk of transmission and severity of potential health risks posed by the pathogenicity of the agent [45]. For example, *Ureaplasma urealyticum*, a mycoplasma bacterium associated with urethritis, is considered to have a low pathogenicity and therefore would not singularly disqualify a potential donor. However, West Nile Virus, which is highly transmissible and can result in permanent impairment or damage to the body, would certainly render a donor unsuitable.

The final determination of eligibility must be made by a medical professional who reviews the potential donor's medical history and confirms screening tests were performed by appropriately trained personnel following standard operating procedures [49]. A summary of these records and the donor eligibility determination must be anonymized and subsequently provided to facilities that acquire donated cell or tissue products. Further, records must be kept until 10 years after the last administration of product obtained from donor cells [50].

If generating a de novo cell platform is not feasible or desired, there are several clinical-grade hESC lines available [51]. These lines were produced under cGMP and have been fully characterized for genetic normality and sterility, as well as being compliant with informed consent and donor eligibility requirements. A license is required to use these lines for research or commercial purposes, and this may prove an expensive aspect to initial manufacturing. However, there are significant benefits to using an established line including increasing availability of both performance and safety data. Additionally, characterization of cell lines is a costly venture, and satisfactory results are not guaranteed. Therefore, using established lines could prove a less risky investment for manufacturing hESC-based HCT/Ps. To date, there are no iPSC lines approved in the USA for generating clinical products.

7.2.3 Cell Mobilization and Collection

When cells are harvested from a patient, whether for autologous or allogeneic uses, it is necessary that cells for donation and donor testing be recovered at the same time, or at least within 7 days of the initial recovery of cells [47]. Cells mobilized and collected prior to determining donor eligibility must be quarantined until a determination has been made [52]. These protections are put in place to ensure screening results are properly obtained, analyzed, and reviewed prior to making a donor eligibility determination.

The procedure by which cells are collected (e.g., surgery, leukapheresis) must be fully documented and performed by a qualified professional. If cells are mobilized or activated in any manner prior to acquisition, this procedure and its administration must also be documented [47]. If acquired cells are shipped to another facility for further manufacturing, transport conditions must be also recorded [53]. Further, donors of hematopoietic stem cells (HSCs) obtained from blood or bone marrow need to be qualified for eligibility prior to receiving any type of myeloablative chemotherapy.

To assist in qualification of allogenic donors in extraordinary circumstances, test specimens may be collected up to 30 days prior to donation [47]. There are additional recommendations for testing donors who have received transfusions and infusions in the weeks prior to donation due to the possibility of plasma dilution rendering donor screening test results unreliable [48]. As requirements may vary greatly, consultation with FDA regulators is recommended for nonstandard cell collections or harvesting procedures.

7.2.4 Cell Bank Systems

Once a cell line is generated and chosen as the platform for manufacturing products, it is highly recommended that a cell bank system be generated. The structure of the bank will depend on whether cells are obtained for autologous or allogeneic uses as shown in Fig. 7.1 [54]. Whether purchased or prepared in-house, manufacturers are responsible for ensuring the quality of each lot of cells used in the production of HCT/Ps [55]. For each ampoule of cells used in manufacturing investigational products, a description of the history, source, derivation, and characterization must be included in the CMC [56].

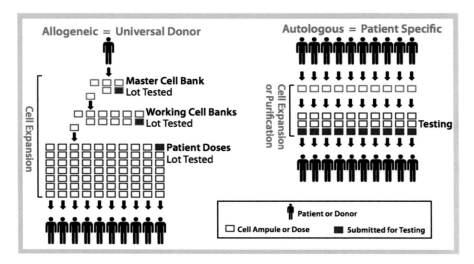

Fig. 7.1 Generalized schemes for allogeneic and patient cell banks [68]

Banking cells allow a common source material for manufacturing production lots and ensures a sufficient supply of cell material over the lifespan of a product. Product manufacturers should describe their intended strategy to provide a continual source of cells in the CMC section, including projections for utilization rate and expected intervals between generation of new cell banks [57]. Additionally, descriptions of vessels in which cells will be stored, labeling standards, and cryopreservation procedures must be detailed in full. As with all manufacturing processes, each test and procedure to establish cell banks must be reported and conducted in accordance with 21 CFR 58, otherwise known as Good Laboratory Practice (GLP) [7].

For all cell banks, it is strongly suggested to establish redundant storage facilities to protect from catastrophic events that could render an entire supply or cell bank unusable. Such events may include power outages, natural disasters, or even human error [57]. Manufacturers should describe all precautions taken to ensure uninterrupted production in the CMC, including use of automated liquid nitrogen systems, secondary power sources, and offsite storage. In this regard, there are several companies that offer guaranteed, monitored storage facilities that are suitable for back-up storage of any amount of clinical cell products, from a few vials to entire cell banks.

7.2.4.1   Master Cell Bank

The master cell bank (MCB) is defined as a collection of cells of uniform composition derived from a single tissue or cell. Generally, the MCB is stored in cryopreserved aliquots in the liquid or vapor phase of liquid nitrogen. The location and identity of individually labeled ampoules of cells should be thoroughly documented, as should the daily temperature of the storage environment. The process to fully characterize cells is fairly expensive and time-consuming, therefore developing and protecting the MCB is key to sustaining long-term production capabilities. Characterization testing should be performed once on aliquots or cell cultures derived directly from the MCB and should include all tests required to establish significant properties of the cells [58].

Characteristics of the MCB that must be established in the CMC include indications of the safety, identity, purity, and stability of cells [57, 59]. The identity of the MCB should be verified by confirming the species of origin through isoenzyme analysis or the presence of unique cell line markers using phenotypic and/or genotypic analysis [57]. After initial verification, identity should be confirmed periodically to ensure robustness of inventory procedures and cell line stability [56, 57]. It is important to note that while isoenzyme analysis allows for quick identification of human rather than animal cells, this type of analysis is only adequate to qualify product identity for facilities producing a single product from a single source. When multiple human products or cell sources are used in a single facility, more extensive identification procedures will be required.

Examining DNA sequences and polymorphisms are far more accurate methods for assessing the identity of cells. To verify the identity or chromosomal stability of PSC lines or products over time, karyotypic or cytogenetic analyses (e.g., fluorescent in situ hybridization techniques (FISH) or DNA microarray) may also be used [60]. Not only can these methods identify specific donors or sources, they are useful for detecting chromosomal aberrations that may occur in cells during culture. While these analyses are initially required to establish the character of the cell line, performing these procedures is time-consuming and expensive in comparison to isoenzyme analyses making their use for routine identity analysis unfeasible, especially for in-process testing.

Determining the purity of the MCB includes identifying and quantifying all existing cells as well as any contaminating elements. These tests are often performed using an array of biomarkers that target desired cell types as well as specific and nonspecific contaminants. Tests for specific pathogens will be required including CMV, HIV-1, HIV-2, EBV, Parvovirus B19, HBV, and HCV as well as tests for adventitious agents including bacteria, fungi, mycoplasma, and viruses [46, 61] (Testing procedures will be discussed later in the chapter).

In addition to documenting characterization test results, all culture conditions should be recorded throughout cell banking and testing processes. This includes recording all media and reagents used during production, as well as the storage condition of cells and materials. Additionally, results of phenotypic and genotypic stability tests must confirm viability of cells before and after any cryopreservation procedure [10].

7.2.4.2   Working Cell Bank

A tiered cell bank that contains both a MCB and working cell bank (WCB) will decrease the potential for unidentified cross-contamination and catastrophic events involving adventitious agent contamination. Generally, cells expanded for production purposes are from the WCB, and the MCB is held in reserve. The WCB should be derived from the MCB and propagated for an approved number of passages in culture before cryopreservation banking.

While limited identity testing is required, WCB cells should be investigated for contaminants that may have been introduced during the culture process prior to banking. Recommended tests include cell line authenticity to check for cross-contamination, as well as bacterial and fungal sterility, mycoplasma, and in vitro adventitious viral agent testing [56].

7.2.4.3 Patient Cell Bank

For autologous cell therapies, it may be beneficial to generate a patient cell bank. Donor eligibility testing requirements may be waived [24]; however, testing of patient cells and tissues for communicable diseases is still recommended for the safety of staff handling these materials during production processes. Otherwise, cell characterization and documentation requirements are similar to the development of allogeneic cell banks.

Performing identity testing to differentiate numerous specific individuals would entail time-consuming DNA sequencing procedures. Therefore, successful implementation of labeling procedures, inventory accountability methods, and product segregation strategies are key to demonstrating and maintaining control for facilities that will host numerous patient cell samples. To ensure patient safety and guaranteed administration of autologous cells, the CMC section should detail the various production processes, documentation procedures, and physical aids designed to provide multiple levels of assurance in this regard [9].

There are considerations for patient privacy that must be addressed along with additional labeling regulations for autologous cell banks. Individual autologous specimens must be labeled "For Autologous Use Only" and "Not Evaluated for Infectious Substances," unless screening and testing have been performed as previously described [46–48]. An individual's personal health information must be protected, and this includes information and documentation regarding stored cells or tissues [62]. As drug developers move into later stages of clinical trials, there is an increased burden to provide protected data services for a larger number of patients. Many HIPAA and CFR-compliant software options are currently available for managing clinical trial data and patient information; however, their cost may be prohibitive during preclinical and early phases.

7.3 Materials Management

To demonstrate control over manufacturing processes, it is necessary to verify and document that supplies and reagents "meet specifications designed to prevent circumstances that increase the risk of the introduction, transmission, or spread of communicable diseases." In more general terms, all reagents used in the manufacture of investigational HCT/Ps must be sterile, and systems must be established to reasonably ensure all product components maintain sterility before, during, and after the manufacturing process through to final product administration [63].

Every chemical entity that "touches" the final product at any point during the manufacturing process must be accounted for in CMC documentation in order to be compliant with FDA regulations regarding cGMP. Further, each of these manufacturing components must be suitable for clinical use and/or proven to be removed from the final product in order to gain final regulatory approval [64]. Because of these limitations, there are many factors beyond supply and cost to consider when sourcing product components and materials for clinical cell therapeutics.

For example, how and where a material is produced will often play a key role in defining clinical suitability. Further, HCT/P manufacturers must be able to document that all materials were obtained from legitimate and approved sources [65]. For most product components, this is achieved by requesting product documentation from the distributor regarding the origin of manufacture as well as any product quality testing that has been performed. However, for HCT/P components of human origin, this process must include ensuring products are thoroughly tested for communicable diseases and were acquired with proper informed consent documentation, as previously described.

In this section, the regulatory considerations and documentation required for production of PSC-based cell therapy products will be described. This will include cells, reagents, and other materials necessary to manufacture products, but that may or may not be included as part of the final product formulation. As with many aspects of PSC-based cell therapeutic manufacturing, meticulous documentation demonstrates control of manufacturing processes and substantiates regulatory compliance.

7.3.1   Tabulation of Reagents Used in Manufacturing

For all reagents used to manufacture an HCT/P, there must exist a tabulation of reagent concentrations, vendors/suppliers, and their source. The reagent concentration must be listed for each manufacturing step in which it is used. For traceability of reagents, it is necessary to record the name of the vendor materials were purchased from as well as the supplier that actually manufactured the material [66]. In certain cases, the supplier and vendor are the same entity, but often reagents are distributed through larger vendors and it is important to establish the exact origin of all product components.

Identifying the source of a reagent requires discerning the geographical origin and physical nature of components. Generally, items are natural or synthetic chemicals, plant-derived materials, or are of microbiological, animal, or human origin. For chemical or plant-derived entities, toxicity and safety profiles should be examined to ensure suitability for clinical use in humans. If product components are made from human- or animal-derived substrates of any type, this must be identified, documented, and approved. If any component will be removed during the manufacturing process, it is considered and excipient—a topic that will be discussed in a later section of this chapter.

The use of animal products should be avoided when human equivalents are available as xeno-free culture conditions are best in the eyes of FDA regulators. However, there may be extraordinary situations in which an animal product is the only option. In these circumstances, porcine products may be used when Certificate of Analysis (COA) documentation is available that certifies products are free of porcine parvovirus. Additionally, bovine products may be substituted, but due to the existence of prion diseases, there are additional documentation requirements [67] including full verification of each bovine specimen's history from birth to slaughter [68].

Where human serum products are used, there must be procedures in place to ensure lots used during manufacture were not previously or subsequently recalled. Additionally, the use of human serum is limited to licensed products obtained from approved blood banks that guarantee donor testing and eligibility. For all other human-derived products, it should be determined if the product is licensed and clinical-grade prior to production [63]. This is often determined by examining the COA and Certificate of Origin documentation, a quality system process required for compliance with cGMP.

There are a number of clinical-grade products that facilitate conformance with cGMP and, therefore, streamline product development. These include disposable, presterilized consumables and process aids that reduce the burden of cleaning and potential for contamination. Prepackaged and presterilized materials such as Water For Injection also reduce the need for sterilization equipment and are prequalified for medical uses [2]. Generally, where FDA preapproved products are available and suitable, they will be highly recommended or required by FDA regulators.

In the extraordinary circumstance that no clinical-grade human or animal product is available and a research-grade product must be used in the manufacturing process, a qualification program exists. Product developers may submit testing data to FDA regulators who will then review testing and vendor information to determine if the product is suitable for use in manufacturing. Analysis must verify the source, safety, and performance of the reagent in order to qualify, which minimally include sterility, endotoxin, mycoplasma, and adventitious agent testing. Additional functional analysis, including purity testing and assays demonstrating the absence of potentially harmful substances, must also be completed as part of qualification [59]. If questions exist about a particular product component, there are procedures and FDA committees in place to assist product manufacturers select suitable products.

7.3.2 Qualification and Oversight of Product Components

The oversight and management of materials used in the production of HCT/Ps is an important facet of cGMP and, therefore, should be thoroughly described in the CMC section. Management of materials begins with procurement, where there are specific requirements for maintaining data and establishing appropriateness of suppliers [69]. The next step is receiving materials, which should be inspected for package integrity and stored at a proper temperature immediately upon arrival to manufacturing facilities. These materials must be evaluated for appropriate product documentation and conformance to manufacturer's specifications before being released for use. Depending on the nature of the product, additional testing beyond visual examination (e.g., functional tests) may be required to define if individual lots are acceptable for use in manufacturing investigational products.

Federal regulations regarding the verification of reagents used in manufacturing investigational PSC-based HCT/Ps can be found in 21 CFR 211.84 "Testing and approval or rejection of drug components, drug product containers, and closures" [55] and 21 CFR 1271.210(a,b) [63]. Strategies for materials management and oversight must be outlined in the CMC section and should be individualized for the facilities and personnel available during early clinical phases. Establishing oversight of all product components should be instituted as early as possible since expectations for robustness of these systems will increase as a clinical trial progresses.

7.3.2.1 Specifications and Acceptance Criteria

For each product component, specifications must exist that include acceptance criteria for qualifying each lot and aliquot safe for use in manufacturing [56]. Written procedures should be established to describe the handling, review, acceptance/rejection, and control of all materials used in manufacturing investigational product. This includes examining all packaging and closures to ensure product integrity as well as conformance to the manufacturer's description of product. Prior to examination by quality control, all materials should be quarantined and stored according to the manufacturer's directions until such time they can be examined and released for use in manufacturing [2]. Labeling systems should be implemented such that quarantined and released items are clearly distinguishable from one another in addition to physical requirements for separation of these items.

It is the responsibility of the manufacturer to be able to identify and trace all materials used in the manufacture of investigational products from receipt to use in individual batches [69]. Therefore, it is recommended that a log containing the date of receipt, quantity, supplier's name, material lot number, storage conditions, and corresponding expiration date be rigorously kept [2, 63]. A comprehensive strategy for examining, qualifying, and releasing materials for use in manufacturing should be developed early to provide consistent oversight of these functions as production increases.

Acceptance criteria should be established for each material used in manufacturing investigational product as well as for final products. These criteria should include relevant attributes such as color, consistency (e.g., powder or liquid) and description of properly secured container. Many acceptance criteria can be inferred from the supplier's product description and/or Certificate of Analysis, while others must be determined through more extensive testing. For example, Becton

Dickinson's (Franklin Lakes, NJ) Matrigel™ is basement-matrix product commonly used in stem cell culture to propagate adherent cells. However, individual lots of this product may vary significantly in levels of protein and endotoxins, making in-house testing of each lot necessary prior to production to ensure consistency of investigational product. (In-process and final product testing will be described in a later section of this chapter.)

7.3.2.2 Certificates of Analysis and Certificates of Origin

Product suppliers and vendors that distribute clinical-grade products should provide documentation supporting analysis of product composition and safety. These documents are usually in the form of COA and COO, which must state individual lot numbers to be considered valid. These documents must be obtained and retained for each lot of product used to manufacture investigational materials.

It is imperative that manufacturers of investigational products examine each and every COA and COO to ensure that materials meet established acceptance criteria for defined attributes [70]. If materials are human or animal-derived, documentation supporting the source/origin of the material as well as tests for adventitious agents must be included in records. Additionally, if documentation confirming a specified attribute is unavailable, it is recommended that the HCT/P manufacturer test the material to evaluate its conformance to acceptance criteria when feasible.

7.3.3 Excipients

Final products must be tested for residual manufacturing agents with known or potential toxicities. Performance of test procedures used to make these determinations must be documented with appropriate standard operating procedures included in the CMC [56]. Residual contaminants for PSC-based HCT/Ps may include stem cells, other mature cell types, chemicals, viruses, proteins, and other agents. While efforts should be made to avoid using materials that will need subsequent removal when developing manufacturing processes, there will inevitably be purification steps included as part of production.

Additional concerns exist about specific products that, while commonly used in cell culture research, may pose safety risks in a clinical setting. If these products are to be included in the final product, they are considered excipients and must be documented and approved prior to distribution. Examples of such excipients include antibiotics, serum products, and other chemical entities necessary for manufacturing procedures such as selection or cryopreservation. Using potentially hazardous materials should be avoided or minimized wherever possible during the manufacturing of HCT/Ps; however, when such materials must be included, their use, concentration, and/or procedures for removal from final product must be thoroughly documented [55, 56].

7.3.3.1 Antibiotics

Commonly, patients are sensitive to penicillin-based antibiotics. For this reason, it is recommended that the use of antibiotics be avoided during the manufacturing of therapeutic products for human use. However, if the use of beta-lactams is required for the production of an HCT/P, a rationale for their use must be submitted and approved by FDA regulators. This rationale should include precautions taken to prevent hypersensitivity reactions during downstream use and reasoning for why an appropriate substitute could not be identified. If antibiotics of any class are used during production, bacteriostasis and fungistasis data must substantiate that suppression of these entities has not occurred during culture [71].

7.3.3.2 Dimethyl Sulfoxide (DMSO) and Other Cryoprotectants

Dimethyl sulfoxide (DMSO) is a commonly used cryoprotectant for preserving human cells. It permeates cells to prevent the formation of ice crystals that are severely detrimental to cell health and survival. Other cryopreservation agents that act by permeating cells include glycerol [72], sucrose [73], and trehalose [74], though some tests have demonstrated poor survival rates when used to cryopreserve stem cells as compared with DMSO. Other strategies for cryoprotection include high molecular weight polymers that are non-permeating such as polyvinylpyrrollidone (PVP) and hydroxylethyl starch [75]. Recently, a group from Japan reported using a carboxylated poly-L-lysine construct to successfully cryopreserve human bone marrow-derived mesenchymal stem cells without the use of proteins or DMSO [76].

Issues with cryoprotective agents may arise from their use at potentially toxic molar levels to cryopreserve cells, which complicates the ability to use products directly from a thawed state. Further, evidence suggests that addition and removal of cryoprotectants such as DMSO may lead to osmotic shock [77] or other detrimental processes in cells, which may compromise the integrity of final products and/or impact the safety of patients [78–82]. Logistically, the ability to ship frozen products that can be thawed with high viability for immediate use is an ideal practice for providing HCT/Ps to investigational study sites, not to mention a strategy that is feasible for full-scale product distribution.

Identifying cryoprotective agents that are xeno-free and nontoxic is a priority to many manufacturers of HCT/Ps since developing optimal cryopreservation procedures can enhance the usefulness and performance of cell products. Fortunately, options for cryopreserving human cells are expanding as the number of investigational products being researched increases. For further information, we suggest an excellent review by Dr. Mirijana Pavlovic and Dr. Bela Balint that details the many considerations that must be taken into account when developing standard cryopreservation procedures for stem cells, many of which are equally applicable to stem cell-derived HCT/Ps [83].

7.3.3.3 Human Serum Albumin

As previously described, the use of human products in manufacturing HCT/Ps entails ensuring products are free of communicable diseases and other adventitious agents. Thus, the use of human serum albumin (HSA) or other serum products necessitates some additional consideration when developing suitable CMC. While it is possible to use autologous serum for the production of autologous HCT/Ps [84], this process is costly, requires preoperative blood donation [85], and is not easily scalable for large-scale production purposes [86]. Commercially, there are a limited number of sources for clinical-grade, FDA-approved human serum albumin. Currently, only Octapharma USA, Inc. (Centreville, VA) and Baxter Pharmaceutical (Deerfield, IL) market HSA as Buminate for clinical use.

As HSA or some form of serum-replacement is often required for the production of HCT/Ps or storage of product components, it is important to identify reliable, clinical-grade products. However, with limited options, many manufacturers have worked to identify serum replacements. There are currently several alternative serum products on the market including Gibco's Knockout™ Serum Replacement distributed by Life Technologies (Carlsbad, CA). This defined formulation has been incorporated into numerous serum-free media recipes for the growth, maintenance, and cryopreservation of various types of stem cells including iPSCs [87]. Similar products include PluriQ™ Serum Replacement offered by Globalstem® (Rockville, MD) and HyClone AdvanceSTEM™ Serum Replacement distributed by ThermoScientific™ (Waltham, MA).

7.3.3.4 Combination Products

For some PSC-based therapies, it may be advantageous to combine products with existing delivery systems or other drug entities to enhance performance or stability. For example, one strategy for delivering cell therapies includes encapsulation in biopolymers [88]. Another PSC-based cell therapy may require a new surgical instrument or device to implant cells within the body. Because the nature and type of devices that may be combined with an HCT/P vary significantly, providing specific regulations are outside the scope of this chapter.

However, generally, if the drug or device component already has FDA marketing approval and will be used in a similar manner, then the CMC previously filed for this entity may be referenced in the new CMC for the combination therapy. Otherwise, the regulatory status of the drug or device for the new intended use must be established through a New Drug Application (NDA), PreMarket Approval application (PMA), or a premarket notification (510 (k)) depending on the exact nature of the combination of products. A consultative review with FDA regulators is suggested early in the process for potential combination products to clarify the best path for regulatory application and oversight.

7.4 Manufacturing

Sponsors of PSC-based cell therapies must consider a range of activities involved in generating, evaluating, and distributing clinical-grade products. A schematic representation of steps required to develop a PSC-based product is shown in Fig. 7.2 [89]. Cell banking and early features of process development have been described previously. In this section, we will focus upon regulations applicable to manufacturing facilities and processes used to generate PSC-based clinical materials for investigational use.

Manufacturing PSC-based cell therapies includes generating and testing products as well as logistics such as packaging, labeling, and transport. Depending on the complexity and intended use of a therapeutic, significant optimization may need to occur in order to translate a research-grade entity into a safe, commercially viable product. In addition to procedures to create products or components, the environment, equipment, and personnel involved must be thoroughly described within the CMC section.

Further, manufacturers of investigational product must demonstrate that the manufacturing environment and equipment contained within are adequately controlled for the generation of sterile HCT/Ps in accordance with cGMP [90, 91] and

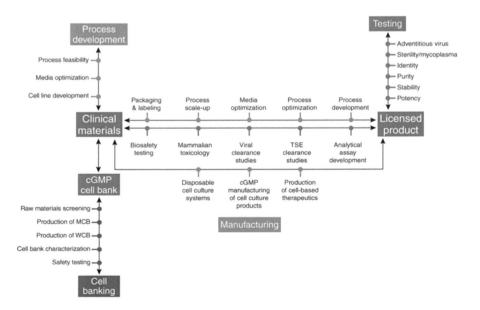

Fig. 7.2 Schematic of processing and manufacturing steps involved in development of a PSC-based cell therapy product [101]

as described in a 2004 guidance entitled, "Sterile Drug Products Produced by Aseptic Processing—Current Good Manufacturing Practice" [92]. While full adherence to all facets of cGMP is not expected at the IND and Phase 1 level of clinical studies, a detailed description of all procedures used during the collection, production, and purification of PSC-based cell therapeutics must be provided.

A comprehensive and systematic evaluation of all manufacturing facilities, processes, equipment, materials, and personnel to identify potential hazards is the first step in developing a robust strategy to eliminate or mitigate possible threats [2]. As with all clinical regulations, expectations for quality increase as trials progress. Taking future manufacturing needs into consideration early will assist in the development of scalable systems for quality by design—an important principle we will discuss later in the chapter.

7.4.1 The Manufacturing Environment

Production of sterile products is one of the most critical elements to ensuring human subject safety in clinical trials; therefore, a thorough plan is required as to how aseptic manufacturing processes and controls will be implemented throughout production and packaging of materials. Aseptic processing begins with proper facilities, equipment, and personnel to conduct the scope of required aseptic manipulations and is further maintained with proper disinfection regimes and rigorous quality control [2]. For PSC-based therapies, Good Tissue Practice (GTP) is at the core of maintaining a manufacturing environment that produces sterile, quality products. The goals and scope of these regulations are outlined in 21 CFR 1271.190(a, b); however, individual regulations may be found throughout 21 CFR 1271, Subpart D [10].

7.4.1.1 Facilities

Facilities occupied by manufacturers of HCT/Ps do not have to register with CBER until a product is given approval [93]. However, site inspections by FDA regulators will occur prior to initiation of Phase 1 studies to verify suitability for intended production. In several states (including California and Massachusetts), facilities hosting and processing human cells and tissues must register these activities with appropriate state authorities. A list of state-specific regulations is located on The American Association of Tissue Bank's website [94].

Regardless of location, suitable facilities should provide sufficient space, lighting, ventilation, heating, cooling, water, and sanitation for intended production processes. Documentation supporting these features should include a diagram of facilities indicating designated areas for sterile manufacturing and processing [2]. One example of a clean room facility schematic is shown in Fig. 7.3 [95]. Notice that the two diagrams are for the same manufacturing space; however, the diagram

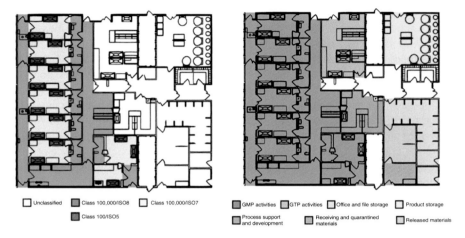

Unclassified   Class 100,000/ISO8   Class 100,000/ISO7

Class 100/ISO5

GMP activities   GTP activities   Office and file storage   Product storage

Process support   Receiving and quarantined   Released materials
and development   materials

Fig. 7.3   Example of floor plan for clean room production of human cell therapies [105]

on the right delineates separation of production activities while the left indicates "clean" areas by International Organization for Standardization (ISO) designations for particle limits.

It is important to evaluate the usage of space within manufacturing environment for susceptibility to contamination with biological substances, including adventitious agents such as bacteria, fungi, viruses, and mycoplasma. Separation and isolation of "clean" activities is key to maintaining an environment suitable for clinical manufacturing. The potential for the presence of adventitious agents should also be minimized through coordinated cleaning and decontamination activities conducted throughout the manufacturing process [69, 96].

Air-handling systems that prevent contamination and cross-contamination are specifically required for the production of sterile products. Full regulations regarding the environmental controls that should be put in place are outlined in 21 CFR 1271.195 [97]. If production of a PSC-based HCT/P involves the use of pathogenic or spore-forming microorganisms, live viruses, or gene therapy vectors, there may be additional containment considerations that must be discussed with the appropriate Center within the FDA prior to manufacturing product [2].

Facilities manufacturing multiple products are required to provide strategies for preventing cross-contamination in addition to contamination by adventitious agents. Within current regulatory standards, stem cells cultured and maintained in a pluripotent state for production are a separate product entity from cells that have entered specific differentiation or manufacturing processes that render them no longer "stem cells." Therefore, environmental control strategies for separating these two product entities must be implemented to prevent potential cross-contamination of final product with highly oncogenic stem cells [93, 98]. Prevention strategies should be outlined in the CMC as they apply to specific elements of the manufacturing process (e.g., multiple incubators as an equipment-based strategy or multiple rooms as a physical barrier).

In the circumstance that manufacturing activities will take place at more than one facility, it is necessary to include descriptions for all procedures and safety measures for both facilities as part of the IND submission. Additionally, intermediate shipment processes to transfer product components or in-process materials between facilities should be outlined in the CMC. Strategies to control and oversee consistency of production at multiple facilities should also be described.

Generally, the US Department of Health and Human Services requires an environmental impact study for facilities manufacturing new products [99]. However, a categorical exclusion exists for sponsors of IND studies for human drugs and biologics under 21 CFR 25.31(e) [100]. Unless there are extraordinary circumstances, this exemption should be applicable for the generation of PSC-based therapeutics.

7.4.1.2 Equipment

A diverse range of equipment will be required to produce, analyze, and store PSC-based HCT/Ps. It is strongly suggested that dedicated equipment be employed in manufacturing investigational product whenever possible. All equipment must be calibrated yearly, or more frequently if recommended by equipment manufacturer. Additionally, all equipment employed for safety-related functions (e.g., viral clearance or bacterial detoxification) must be regularly tested to ensure proper performance with a "high degree of assurance." Expectations regarding equipment for cGMP during Phase 1 studies are outlined in a July 2008 FDA Guidance [2] as well as 21 CFR 1271.200 [101].

The 2008 guidance includes specific recommendations for aseptic workstations employed in the manufacture of cell therapy products. First, the placement and installation of workstations must allow sufficient airflow around the apparatus. Next, the placement of all items in and around the workstation should not disrupt the unidirectional airflow necessary for sterile processing applications. Prior to manufacturing investigational product, the workstation should be calibrated to verify proper function and performance by a certified professional [2]. The inside of the workstation should be disinfected with sterile disinfectant solution frequently, and a thorough decontamination procedure should be applied yearly or after any significant contamination event. All use and maintenance of workstations for production of HCT/Ps should be documented to support control of the manufacturing process [97].

Generally, cell manipulations and all other sterile processing (e.g., media preparation) should be performed only under laminar airflow conditions that meet ISO 5 standards. The ISO clean room classification system is related to the number and size of airborne particles allowed within a contained environment. Current ISO classes are outlined in Table 7.2 [102]. Commonly, biosafety cabinets are incorporated into manufacturing procedures to meet FDA requirements regarding sterile production activities, as only the environment to which product will be exposed must meet this standard.

Table 7.2 ISO 14644 Cleanroom Classes for Particle Monitoring [111]

| ISO Class | Maximum Particles per Cubic Meter (m³) | | | | | |
	0.1 μm	0.2 μm	0.3 μm	0.5 μm	1.0 μm	5.0 μm
1	10	2	---	---	---	---
2	100	24	10	4	---	---
3	1,000	237	102	35	8	---
4	10,000	2,370	1,020	352	83	---
5	100,000	23,700	10,200	3,520	832	29
6	1,000,000	237,000	102,000	35,200	8,320	293
7	---	---	---	352,000	83,200	2,930
8	---	---	---	3,520,000	832,000	29,300
9	---	---	---	35,200,000	8,230,000	293,000

Sterility of the production environment can be confirmed using touch and settling plates containing bacterial growth media and/or active air monitoring to demonstrate the absence of adventitious agents on surfaces or in the air [10]. The viability of all particles is calculated by dividing the number of colonies grown on plates by the total number of particles as measured by a calibrated instrument designed specifically for this purpose. Routines and methods for performing this testing during production, as well as strategies to document and review results should be outlined in the CMC section.

Temperature appropriate storage must be dedicated for HCT/Ps as well as various product components [103]. This equipment will likely include a range of incubators, refrigerators, freezers, and possibly cryogenic storage units that must be monitored for proper function and ability to maintain an appropriate temperature range. Automatic monitoring or user-generated logs are acceptable for recording daily temperatures; however, alarm systems should also be implemented to ensure the integrity of items contained within. Additionally, this equipment must be connected to adequate power sources, and alternative sources of energy should be identified for equipment critical to maintaining production capabilities (e.g., cryogenic storage unit containing MCB). Storage logistics must be outlined for the entire production cycle including intermediate and final product storage conditions and locations.

7.4.2 Procedures and Personnel

Personnel conducting manufacturing activities must be trained and experienced in aseptic technique in addition to standard operating procedures used to generate and test various product components. Documentation of applicable training should include the initial qualifications and experience of the trainee, a description of the type of training, name of trainer, as well as the dates training was performed and completed. These records should be reviewed by quality assurance periodically to ensure training was appropriate and complete for each employee and process.

As a cell or tissue-based product, the methods used to produce PSC-based therapeutics should consistently be in alignment with the principles of GTP [10]. Therefore, in addition to training, personnel should be familiar with the principles of quality control, cGMP and GTP. In each step of product preparation, extreme care must be given to both aseptic processing techniques and adherence to standard operating procedures, as maintenance of sterility and proper manufacturing documentation are key to generating materials suitable for clinical use. Quality assurance must verify investigational product lots were manufactured by trained personnel, in addition to ensuring aseptic technique and standard operating procedures were adhered to at all times [2].

Manufacturing documentation must include standardized procedures for producing each component of the product as well as any adjunct materials used during the manufacturing process (e.g., media for maintaining stem cells, substrate coating solutions for culture ware, etc.). The novel and complex nature of methods used to generate PSC-based HCT/Ps necessitates meticulous manufacturing procedures and records.

Batch records must be designed specifically for each task to include space for recording lot numbers of materials, calculations performed, measurements made (e.g., pipetting a volume of liquid or weighing a mass of powder) as well as the signature of the operator(s) and verifier. Procedural records for all maintenance and differentiation of cells must be explicitly detailed (e.g., exact number of minutes for an enzyme treatment or exact number of cell triturations). These records must be reviewed in addition to quality control testing prior to release of products for clinical use by the quality assurance unit.

7.4.2.1 Cell Collection, Culture Conditions, and Final Harvest

The methodology and conditions used to collect and culture cells for the production of an HCT/P must be presented as an outline of precise procedures that are replicable and easily understood by FDA regulators. This "product life cycle" should include all of the processes involved in both manufacturing and evaluating therapeutic products as well as indicate specific control points at which pass/fail determinations are made as to whether individual production batches proceed forward in the manufacturing process [56].

A product life-cycle schematic should also include descriptions of the timing required for specific steps and overall duration, as well as controls used to ensure aseptic processing [57]. The entire manufacturing process may take place over the course of days, weeks, or even months making it important to elucidate precise control points within the manufacturing process that are indicative of the success and/or sterility of each section of production.

Since terminal sterilization is not appropriate for generating PSC-based therapies, a closed system and/or aseptic processing methods are required [92]. A detailed description of culture systems, such as flasks, bags, or large-scale bioreactors, must be included as part of the CMC. These descriptions must relate to the production of clinically relevant sized batches, that is, not for small-scale batches with the intent to simply "scale up" when demand increases [56]. Where consumable platforms are used, as is this case for the majority of culture flasks or bags, it is generally a good idea to identify and validate alternative suppliers. Beyond establishing flexibility within control of the manufacturing process, it is beneficial to demonstrate that reliance on specific components will not be hindered by changes in availability during full-scale production runs.

While specific details included in the description of culture conditions will vary greatly between HCT/Ps, there are several key features that must be addressed because of their possible impact on final products. These features may include mechanical manipulations of cells (such as trituration or sorting) or use of enzymatic digestion procedures that are common for passaging of many cell types [92]. Additionally, any procedures to isolate or sort cells, including separation devices, density gradients, magnetic beads, or fluorescent activated cell sorting (FACS), should be described in full with substantial data supporting the robustness of these methods.

A detailed description of the procedures used to perform final harvesting of cells must also be included as part of the CMC. As previously mentioned, enzymatic applications or mechanical techniques, such as centrifugation, must be described fully including any wash conditions or media used. If cells are cryopreserved, this must be documented and performed using standardized equipment such as controlled rate freezers that regulate and record the rate at which items are frozen. Cryopreservation procedures should be tested and validated for suitability prior to freezing lots of product intended for clinical use. Finally, the storage conditions for harvested cells, including whether or not they are cryopreserved, should be included in the description of culture conditions [56].

7.4.2.2 Ex Vivo Gene Modification and Irradiation of Modified Cells

For certain PSC-based therapies, it may be necessary to utilize ex vivo gene manipulation strategies to generate final products. These modifications may include infection or transfection of cells with specific genetic modifiers in the form of viral or plasmid vectors. In these circumstances, details should be included regarding methods, devices, and reagents used in the selection and modification of cells. Where cells are placed back into culture after genetic modification, details of culture conditions, including media components and duration, must also be described. Further, there are specific safety issues that must be addressed when incorporating these techniques into the production of an HCT/P, including insertional mutagenesis, oncogenesis, immunogenicity, and the potential for inadvertent transmission and expression in non-target cells.

If a genetically modified allogeneic cell product is subjected to irradiation prior to injection, substantial data is required to demonstrate that cell products maintain desired characteristics, but are replication-incompetent. Further, the device used to perform cell irradiation must be regularly calibrated with a maintenance scheme outlined within the CMC. For additional information regarding regulations on the use of gene therapy techniques for the development of clinical products, consult a 2008 FDA Guidance titled "Content and Review of Chemistry, Manufacturing, and Control (CMC) Information for Human Gene Therapy Investigational New Drug Applications (INDs)" [104].

7.4.2.3   Final Product Formulation, Packaging, and Labeling

A finalized product formulation that includes the HCT/P, as well as any delivery or storage systems that are used to preserve, transport, or generate final dose preparations must be described in detail with schematic diagrams, as appropriate. This description should include the unit dosage amount, total number of units produced per manufacturing run, and appropriate storage methods for products. Whether products are delivered fresh or in a cryopreserved state, instructions for clinical site storage and preparation should also be included [56]. It is important to note that the formulation is subject to change as more safety and efficacy data become available. However, an initial formulation for intended Phase 1 studies must be included in the IND application as a starting point for communication with regulators.

The types of containers and closures used for packaging HCT/Ps must be reliable, compatible with product and storage conditions, as well as compliant with all regulatory standards. Full requirements for cGMP control of drug product containers and closures are included in 21 CFR 210 and 211 with additional USP requirements listed in the USP/NF [64, 90]. For submission of an IND, a description of intermediate and final product packaging should be described, as well as any precautions needed to ensure the protection and integrity of HCT/Ps during use in clinical trials [105].

For many HCT/Ps, intended administration is by injection, whether by intravenous, intrathecal, or other routes. This fact renders the packaging of products of highest concern to regulatory bodies, who will judge the quality and suitability of individual product packaging based on its ability to keep the product sterile and protected from environmental factors that may affect the overall quality, safety, or potency of contents [55]. Common causes of product degradation include exposure to light or reactive gases (e.g., oxygen), loss of solvent, absorption of water vapor, or microbial contamination.

All investigational drug products and HCT/Ps must be labeled "Caution: New Drug—Limited by Federal Law to Investigational Use" [106]. Other labeling requirements specific to HCT/Ps are located in 21 CFR 1271.250 "Labeling Controls" [107] and 21 CFR 1272.370 "Labeling" [108]. As previously described, cell therapy products intended solely for autologous use must contain a label indicating "Autologous Use Only" and, if donor testing and screening is not performed, "Not Evaluated for Infectious Substances" [23]. For non-autologous HCT/Ps, distinct identifiers not containing a donor's personal information must be contained on all product labeling and records pertaining to that HCT/P [50].

Generally, each HCT/P must contain a label that clearly and accurately indicates the product's distinct identification code, description, expiration, and any warning required by 21 CFR 1271.60(d)(2), 1271.65(b)(2), or 1271.90(b), where applicable and physically possible. Additional information is required either on package labeling or in accompanying product information, such as the name and address of the establishment making the HCT/P available for distribution, storage temperature, as well as appropriate warnings and instructions related to the prevention of transmission of communicable disease [52, 109, 110].

Accepted conventions for identification and labeling of final cell therapy products are contained within the International Standard for Blood and Transplant (ISBT) 128 Technical Specifications for Cellular Therapies [111]. Responsibility for implementation and management of these standards is maintained by the International Council for Commonality in Blood Banking Automation (ICCBBA), which ensures that medical products of human origin are assigned a unique identifier that is standardized across international borders.

Quarantined materials defined in 21 CFR 1271.3(q) [7], including HCT/Ps for which donor eligibility testing has not yet been completed, must be easily distinguishable from materials that are available for release and distribution [52]. In the circumstance that donor eligibility testing fails, in addition to physically separating these items from materials released for manufacturing, labeling must indicate "For Nonclinical Use Only" and contain a Biohazard legend [7]. Alternatively, if there is a documented urgent medical need, a provision exists that may allow for use of an otherwise ineligible product. However, in these circumstances, product labeling must state "Warning: Advise patient of communicable disease risks" and "Warning: Reactive test results for (name of specific disease agent or disease)" [109].

7.4.2.4 Product Shipping, Storage, and Handling

The CMC section must include information about systems to identify and track products from the time of initial collection, through production and shipment until administration of final products to individual patients. Documentation must account for the location and environment of HCT/Ps at all times—including during the shipment of products from manufacturing facilities to clinical sites. (Product tracking will be discussed further later in this chapter.)

Exterior packaging must be constructed to prevent contamination of the product and maintain proper temperature and pressure conditions [53]. The fragile and potentially hazardous nature of PSC-based HCT/Ps makes it vital to label packaging with all regulated and relevant information. This includes the general category of products contained in the package, identification of a temperature-controlled product, and proper UN classification codes.

It is vital to demonstrate that products will be stable throughout the duration of short- and long-term storage, as well as any transport processes. This is achieved by analyzing storage and shipment processes to establish acceptance criteria for handling of products. Standard operating procedures must be developed to fully outline the responsibilities of personnel on both shipping and receiving ends of packages. Federal regulations outlining specific responsibilities are located in 21 CFR 1271.265 "Receipt, predistribution shipment, and distribution of an HCT/P" [112].

Before standardized procedures can be developed, however, acceptable temperature limits must be established for storage and shipment of HCT/Ps [103]. Generally, cryopreserved products that are shipped frozen should be maintained at a temperature appropriate to the freezing medium (e.g., −109 °F/−78 °C for dry ice shipments). For non-cryopreserved cell therapy products, validated shipping services and products are available to ensure relative homeostasis during the shipping process. PSC-based products that are shipped "live" should never be exposed to freezing temperatures (below 32 °F/0 °C) or high temperatures (above 75 °F/24 °C) because of the potential for physiological damage and cell death.

In addition to validated shipping services, products are available to accurately record temperatures inside product packaging to verify temperature criteria outlined in product specifications are not breached. The feasibility of transportation strategies should be supported by data obtained from "test" product shipments that accurately reflect condition and duration of shipping process. Shipments should include actual product in order to perform viability and any additional testing. When shipping products to clinical sites for use, environmental monitoring of packages will be required to ensure quality and safety of materials.

7.4.2.5 Process Validation

Any process that cannot be fully verified by subsequent inspection and test must be justified, validated, and approved. Process validations should indicate that manufacturing procedures and equipment are able be used consistently and with a high degree of assurance during production of investigational HCT/Ps. Generally, this can be demonstrated by providing data from three or more productions runs that consistently generate product batches meeting all acceptance criteria. More detailed requirements for process validation are described in 21 CFR 820.75 [113].

7.5 Quality Control

For each lot of product component, adjunct material, or final investigational product, tests must be performed to verify conformance with specifications regarding identity, sterility, purity, and potency [114]. It may be difficult to fully qualify certain attributes of biological products such as PSC-based cell therapeutics, especially during early studies where product knowledge and characterization is limited. For this reason, it is critical to perform extensive product testing and retain samples of each investigational lot for direct comparison to products used in later stages of clinical trials. In this manner, the reproducibility of comparable HCT/Ps for additional clinical or safety studies will be ensured.

To control manufacturing processes, intermediate test procedures should be implemented in addition to final product testing. The timing of intermediate tests should reflect significant points in the manufacturing process such as a change in culture vessel or after the introduction of a new chemical agent [2]. All product and intermediate testing should be performed in a controlled environment under GLP, and significant consideration should be given to transfer of test samples from manufacturing areas to quality control areas to prevent potential contamination during transport.

A written plan that describes quality control functions, personnel, and training should be provided as part of the CMC. In addition to testing and oversight of materials, quality control is responsible for reviewing and approving test procedures, as well as validating acceptance criteria have been met for all released products [115]. While quality should be a goal of every member of the manufacturing department, it is highly recommended that at least one individual be assigned to perform quality control functions independently of manufacturing responsibilities. Generally, it is not cGMP-compliant for quality control personnel to review work they performed or product they produced.

In some instances where resources are limited, it may be necessary for a single individual to perform manufacturing and quality control duties. In such circumstances, it is recommended that another qualified individual not involved in manufacturing conduct periodic reviews of both manufacturing and quality control activities [10]. As HCT/P production needs grow, there will be stricter requirements for separation of quality and manufacturing activities. Regulations outlining production and process controls necessary for quality systems are located in 21 CFR 1271.220 [116] and 1271.225 [117] as they apply to Good Tissue Practice (GTP), as well as 21 CFR 820.70 for medical devices [118].

In this section, the various types of testing required to establish product safety and quality will be described. For PSC-based therapies, this includes demonstrating the identity and purity of final products, as well as establishing the presence of viable cells free from potentially infectious agents. Of equal importance are tests indicating the potency of a therapeutic product [119], a process that is far less defined for novel regenerative medicine products.

7.5.1 Identity Testing

In addition to establishing cell bank identity, testing must confirm the unique identity of final HCT/Ps. Quantitative testing by previously described genotypic, phenotypic, and/or biochemical assays is generally required to confirm cell identity and assess heterogeneity of cell populations [120]. All HCT/P lots must have their identity and corresponding labeling verified prior to release for use in investigational studies.

When generating PSC-based therapies, there are often cells present in the facility at many different stages of development. Therefore, identifying cells often means determining their specific maturation state from PSC to final derived cell product. Frequently, this is achieved through the use of biomarkers such as cell-surface receptors, transcription factors, or cytokines to specifically identify cells (e.g., Oct4 transcription factor to identify pluripotent stem cells). As cells evolve, often a panel of markers will be required to fully identify and qualify cell products. For example, to qualify mature oligodendrocytes, one might examine the expression of several oligodendrocyte-specific markers including Olig1/2 [121] along with less specific neuronal markers such as doublecortin.

7.5.2   Sterility Testing

Regulations regarding sterility and microbiological testing are located in 21 CFR 610.12, "Sterility" [122]. Generally, test methods must be validated for use in specific manufacturing paradigms and include standardized controls. Culture-based methods that rely on growth of microorganisms as well as nongrowth-based methods that evaluate microbiological surrogates may be suitable depending on the product and facilities.

Since many PSC-based therapies have a short "shelf life," Rapid Microbiological Methods (RMM) may be implemented to initiate release of final products prior to the standard 14-day microbiological culture requirements for many growth-based detection methods. However, implementing RMM procedures is only allowed when in-process sterility tests are negative, and a rapid method such as a Gram stain is incorporated and found negative in comparison to a known standard as part of final product testing. In principle, these methods can detect either specific microorganisms or other microbiological surrogates downstream of contamination; however, the methods used to establish sterility must be validated and approved by the FDA.

Validation characteristics for sterility testing include the limit of detection, specificity, degree of reproducibility, and robustness of analytical techniques. Specific guidelines for designing and selecting analytical microbiological methods for HCT/Ps are described in detail in an FDA Guidance published in 2008, titled "Validation of Growth-Based Rapid Microbiological Methods for Sterility Testing of Cellular and Gene Therapy Products" [123].

7.5.2.1   USP Sterility Test

Requirements for performing and validating sterility test methods for HCT/Ps are located in the USP 28 and National Formulary 23 General Chapter 71 [71] and 21 CFR 610.12 [122]. These tests must be approved for their ability to confirm either the presence or absence of bacterial or fungal elements in cell-based products. Generally, the first step in investigating product sterility includes performing a

Gram stain to rule out the presence of common Gram-positive bacteria. However, because this method is unsuitable for definitively identifying Gram-negative and Gram-variable bacterial species, other techniques are required for confirming full sterility of clinical-grade products.

For a more comprehensive sterility profile, a methodology incorporating a variety of different growth media is described in USP <71> *Sterility Tests* [71]. This methodology is time consuming due to its manual nature, which has led many researchers to attempt new strategies to comply with sterility testing regulations. A group at NIH's Department of Transfusion Medicine compared methods using Becton Dickinson's (Franklin Lakes, NJ) Bactec™ and bioMerieux's (Marcy-l'Etoile, France) BacT/ALERT® automated testing equipment to the manual CFR/USP method. The results of these experiments found that both of the automated methods produced consistent, reliable results within 7 days when applied to cell therapy products [126, 127]. These methods are currently being reviewed and USP <1223> *Validation of Alternative Microbiological Methods* will be updated in the future to reflect the results of this ongoing research [128].

7.5.2.2 Mycoplasma

Testing for mycoplasma must be performed after any pooling of cell cultures but prior to washing, as this detection method requires media supernatant rather than cells. Liquid samples collected are subjected to PCR-based amplification that allows detection of small amounts of mycoplasma DNA [62]. While many companies provide qualified mycoplasma testing services, currently, there is only a single FDA-approved product on the market for performing rapid detection methods in-house. Approved in the US in 2012, the MycoTOOL® PCR Mycoplasma Detection Kit from Roche Applied Science (Basel, Switzerland) has subsequently been approved by the European Medicines Agency, the Japanese Pharmaceuticals and Medical Devices Agency and over 60 other countries worldwide [129]. There are many other available kits for "unofficial" detection of mycoplasma; however, none of these have been approved for testing clinical-grade investigational products under cGMP.

7.5.2.3 Adventitious Agents

The presence of adventitious agents must be screened for in allogenic cells and tissues used in product manufacturing. Generally, this includes Epstein-Barr virus, cytomegalovirus, hepatitis B, hepatitis C, and human immunodeficiency virus. In addition, tests for hemadsorbing viruses, retroviruses, adenoviruses, and adeno-associated viruses may be required depending on the manufacturing and/or reprogramming processes used during production. These tests are performed using in vitro and in vivo methodologies and must be conducted under GLP conditions. Detailed descriptions of specific FDA-approved test methodologies for each of these requirements are available on the FDA website [130].

The facilities and diverse expertise required to perform this level of testing are frequently outside the scope of a product manufacturer's abilities and resources. Because of the quality, number, and diversity of tests that must be performed, it is highly recommended that these particular characterization tests be contracted out to facilities that are specifically equipped to perform these analyses. In this regard, there are many organizations that will perform adventitious agent testing in compliance with federal regulations including BioReliance (Rockville, MD), Molecular Diagnostic Services, Inc. (San Diego, CA), Genewiz (South Plainfield, NJ), Texcell (Frederick, MD/Cedex, France), and SGS (Geneva, Switzerland). While contracting this testing out may be an expensive prospect, using validated services will ensure initial product characterizations are of adequate quality for rigorous FDA standards.

7.5.3 Purity

Prior to release of an HCT/P, testing must be performed to demonstrate the final product is free from undesirable materials potentially introduced during the manufacturing process [112]. This includes residual contaminants specifically used during production as well as pyrogenic/endotoxin contaminants that may result from process impurities. For PSC-based therapies, this also includes establishing all the cells present in the final product are desired and approximately equal with regard to quality.

7.5.3.1 Detection of Cellular Contaminants

The most obvious cellular contaminant may be stem cells, but there are many other potential types of cellular contamination that could be present in PSC-based products. This is due to the fact that numerous types of cells may be differentiated from a single population of progenitor cells. Methodologies to identify cellular contaminants frequently rely on the use of cell-surface markers because coordinated panels can be used to identify discrete cells types or characteristics. An example of this type of analysis is shown for hematopoietic progenitors derived from PSCs in Fig. 7.4 [131].

Two chemicals, 4′, 6-diamidino-2-phenylindole (commonly referred to as DAPI) and Hoechst stain, are commonly used to identify all cells present in a sample because of their ability to permeate cells and emit blue fluorescence once bound to double-stranded DNA. This allows for a count of all cells that can then be used to determine relative percentages of other cells present, although it should be noted that Hoechst is only useful in identifying cells that are alive. Fluorescent microscopy is a relatively low throughput method to analyze cell-surface markers that are reflective of the cellular content of final products.

Flow cytometry is a higher throughput method that quickly profiles cells using panels of fluorescent biomarkers simultaneously. Guidelines for using flow cytometry as an analytical technique are outlined in USP <1027> *Flow Cytometry* [132].

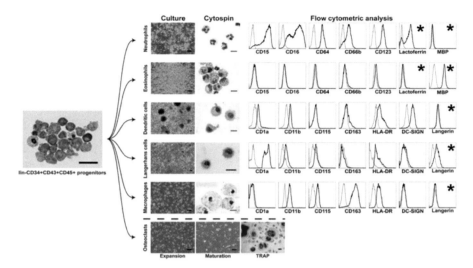

Fig. 7.4 Flow cytometry analysis of cells differentiated from hPSC-derived blood progenitors [137]

Benefits of using flow cytometry include the ability to quickly evaluate cell viability using propidium iodide and compare all generated sample profiles to standards as part of test validation. Generally, calibration and performance of flow cytometry is a more exact quantitative process than microscopic visualization; thus, it is highly recommended this method be incorporated into the production of investigational products where it is suitable to promote production capacity and consistency.

7.5.3.2 Identifying Residual Contamination

There are many potential types of residual contamination in cell-based products including proteins, DNA, RNA, solvents, cytokines, growth factors, antibodies, and serum. Accordingly, a range of tests may be required to verify levels of residual contamination present in the final product. Though procedures to identify common residual contaminants are mandatory, the selection of additional tests depends on the exact nature of the manufacturing process and specific HCT/P.

Immunological assays (e.g., ELISA) may be used to quantify residual protein contamination in products. RT-PCR methods are frequently utilized to qualify and quantify the presence of contaminating DNA. A cytokine profile assay may be useful to detect the presence of cytokines left over from the manufacturing process as well as specific contaminating cell types potentially present in the final product [56]. For example, to evaluate the presence of fibroblast or keratinocyte cells in the final product of an iPSC-based therapy, the level of IL-1, PDGFα, and TGFβ1 may be examined. Finally, to identify contaminating small molecules and peptides, analytical methods such as high performance liquid chromatography (HPLC) and mass spectrometry are useful and robust.

7.5.3.3 Pyrogenicity/Endotoxin Testing

A pyrogen is defined as any substance that has the potential to raise body temperature to physiologically high levels and includes endotoxins found in the cell wall of Gram-negative bacteria. Traditional in vivo tests for pyrogenic activity use intravenous injection of a drug product into rabbits. Due to the more complicated logistics of performing this in vivo method, acceptable in vitro alternatives have been developed and approved for use. Limulus amebocyte lysate (LAL)-based methods detect Gram-negative bacteria using the coagulation reaction that occurs between LAL and any endotoxin present in samples.

Quantification of this reaction is achieved using either a gel-clot, turbidimetric, or chromogenic technique to compare samples to known standards. The measurements made are expressed in Endotoxin Units (EU) per milliliter (mL) where one EU is approximately 100 picograms of *E. Coli* lipopolysaccharide representing an estimated 10^5 bacteria. Gel-clot techniques are simple to perform; however, photometric techniques allow for both end-point and kinetic measurements with detection levels as low as 0.001 EU/mL. Unlike testing for mycoplasma, there are many available products and kits to perform LAL-based endotoxin analyses in-house as well as an abundance of contract testing services.

The regulations for using LAL-based methodologies are described in the USP <85> *Bacterial Endotoxins Test* [124] and, where an applicable device is involved, USP <161> *Transfusion and Infusion Assemblies and Similar Medical Devices* [134]. The FDA has set a maximum limit of 5 EU per kilogram (kg) for non-intrathecal drugs and 0.2 EU/kg for drugs that will be administered intrathecally. These values are expressed in EU/kg to account for variation in dosage volumes, such that a 10 mL/kg maximum dose of a drug should contain no more than 0.5 EU/mL [135].

Reducing the level of endotoxin present in final HCT/Ps starts with utilizing quality, sterile components that are low in endotoxin during the manufacturing process. For example, Water for Injection (WFI) is required to have an endotoxin level less of than 0.25 EU/mL, which makes it a better choice for cGMP preparation of media or culture reagents compared to most other sterile waters. Thus, it is important to examine each step of the production process, especially product components and packaging materials, to prevent the introduction of potential pyrogenic agents into final products.

7.5.3.4 Tumorigenicity

During the IND phase, production processes, especially use of growth factors and purification procedures, will need to be evaluated and verified for overall robustness in removing oncogenic elements such as stem cells. Factors influencing the overall tumorigenicity of PSC-based products include the differentiation status of cells within the final formulation, the extent and nature of cell manipulations, profile of expressed transgenes in transduced cells, previous demonstrations of tumorigenic potential in preclinical studies and the target patient population.

The use of PSCs in manufacturing definitively necessitates tumorigenicity testing for each lot of final product to ensure it is free from undifferentiated cells or other oncogenic elements [60]. Traditional in vivo methods for assessing tumorigenicity, dysplasia, and hyperplasia include the use of immunocompromised rodents inoculated with cell products and observed for the formation of nodules over an appropriate time period. Routine histological examinations confirming either the presence of absence of tumor cells support the overall determination of tumorigenicity for a product. Since this lengthy process requires approval and facilities to host animals, which may not be feasible for small manufacturers, there are many contract research organizations that will provide this type of preclinical safety testing.

However, because this methodology is time-consuming and expensive, it generally unsuitable for in-process and final testing of products. Thus, flow cytometry methods incorporating the use of recognized stem cell markers (e.g., Oct4) might also be deemed acceptable for demonstrating freedom of final products from stem cell/tumorigenic contamination on a routine basis. However, these methods must be validated for their sensitivity, a process that involves "spiking" product samples with increasingly dilute amounts of stem cells to determine at what point stem cells are undetectable. Regulated limits for stem cells and other oncogenic contaminants are very low, but individual acceptance criteria should be based on preclinical data and discussed with FDA reviewers in consideration of specific products and applications.

7.5.4 Potency

Tests for the potency of an HCT/P should demonstrate the specific characteristics of transplantable cells that will produce desired clinical outcomes. Complicating matters, some HCT/Ps require transplantation of progenitor cells for engraftment rather than cells with mature phenotypes. It is often challenging to qualify the ability of progenitor cells to mature into fully functional cells in vivo at pre-engraftment/early developmental stages. Thus in vitro or in vivo assays, or a combination of both, must be specifically tailored to indicate the strength of each batch of HCT/P produced. These tests should qualitatively and quantitatively demonstrate specific desired characteristics of final products that are highly specific for each HCT/P.

Frequently, potency tests occur through verification of a panel of biological indicators including cell-surface markers or functional analysis. As cells differentiate and mature, different cell type-specific markers must be utilized to qualify the exact identity and maturity of cell products. For example, to qualify dopaminergic neurons, one could examine the expression of tyrosine hydroxylase on the cell surface using microscopy or flow cytometry. Additionally, assays could be performed to determine if cells are functional by appropriate electrophysiological profile or measurement of dopamine release upon stimulation.

In vivo tests for engraftability of cells may include examining human biomarkers in animals that were implanted with cells to assess viability and integration. While a range of accepted methodologies exist to evaluate specific features and functions of cells, assays indicating potency of PSC-based therapies and their capacity for successful engraftment must be specifically justified for each HCT/P. The complex nature of developing potency tests is covered in detail in a later chapter of this book.

7.5.5 Stability

For PSC-based therapies, there are two phases of product that need to be analyzed for stability. First, cells that are used in production need to be evaluated for genotypic and phenotypic stability as well as suitability for specific production processes as defined in acceptance criteria. Second, a stability assessment must be performed on end of production cells as a one-time test to assess final product safety. While these tests may be similar in nature, qualifying the stability of stem cells as pluripotent entities is a vastly different exercise from demonstrating the stability of mature cell types.

The stability of cell banks under defined storage conditions must be established for short-term production (less than a year) and generated continually as clinical trials proceed. The CMC should include a proposal of how the MCB and WCB will be monitored throughout production as well as indications for newly thawed containers. If production or thawing of cells from either the MCB or WCB does not take place for significant periods of time, viability testing should be performed at predefined intervals described in the CMC. Establishing consistency and stability of production activities through the redundancy of back-up systems is highly recommended.

To evaluate the stability of cell lines as they are repetitively cultivated for production, time points should be examined to determine the minimum and maximum number of subcultivations during which cells are adequate for production. For cell lines that contain recombinant DNA constructs, the coding sequence of the construct should also be verified within these limits by either nucleic acid testing or product analysis. For circumstances in which products cannot be analyzed by such methods, other traits may be useful for assessment of stability including morphological or growth characteristics, biochemical or immunological markers, as well as other relevant genotypic or phenotypic markers suitable for the HCT/P [56].

Often, the first criterion for release of final products is cell viability at the end of production. The survival rate of cells must be established at the time when product is released for distribution and include additional tests for all storage and preparation conditions prior to clinical administration. When qualifying HCT/Ps, it is important to demonstrate consistency with regard to viability as this indicates control over the entire process of production and substantiates product acceptance criteria.

Establishing absolute minimum release criteria for viable cells will greatly depend on potential negative effects of delivering dead or unhealthy cells into the body, methods for postproduction HCT/P preparation at clinical sites, and the minimum dose requirement for engraftable cells. If cells are purified on-site by some mechanism that guarantees the delivery of nearly 100 % viable cells, the first issue is largely negated. However, the robustness of mechanisms used to generate consistent viability should be clearly established in the CMC.

When cells are cryopreserved, viability data must support adequate survival of preserved cells once they are reconstituted for clinical production purposes. If cells demonstrate high viability after the preservation process, further testing is not generally required. However, if low viability is observed and considered an acceptable part of the manufacturing process, this must be justified and supported by data showing viable cells are fully functional and capable of producing an appropriate amount of clinical product [57].

While establishing product stability is a key part of a successful CMC, it is emphasized by the FDA that the amount of information required is dependent on the scope and length of the clinical investigation [64]. For PSC-based HCT/Ps, this entails demonstrating cells are within acceptable chemical and physical limits for the duration of production and any storage or transport processes until administration of the cell therapy product to a patient. Significant consideration must be given to postproduction preparation methods (e.g., syringe loading), and expiration limits for each step of dose preparation must be established in the CMC. Additionally, where cryopreservation procedures are used to store cellular product components, in-process materials, or final products, the stability of each individual cryopreservation step should be assessed at appropriate time intervals.

7.5.6 In-Process Testing

Testing to evaluate in-process quality or stability may include assays to monitor all product aspects including sterility, identity, purity, viability, and potency. For each test to be included in the stability panel, a description of the test method, sampling time point, and composition of test article should be included (e.g., cell pellet or supernatant). Sampling time points should be outlined in the CMC and determined by significant events in the production process (e.g., thawing of cells or change in culture vessel) or, alternatively, by anticipated developmental level of product.

Many of the tests previously described in this chapter could be used to establish acceptance criteria for in-process materials. For example, stem cell lines can be examined using karyotypic or cytogenetic analyses (e.g., FISH) to evaluate chromosomal aberrations that may have occurred over longer periods of time in culture. Additionally, the use of intermediate biomarkers that are not included in the assessment of final products may prove useful for developing in-process specifications or control points.

Specifications establishing acceptance criteria for release of in-process materials into the next phase of manufacturing must be generated and adhered to by operators [136]. The length of time to achieve test results for in-process testing should be considered carefully such that contaminated or otherwise unsuitable in-process materials do not move forward in manufacturing while waiting on test results. Delayed results could lead to wasting of valuable resources and/or contamination of other in-process materials.

For PSC therapies, it is often useful to evaluate spent media rather than testing cells directly where appropriate. This is especially true for in-process testing of adherent cell populations and should be taken into consideration when developing product test procedures. Indirect assessments of HCT/Ps should be documented and performed in accordance with specific test requirements and justified within CMC documentation.

7.5.7 Establishing Product Expiration

Though final product specifications are not expected in an IND application, methods to optimize specifications and in-process testing should be developed to advance Quality by Design—a topic discussed in the next section. Finalized expiration dates may not be available if products are cryopreserved because available data suggests that cells are viable even after long-term storage. However, products used for Phase 1 studies must have qualifying data that demonstrates they are used within acceptable time limits according to all available stability data. The expiration date of a product will be based on the specific HCT/P, processing methods, storage condition, and type of product packaging [137]. Viability and functionality of long-term storage of products should be evaluated on a yearly basis using appropriate methods defined for final product testing.

The expiration of final products must also be assessed for usage conditions. Clinical protocols defining the preparation and administration of final products with specific expiration limits for products once they have arrived on-site or been opened for use must be outlined in the CMC [138]. Determining the usage expiration of PSC-based final products, generally, is a shorter objective than for lifetime expiration because in vitro cells are typically only viable in culture for a matter of days or weeks. Expiration criteria may be based on viability and/or the loss of a specific biomarker or function as previously described for final product testing.

While final product expiration dates are a fairly concrete concept, the notion of in-process materials "expiring" may not be as obvious. For individual product preparation steps, there may be a specific point when the product is no longer suitable for continuation in the manufacturing process. For example, if a PSC-based therapy is dependent on cells being present at a particular maturation state to initiate the next phase of the manufacturing process, but cells did not reach this stage within their usual expansion profile, these cells may not be of the same quality as cells previously used to generate product. Beyond documenting any aberrations from standard operating procedures or specifications, routine testing should measure in-process materials to ensure the consistency and safety of all manufactured final products.

7.6 Quality Assurance

The full scope of quality system regulation required for cGMP is described in 21 CFR 820 [71]. At the IND level, expectations for quality systems are less stringent than for later phases of clinical trials. However, a quality control unit must be established prior to manufacturing product or components in order to assess materials management, production processes, and quality of final batches [115]. The oversight of all quality and production processes is expected to expand progressively with each level of clinical production. In this section, the important facets of quality systems and principles of Quality by Design as they apply to PSC-based therapies will be presented.

Quality by Design is a concept that relates quality of product design, materials, and processes to clinical performance of manufactured therapeutics. This occurs through review of data contained within manufacturing and clinical documentation to assimilate knowledge and identify areas for improvement of either processes or product. A diagram from the FDA demonstrating Quality by Design as it applies to manufacturing cell therapy products is shown in Fig. 7.5 [139]. The cyclical nature of quality processes reflects continual optimization of products that occurs as clinical studies progress.

7.6.1 Documentation Review

One of the primary roles of quality assurance is to review all documents generated for materials management, facilities and equipment, safety procedures, manufacturing, and quality control. Records must be examined for completion and to identify any aberrations or errors [140]. Any deviation from standard operating procedures must be investigated for potential effect on product safety and/or quality. Investigations should be overseen by quality assurance, but involve all appropriate departments in order to determine an appropriate course of action.

In addition to reviewing records, quality assurance is responsible for integrating data from all departments in order to identify relevant trends. Trends may be positive or negative in nature, such that improvements are identified through either refinement or corrective processes. For example, a positive trend might be established through correlation of clinical benefit with specific product characteristics demonstrated during product quality or potency testing. In an IND application,

Fig. 7.5 FDA graphic illustrating principles of quality by design [142]

product acceptance criteria will largely be based on preclinical studies performed on nonhuman subjects. Therefore, it is important to identify lots that perform best in human subjects during early clinical studies in order to refine acceptance criteria for release of future products.

Conversely, if lots of a product component or product made by a particular operator have been of consistently poorer quality than lots made by other operators, an investigation of this potentially negative trend would be warranted. The goal of investigations into negative product trends is to determine proper corrective and preventative action (CAPA) to ensure improved quality of future lots. An examination of manufacturing documentation for variations in operator technique, materials, or use of equipment may determine additional training or a change in materials or equipment would be corrective. Quality assurance must maintain records of all investigations and CAPA, as well as oversee implementation of CAPA in coordination with appropriate departments.

7.6.2 Document Controls

As previously described, documents outlining all procedures for materials management, manufacturing, quality control, and quality assurance should be standardized and include areas for appropriate signatures and verifications. Documents for new procedures or changes to existing documents must be reviewed and approved by a designated individual who is qualified to determine if changes are significant enough to require validation studies. This process ensures that all production of investigational products is being conducted in exactly the same manner to promote product consistency and validity of acquired clinical data.

All documents should be dated, and a history of changes to standardized documents should be meticulously recorded [141]. Previous versions of documents must be archived in order to fully trace all changes to product manufacturing, regardless of apparent significance of changes. Under 21 CFR 312.57, sponsors must retain all supporting records for at least 2 years after a marketing application is approved for the drug [142]. If a marketing application is not approved for the drug, records must be kept until 2 years after shipment and delivery of the drug for investigational use is discontinued and the FDA is notified [2].

7.6.3 Non-conforming Products and Adverse Events

Under 21 CFR 211.192, the quality control unit is endowed with the responsibility to review and approve production and control records to determine compliance before a batch is released or distributed [140]. This individual or unit is also responsible for investigating any unexplained discrepancies or deviations in standard operating procedures or the failure of a batch or any of its components to meet specifications, regardless if the batch is distributed.

Even early in the clinical process, adverse event reporting is required by federal regulations. An adverse reaction to a drug or drug product occurs when any noxious and unintended response is observed following the handling or administration of said drug, for which there is a reasonable possibility that the HTC/P is the cause. Any adverse reaction involving a communicable disease must be investigated to determine potential exposure, especially in cases where infected batches were made available for distribution. Additional reporting is required when the adverse reaction involves life-threatening injuries, permanent impairment, necessitates medical or surgical intervention, or is fatal [143].

An investigation into a failed or non-conforming product, component, or adverse event must be extended to include other batches of the same or other drug products associated with the specific failure or discrepancy [140]. For this reason, it is vital to pursue initial investigations until a root cause can be identified with a high degree of certainty. A written record of the investigation including conclusions and summary of follow-up actions is required for any potentially significant incident. Follow-up includes outlining any CAPA taken to ensure similar events do not take place in the future [143, 144].

7.6.4 Product Tracking

Quality assurance must ensure each HCT/P is tracked during all phases of manufacturing, from donor to final disposition, in order to facilitate investigation of any actual or suspected transmission event involving a communicable disease or adventitious agent [145]. The CMC should describe the system established to track and segregate investigational products, including the assignment of distinct identification codes that relate the HCT/P to the original donor and all associated records [56]. As previously mentioned, identifier codes must not include any personal information relating to the donor, such as name, social security number, or other medical record number [45].

Prior to distribution of an HCT/P to a consignee, the consignee must be informed in writing of the requirements for specific tracking systems to be implemented for that product. Compliance with and maintenance of established tracking systems is imperative and required by FDA regulatory authorities [145]. Strategies for facilitating communication between consignee and distributor should be included in the description of tracking system.

7.6.5 Additional Regulatory Guidance for Product Developers

The FDA and CBER offer many different forms of assistance for developers of new drug products including consultation with a variety of scientific and medical professionals. During early phases of development, there is support to ensure preclinical experimental design and analysis will yield sufficient and relevant data necessary to progress a drug candidate forward to the next clinical phase. In later phases, similar assistance is available for the design and execution of clinical experiments, as well as guidance to develop optimized product testing and specifications appropriate to each product.

Because every HCT/P is unique in its origin, production, and intended use, the regulatory approval process is highly individualized. A lengthy dialogue between researchers and regulatory officials begins with a pre-IND consultation intended to foster communication and clarification between sponsors/manufacturers and regulators. For PSC-based therapies, there remain many technical and regulatory challenges to overcome before products become commercially available. The novelty of these products requires IND applications that demonstrate a well-considered plan with rigorous process and safety controls. Therefore, a thorough CMC section is vital to a successful IND application and advancement of an HCT/P to Phase 1 clinical trials.

References

1. US National Archives and Records Administration. Code of federal regulations. IND Content and format. 2013;Title 21, Part 312.23.
2. USFDA Center for Biologics Evaluation and Research. FDA guidance for industry: cGMP for phase 1 investigational drugs. Jul 2008. Available from: http://www.fda.gov/downloads/Drugs/GuidanceComplianceRegulatoryInformation/Guidances/ucm070273.pdf
3. US National Archives and Records Administration. Code of federal regulations. Human cells, tissues, and cellular and tissue-based products. 2013;Title 21, Part 1271.
4. US Code. Public Health Service Act. Regulation of biological products. 1999;Title 42, Part 262, Section 351 [cited 4 Aug 2013]. Available from: http://www.fda.gov/RegulatoryInformation/Legislation/ucm149278.htm
5. US Code. Public Health Service Act. Regulations to control communicable diseases. 1999;Title 42, Part 264, Section 361 [cited 4 Aug 2013]. Available from: http://www.fda.gov/RegulatoryInformation/Legislation/ucm149429.htm
6. US National Archives and Records Administration. Code of federal regulations. General provisions. 2013;Title 21, Part 1271, Subpart A.
7. US National Archives and Records Administration. Code of federal regulations. Minimal manipulation means. 2013;Title 21, Part 1271.3.
8. DeFrancesco L. FDA prevails in stem cell trial. Nat Biotechnol. 2012;30(10):906.
9. US National Archives and Records Administration. Code of federal regulations. Good laboratory practice for nonclinical laboratory studies. 2013;Title 21, Part 58.
10. US National Archives and Records Administration. Code of federal regulations. Current good tissue practice. 2013;Title 21, Part 1271, Subpart D.
11. Gutierrez-Aranda I, Ramos-Mejia V, Munoz-Lopez M, Real PJ, Mácia A, Sanchez L, Ligero G, Garcia-Parez JL, Menendez P. Human induced pluripotent stem cells develop teratoma more efficiently and faster than human embryonic stem cells regardless the site of injection. Stem Cells. 2010;28(9):1568–70.
12. Okita K, Ichisaka T, Yamanaka S. Generation of germline-competent induced pluripotent stem cells. Nature. 2007;448(7151):313–7.
13. Yu J, Hu K, Smuga-Otto K, Tian S, Stewart R, Slukvin II, Thomson J. Human induced pluripotent stem cells free of vector and transgene sequences. Science. 2009;324(5928):797–801.
14. Kim D, Kim CH, Moon JI, Chung YG, Chang MY, Han BS, et al. Generation of human induced pluripotent stem cells by direct delivery of reprogramming proteins. Cell Stem Cell. 2009;4(6):472–6.
15. Fusaki N, Ban H, Nishiyama A, Saeki K, Hasegawa M. Efficient induction of transgene-free human pluripotent stem cells using a vector based on Sendai virus, an RNA virus that does not integrate into the host genome. Proc Jpn Acad Ser B Phys Biol Sci. 2009;85(8):348–62.
16. Jia F, Wilson KD, Sun N, Gupta DM, Huang M, Li Z, et al. A nonviral minicircle vector for deriving human iPS cells. Nat Methods. 2010;7(3):197–9.
17. Warren L, Manos PD, Ahfeldt T, Loh YH, Li H, Lau F, et al. Highly efficient reprogramming to pluripotency and directed differentiation of human cells with synthetic modified mRNA. Cell Stem Cell. 2010;7(5):618–30.
18. Anokye-Danso F, Trivedi CM, Juhr D, Gupta M, Cui Z, Tian Y, et al. Highly efficient miRNA-mediated reprogramming of mouse and human somatic cells to pluripotency. Cell Stem Cell. 2011;8(4):376–88.
19. Miyoshi N, Ishii H, Nagano H, Haraguchi N, Dewi DL, Kano Y, et al. Reprogramming of mouse and human cells to pluripotency using mature microRNAs. Cell Stem Cell. 2011;8(6):633–8.
20. Awe JP, Lee PC, Ramathal C, Vega-Crespo A, Durruthy-Durruthy J, Cooper A, et al. Stem Cell Res Ther. 2013;4(4):87.
21. US National Archives and Records Administration. Code of federal regulations. General provisions. 2013;Title 21, Part 1271, Subpart B.

22. US National Archives and Records Administration. Code of federal regulations. Occupational safety and health standards. 2013;Title 29, Part 1910.
23. US National Archives and Records Administration. Code of federal regulations. Donor-eligibility determination not required. 2013;Title 21, Part 1271.90(a).
24. Ikehata H, Kudo H, Masuda T, Ono T. UVA induces C -> T transitions at methyl-CpG-associated dipyrimidine sites in mouse skin epidermis more frequently than UVB. Mutagenesis. 2003;18(6):511–9.
25. Yokoyama Y. Hematopoietic stem cells and mature blood cells from pluripotent stem cells. Nihon Rinsho [Japanese]. 2011;69(12):2137–41.
26. Red Cross. What we do: lifesaving blood. 2013 [cited 21 Aug 2013]. Available from: http://www.redcross.org/what-we-do/blood-donation
27. Merling RK, et al. Transgene-free iPSCs generated from small volume peripheral blood non-mobilized CD34+ cells. Blood. 2013;121(14):e98–107.
28. Dowey SN, Huang X, Chou BK, Ye Z, Cheng L. Generation of integration-free human induced pluripotent stem cells from postnatal blood mononuclear cells by, plasmid vector expression. Nat Protoc. 2012;7(11):2013–21.
29. Ramos-Mejía V, Montes R, Bueno C, Ayllón V, Real PJ, Rodriguez R, Menedez P. Residual expression of the reprogramming factors prevents differentiation of iPSC generated from human fibroblasts and cord blood CD34+ progenitors. PLoS ONE. 2012;7(4), e35824.
30. Ye L, Muench MO, Fusaki N, Beyer AI, Wang J, Qi Z, et al. Blood cell-derived induced pluripotent stem cells free of reprogramming factors generated by Sendai viral vectors. Stem Cells Transl Med. 2013;2(8):558–66.
31. Zuk PA, Zhu M, Ashjian P, De Ugarte DA, Huang JI, Mizuno H, et al. Human adipose tissue is a source of multipotent stem cells. Mol Biol Cell. 2002;13(12):4279–95.
32. Sugi IS, Kida Y, Kawamura T, Suzuki J, Vassena R, Yin YQ, et al. Human and mouse adipose-derived cells support feeder-independent induction of pluripotent stem cell. Proc Natl Acad Sci U S A. 2010;107(8):3558–63.
33. Sun N, Panetta NJ, Gupta DM, Wilson KD, Lee A, Jia F, et al. Feeder-free derivation of induced pluripotent stem cells from adult human adipose stem cells. Proc Natl Acad Sci U S A. 2009;106(37):15720–5.
34. Narsinh KH, Jia F, Robbins RC, Kay MA, Lonake MT, Wu JC. Generation of adult human induced pluripotent stem cells using nonviral minicircle DNA vectors. Nat Protoc. 2011;6(1):78–88.
35. US National Archives and Records Administration. Code of federal regulations. Scope. 2013;Title 21, Part 1270.1(c).
36. Malarkey M. Letter from Mary A. Malarkey, Director of Compliance and Biologics Quality, USFDA, to Christopher J. Centeno, Medical Director, Regenerative Sciences, Inc. 25 Jul 2008. Available from: http://www.fda.gov/BiologicsBloodVaccines/Guidance ComplianceRegulatoryInformation/ComplianceActivities/Enforcement/UntitledLetters/ ucm091991.htm
37. Regenexx Procedures Family: Advanced Stem Cell Procedures. 2013. [cited 28 Aug 2013]. Available from: http://www.regenexx.com/regenexx-procedures-family/
38. Chirba MA and Nobel AA. Our Bodies, our cells: FDA regulation of autologous adult stem cell therapies. Bill of Health. June 2013. Available from: http://works.bepress.com/ maryann_chirba/38
39. Malarkey M. Letter from Mary A. Malarkey, Director of Compliance and Biologics Quality, USFDA, to David Eller, CEO, CellTex Therapeutics Corporation. 24 Sept 2012 [cited 8 Aug 2013]. Available from: http://www.fda.gov/ICECI/EnforcementActions/Warning Letters/ 2012/ucm323853.htm
40. Cyranoski D. Controversial stem-cell company moves treatment out of the United States. Nat News. 20 Jan 2013. Available from: http://www.nature.com/new/controversial-stem-cell-company-moves-treatment-out-of-the-united-states-1.12332

41. USFDA Consumer Health Information. FDA warns about stem cell claims. Jan 2012 [cited 5 Aug 2013]. Available from: http://www.fda.gov/downloads/ForConsumers/Consumer Updates/UCM286213.pdf
42. USFDA Basics. What are stem cells? How are they regulated? 7 Jan 2012 [cited 28 Aug 2013]. Available from: http://www.fda.gov/AboutFDA/Transparency/Basics/ucm194655.htm
43. US National Archives and Records Administration. Code of federal regulations. Donor eligibility. 2013;Title 21, Part 1271, Subpart C.
44. US National Archives and Records Administration. Code of federal regulations. General requirements for informed consent. 2013;Title 21, Part 50.20.
45. USFDA Center for Biologics Evaluation and Research. FDA guidance for industry: eligibility determination for donors of human cells, tissues, and cellular and tissue-based products (HCT/Ps) small entity compliance guide. Aug 2007. Available from: http://www.fda.gov/ biologicsbloodvaccines/GuidanceComplianceRegulatoryInformation/Guidances/tissue/ ucm073964.pdf
46. US National Archives and Records Administration. Code of federal regulations. How do I screen a donor? 2013;Title 21, Part 1271.75.
47. US National Archives and Records Administration. Code of federal regulations. What are the general requirements for donor testing? 2013;Title 21, Part 1271.80.
48. US National Archives and Records Administration. Code of federal regulations. What donor testing is required for different types of cells and tissues? 2013;Title 21, Part 1271.85.
49. US National Archives and Records Administration. Code of federal regulations. How do I determine whether a donor is eligible? 2013;Title 21, Part 1271.50.
50. US National Archives and Records Administration. Code of federal regulations. Record retention requirements. 2013;Title 21, Part 1271.55(d).
51. ESI Bio Corporate Website. ES cell lines. 2014. [cited 2 June 2014]. Available from: http:// www.esibio.com/products/product-category/cell-lines/
52. US National Archives and Records Administration. Code of federal regulations. What quarantine and other requirements apply before the donor-eligibility determination is complete? 2013;Title 21, Part 1271.60.
53. US National Archives and Records Administration. Code of federal regulations. Receipt, predistribution shipment, and distribution of an HCT/P. 2013;Title 21, Part 1271.265.
54. Brandenberger R, Burger S, Campbell A, Fong T, Lapinskas E, Rowley JA. Cell therapy bioprocessing: integrating process and product development for the next generation of biotherapeutics. BioProcess Int. 2011;9(S1):30–7. Available from: http://www.bioprocessintl. com/manufacturing/cell-therapies/cell-therapy-bioprocessing-314870/.
55. US National Archives and Records Administration. Code of federal regulations. Testing and approval or rejection of components, drug product containers, and closures. 2013;Title 21, Part 211.84.
56. USFDA Center for Biologics Evaluation and Research. Guidance for FDA reviewers and sponsors: content and review of chemistry, manufacturing and control (CMC) information for human somatic cell therapy investigational new drug applications. Apr 2008. Available from: http://www.fda.gov/biologicsbloodvaccines/GuidanceComplianceRegulatoryInformation/ Guidances/tissue/ucm073964.pdf
57. International Conference on Harmonisation. Guidance on quality of biotechnological/biological products: derivation and characterization of cell substrates used for production of biotechnological/biological; availability. Fed Regist. 1998;63(182):50244–9.
58. USFDA Center for Biologics Evaluation and Research. Draft of points to consider in the characterization of cell lines used to produce biologicals. Jul 1993. Available from: http:// www.fda.gov/biologicsbloodvaccines/safetyavailability/ucm162863.pdf
59. US National Archives and Records Administration. Code of federal regulations. General provisions. 2013;Title 21, Part 610, Subpart B.
60. USFDA Center for Biologics Evaluation and Research. Draft guidance for industry: preclinical assessment of investigational cellular and gene therapy products. Nov 2012. Available from:

http://www.fda.gov/biologicsbloodvaccines/GuidanceComplianceRegulatoryInformation/
Guidances/CellularandGeneTerapy/ucm329861.pdf

61. US National Archives and Records Administration. Code of federal regulations. General provisions. 2013;Title 21, Part 610, Subpart E.

62. Health Information Privacy and Portability Act of 1996. United States Statutes at large. (Pub. L. no. 104–191). 1996:100 Stat. 2548. Available from: http://www.gpo.gov/fdsys/pkg/PLAW-104pub;191/content-detail.html

63. US National Archives and Records Administration. Code of federal regulations. Reagents. 2013: Title 21, Part 1271.210(a, b).

64. US National Archives and Records Administration. Code of federal regulations. Current good manufacturing practice in manufacturing, processing, packing, or holding of drugs; general. 2013;Title 21, Part 210.

65. US National Archives and Records Administration. Code of federal regulations. Current good tissue practice. 2013;Title 21, Part 1271.10.

66. US National Archives and Records Administration. Code of federal regulations. Reagents. 2013;Title 21, Part 1271.210(d).

67. US National Archives and Records Administration. Code of federal regulations. Requirements for ingredients of animal origin used for product of biologics. 2013;Title 9, Part 113.53.

68. Use of Materials Derived from Cattle in Medical Products Intended for Use in Humans and Drugs Intended for Use in Ruminants, Proposed Rule. Federal register 72. 12 Jan 2007:1581.

69. US National Archives and Records Administration. Code of federal regulations. Purchasing controls. 2013;Title 21, Part 820.50.

70. US National Archives and Records Administration. Code of federal regulations. Supplies and reagents. 2013;Title 21, Part 1271.210.

71. United States Pharmacopeia and National Formulary (USP36-NF31). Sterility tests. Rockville, MD: United States Pharmacopeia Convention; 2012. Chapter 71.

72. Grein TA, Freimark D, Weber C, Hudel K, Wallrapp C, Czermark P. Alternative to dimethylsulfoxide for serum-free cryopreservation of human mesenchymal cells. Int J Artif Organs. 2010;33(6):370–80.

73. Balci D, Can A. The assessment of cryopreservation conditions for human umbilical cord stroma-derived mesenchymal stem cells towards a potential use for stem cell banking. Curr Stem Cell Res Ther. 2013;8(1):60–72.

74. Buchanan SS, Gross SA, Acker JP, Toner M, Carpenter JF, Pyatt DW. Cryopreservation of stem cells using trehalose: evaluation of the method using a human hematopoietic cell line. Stem Cells Dev. 2004;13(3):295–305.

75. Fuller BJ. Cryoprotectants: the essential antifreezes to protect life in the frozen state. Cryo Lett. 2004;25(6):375–88.

76. Matsamura K, Hayashi F, Nagashima T, Hyon SH. Long-term cryopreservation of human mesenchymal stem cells using carboxylated poly-L-lysine without the addition of proteins or dimethyl sulfoxide. J Biomater Sci Polym Ed. 2013;24(12):1484–97.

77. Woods EJ, Pollok KE, Byers MA, Perry BC, Purtteman J, Heimfeld S, Gao D. Cor blood stem cell cryopreservation. Trasnfus Med Hemother. 2007;34(4):276–85.

78. Liseth K, Abrahamsen JF, Bjørsvik S, Grøttebø K, Bruserud Ø. The viability of cryopreserved PBPC depends on the DMSO concentration and the concentration of nucleated cells in the graft. Cytotherapy. 2005;7(4):328–33.

79. Benekli M, Anderson B, Wentling D, Bernstein S, Czuczman M, McCarthy P. Severe respiratory depression after dimethylsulphoxide-containing autologous stem cell infusion in a patient with AL amyloidosis. Bone Marrow Transplant. 2000;25(12):1299–301.

80. Higman MA, Port JD, Beauchamp Jr NJ, Chen AR. Reversible leukoencephalopathy associated with re-infusion of DMSO preserved stem cells. Bone Marrow Transplant. 2000; 26(7):797–800.

81. Hequet O, Dumontet C, El Jaafari-Corbin A, Salles G, Espinhouse D, Arnaud P, et al. Epileptic seizures after autologous peripheral blood progenitor infusion in a patient treated with high-dose chemotherapy for myeloma. Bone Marrow Transplant. 2002;29(6):544.

82. Hoyt R, Szer J, Grigg A. Neurological events associated with the infusion of cryopreserved bone marrow and/or peripheral blood progenitor cells. Bone Marrow Transplant. 2000; 25(12):1285–7.

83. Pavlovic M, Balint B. Principle and practice of stem cell cryopreservation. In: Pavlovic M and Balint B, ed. Stem cells and tissue engineering. New York: Springer; 2013. p. 71–81.

84. Matsuo A, Yamazaki Y, Takase C, Aoyagi K, Uchinuma E. Osteogenic potential of cryopreserved human bone marrow-derived mesenchymal stem cells cultured with autologous serum. J Craniofac Surg. 2008;19(3):693–700.

85. Reuther T, Kettmann C, Scheer M, Kochel M, Iida S, Kubler AC. Cryopreservation of osteoblast-like cells: viability and differentiation with replacement of fetal bovine serum in vitro. Cells Tissues Organs. 2006;183(1):32–40.

86. Dimarakis I, Levicar N. Cell culture medium composition and translational adult bone marrow-derived stem cell research. Stem Cells. 2006;24(5):1407–8.

87. Wagner K, Welch D. Cryopreservation and recovering of human iPS cells using complete knockout serum replacement feeder-free medium. J Vis Exp. 2010;41:2237.

88. Hunt NC, Grover LM. Cell encapsulation using biopolymer gels for regenerative medicine. Biotechnol Lett. 2010;32(6):733–42.

89. Carpenter MK, Frey-Vasconcells J, Rao M. Developing safe therapies from human pluripotent stem cells. Nat Biotechnol. 2009;27:606–13. Available from: http://www.nature.com/nbt/journal/v27/n7/fig_tab/nbt0709-606_F1.html.

90. US National Archives and Records Administration. Code of federal regulations. Current good manufacturing practice for finished pharmaceuticals. 2013;Title 21, Part 211.

91. US National Archives and Records Administration. Code of federal regulations. Investigational new drug application. 2013;Title 21, Part 312.

92. USFDA Center for Biologics Evaluation and Research. FDA guidance for industry: sterile drug products produced by aseptic processing – current good manufacturing practice. Sept 2004. Available from: http://www.fda.gov/downloads/drugs/.../Guidances/ucm070342.pdf

93. US National Archives and Records Administration. Code of federal regulations. Scope. 2013;Title 21, Part 1271.1(b).

94. American Association of Tissue Banks Website. State requirement for tissue bank licensure, registration or certification. Nov 2006 [cited 2 Jun 2014]. Available from: http://www.aatb.org/State-Requirements-for-Tissue-Bank-Licensure-Registration-or-Certification

95. Dietz AB, Padley DJ and Gastineau DA. Infrastructure development for human cell therapy translation. Clin Pharmacol Ther. 2007;82:320–4. Available from: http://www.nature.com/clpt/journal/v82/n3/fig_tab/6100288f1.html /#figure-title

96. United States Pharmacopeia and National Formulary (USP36-NF31). Microbiological evaluation of clean rooms and other controlled environments. Rockville, MD: United States Pharmacopeia Convention; 2012. Chapter 1116.

97. US National Archives and Records Administration. Code of federal regulations. Environmental control and monitoring. 2013;Title 21, Part 1271.195.

98. US National Archives and Records Administration. Code of federal regulations. Identification. 2013;Title 21, Part 820.60.

99. US National Archives and Records Administration. Code of federal regulations. Human Drugs and Biologics. 2013;Title 21, Part 25.

100. US National Archives and Records Administration. Code of federal regulations. Action on an IND. 2013;Title 21, Part 25.31(e).

101. US National Archives and Records Administration. Code of federal regulations. Equipment. 2013;Title 21, Part 1271.200.

102. Part 4: Design, Construction and Start-Up. Cleanrooms and associated controlled environments. ISO 14644–4:2001. Available from: http://www.iso.org/iso/catalogue_detail.htm?csnumber=25007

103. US National Archives and Records Administration. Code of federal regulations. Control of storage areas. 2013;Title 21, Part 1271.260(e).

104. USFDA Center for Biologics Evaluation and Research. Guidance for FDA Reviewers and sponsors: content and review of chemistry, manufacturing and control (CMC) information for human gene therapy investigational new drug applications. Apr 2008. Available from: http://www.fda.gov/BiologicsBloodVaccines/GuidanceCom plianceRegulatoryInformation/Guidances/CellularandGeneTherapy/ucm072587.htm
105. USFDA Center for Biologics Evaluation and Research. Guidance for industry: container closure systems for packaging human drugs and biologics. May 1999. Available from: http://www.fda.gov/downloads/Drugs/Guidances/ucm070551.pdf
106. US National Archives and Records Administration. Code of federal regulations. Labeling of an investigational new drug. 2013;Title 21, 312.6.
107. US National Archives and Records Administration. Code of federal regulations. Labeling controls. 2013;Title 21, 1271.250.
108. US National Archives and Records Administration. Code of federal regulations. Labeling. 2013: Title 21, 1271.370.
109. US National Archives and Records Administration. Code of federal regulations. How do I store an HCT/P from a donor determined to be ineligible, and what uses of the HCT/P are not prohibited. 2013;Title 21, 1271.65.
110. US National Archives and Records Administration. Code of federal regulations. Are there exceptions from the requirement of determining donor eligibility and what labeling requirements apply? 2013;Title 21, 1271.90.
111. International Standard for Blood and Transplant (ISBT) 128 Technical Specification for Cellular Therapies. 4th ed. 2013. Available from: http://www.ICCBBA.org
112. US National Archives and Records Administration. Code of federal regulations. Availability for distribution. 2013;Title 21, Part 1271.265(c).
113. US National Archives and Records Administration. Code of federal regulations. Process validation. 2013;Title 21, Part 820.75.
114. US National Archives and Records Administration. Code of Federal regulations. Testing and release for distribution. 2013;Title 21, Part 211.165.
115. US National Archives and Records Administration. Code of federal regulations. Responsibilities of quality control unit. 2013;Title 21, Part 211.22.
116. US National Archives and Records Administration. Code of federal regulations. Processing and process controls. 2013;Title 21, Part 1271.220.
117. US National Archives and Records Administration. Code of federal regulations. Process changes. 2013;Title 21, Part 1271.225.
118. US National Archives and Records Administration. Code of federal regulations. Production and process controls. 2013;Title 21, Part 820.70.
119. US National Archives and Records Administration. Code of federal regulations. Potency. 2013;Title 21, Part 610.10.
120. US National Archives and Records Administration. Code of federal regulations. Identity. 2013;Title 21, Part 610.14.
121. Lu QR, Park JK, Noll E, Chan JA, Alberta J, Yuk D, et al. Oligodendrocyte lineage genes (OLIG) as molecular markers for human glial brain tumors. Proc Natl Acad Sci U S A. 2001;98(19):10851–6.
122. US National Archives and Records Administration. Code of federal regulations. Sterility. 2013;Title 21, Part 610.12.
123. USFDA Center for Biologics Evaluation and Research. Guidance for industry: validation of growth-based rapid microbiological methods for sterility testing of cellular and gene therapy products. Feb 2008. Available from: http://www.fda.gov/biologicsbloodvaccines/GuidanceComplianceRegulatoryInformation/Guidances/CellularandGeneTherapy/ucm072612.pdf
124. United States Pharmacopeia and National Formulary (USP36-NF31). Bacterial endotoxins test. Rockville, MD: United States Pharmacopeia Convention; 2012. Chapter 85.
125. US National Archives and Records Administration. Code of federal regulations. Quality system regulation. 2013;Title 21, Part 820.

126. Khuu HM, Stock F, McGann M, Carter CS, Atkins JW, Murray PR, Read EJ. Comparison of automated culture systems with a CFR/USP-compliant method for sterility testing of cell therapy products. Cytotherapy. 2004;6(3):183–95.
127. Khuu HM, Patel N, Carter CS, Murray PR, Read EJ. Sterility testing of cell therapy products: parallel comparison of automated methods with a CFR-compliant method. Transfusion. 2006;46(12):2071–82.
128. United States Pharmacopeia and National Formulary (USP36-NF31). Cell and gene therapy products. Rockville, MD: United States Pharmacopeia Convention; 2012. Chapter 1047.
129. MycoTool PCR Mycoplasma Detection Kit: Overview. 2013 [cited 14 Aug 2013]. Available from: http://www.roche-applied-science.com/shop/custom-biotech/products/mycotool-pcr-mycoplasma-detection-kit
130. Testing HCT/P Donors for Relevant Communicable Disease Agents and Diseases. 29 Jul 2013 [cited 14 Aug 2013]. Available from: http://www.fda.gov/ biologicsbloodvaccines/safetyavailability/tissuesafety/ucm095440.htm
131. Choi KD, Vodyanik M, Slukvin II. The hematopoietic differentiation and production of mature myeloid cells from human pluripotent stem cells. Nat Protoc. 2011;6(3):296–313. Available from: http://www.ncbi.nlm.nih.gov/pmc/articles/PMC3066067/figure/F2/
132. United States Pharmacopeia and National Formulary (USP36-NF31). Flow cytometry. Rockville, MD: United States Pharmacopeia Convention; 2012. Chapter 1027.
133. United States Pharmacopeia and National Formulary (USP36-NF31). Growth factors and cytokines used in cell therapy manufacturing. Rockville, MD: United States Pharmacopeia Convention; 2012. Chapter 92.
134. United States Pharmacopeia and National Formulary (USP36-NF31). Transfusion and infusion assemblies and similar medical devices. Rockville, MD: United States Pharmacopeia Convention; 2012. Chapter 161.
135. Bacterial Endotoxins/Pyrogens. Inspections, compliance, enforcement, and criminal investigations. 20 Mar 1985 [cited 4 Aug 2013]. Available from: http://www.fda.rov/ICECI/Inspections/InspectionGuides/InspectionTechnicalGuides/ucm072918.htm
136. US National Archives and Records Administration. Code of federal regulations. Management responsibility. 2013;Title 21, Part 820.80.
137. US National Archives and Records Administration. Code of federal regulations. Acceptable temperature limits. 2013;Title 21, Part 1271.260(e).
138. US National Archives and Records Administration. Code of federal regulations. General biological products standards. 2013;Title 21, Part 610.
139. USFDA Department of Health and Human Services. Pharmaceutical quality for the 21st century: a risk-based approach progress report. May 2007. Appendix 19: quality by design graphic. Available from http://www.fda.gov/AboutFDA/CentersOffices/OfficeofMericaProductsandTobacco/CDER/ucm128080.htm#APPENDIX19
140. US National Archives and Records Administration. Code of federal regulations. Production record review. 2013;Title 21, Part 211.192.
141. US National Archives and Records Administration. Code of federal regulations. Document controls. 2013;Title 21, Part 820.40.
142. US National Archives and Records Administration. Code of federal regulations. Recordkeeping and record retention. 2013;Title 21, Part 312.57.
143. US National Archives and Records Administration. Code of federal regulations. Adverse reaction reports. 2013;Title 21, Part 1271.350(a).
144. US National Archives and Records Administration. Code of federal regulations. Corrective and preventative action. 2013;Title 21, Part 820.100.
145. US National Archives and Records Administration. Code of federal regulations. Tracking. 2013;Title 21, 1271.290.
146. United States Pharmacopeia and National Formulary (USP36-NF31). Cell and gene therapy products. Rockville, MD: United States Pharmacopeia Convention; 2012. Chapter 1046.
147. United States Pharmacopeia and National Formulary (USP31-NF26). Validation of alternative microbiological methods. Rockville, MD: United States Pharmacopeia Convention; 2008. Chapter 1223.

Chapter 8
Regenerative Rehabilitation: Synergizing Regenerative Medicine Therapies with Rehabilitation for Improved Muscle Regeneration in Muscle Pathologies

Kristen Stearns-Reider and Fabrisia Ambrosio

8.1 Introduction

With recent advances in the understanding of the molecular basis for tissue regeneration, regenerative medicine therapies for a host of musculoskeletal disorders are becoming available at an ever increasing pace. One promising area for the application of such therapies is toward the regeneration of skeletal muscle tissue. A host of disorders and pathologies contribute to the loss of skeletal muscle, including muscular dystrophies, acute trauma, tumor resection, and age-related sarcopenia. While some of these disorders have a relatively mild impact on the loss of muscle strength and function, others are so severe that they lead to the need for limb amputation or, in the worst cases, death. Therefore, regenerative medicine strategies are critical for the treatment of many musculoskeletal disorders.

While advances in muscle tissue regeneration are occurring at an unprecedented pace, the use of rehabilitation to support such procedures has traditionally received less attention. The cellular- and tissue-level response to mechanical loading has been well described in the musculoskeletal system; however, little is known regarding how this process could be leveraged to facilitate muscle regeneration. In the musculoskeletal system, there is a wealth of knowledge regarding the cellular- and tissue-level response to mechanical loading. However, there is little information as

K. Stearns-Reider
Department of Integrative Biology and Physiology, University of California, Los Angeles, Los Angeles, CA, USA

F. Ambrosio (✉)
Department of Physical Medicine and Rehabilitation, University of Pittsburgh, Pittsburgh, PA, USA

McGowan Institute for Regenerative Medicine, University of Pittsburgh, Pittsburgh, PA, USA
e-mail: ambrosiof@upmc.edu

© Springer Science+Business Media New York 2016
M.K. Childers (ed.), *Regenerative Medicine for Degenerative Muscle Diseases*,
Stem Cell Biology and Regenerative Medicine,
DOI 10.1007/978-1-4939-3228-3_8

to how this process could be leveraged in a targeted and specific manner in order to facilitate tissue remodeling following regenerative medicine applications for muscle regeneration. In the following sections, we will (1) review the most recent advances in regenerative medicine therapies for skeletal muscle regeneration, (2) provide an overview of the principles of mechanotransduction as they apply to the musculoskeletal system, and finally, (3) present early evidence supporting the use of physical rehabilitation as a tool to facilitate muscle regeneration following the application of regenerative medicine technologies. While there are a number of different musculoskeletal pathologies that may benefit from the use of regenerative medicine strategies, this review will focus on two primary applications that have received the majority of research focus: (1) volumetric muscle loss and (2) muscular dystrophy. Although this field is in its infancy, the available evidence supporting the importance of physical rehabilitation in facilitating muscle regeneration provides a foundation that may guide future investigations aimed at treating many severe musculoskeletal pathologies and disorders.

8.2 Regenerative Medicine Therapies for Muscle Pathology

Young, healthy skeletal muscle has a tremendous capability for regeneration following a relatively minor injury. However, this capacity is severely diminished with disease, volumetric muscle loss, and age, all of which can dramatically affect strength and functional capacity. To address these issues, many regenerative medicine therapies are being developed to regenerate muscle tissue and restore strength and functional capacity. These therapies can be generally divided into three areas of research focus: (1) stem cell transplantation, (2) biologic and engineered scaffolds, and (3) a combination approach using both stem cells and scaffolds. The basic principles, current findings, and limitations of each type of therapy are discussed below.

8.2.1 Stem Cell Transplantation

The idea of a stem cell first took form at the turn of the twentieth century when Ernst Haeckel described the presence of *stammzelles* [1]. These were described as primordial cells with the capacity to evolve into all types of cells and multicellular organisms. Since that time, numerous studies have investigated the use of stem cells to treat a plethora of disorders and diseases. Stem cells are unspecialized cells, capable of self-renewal, that have the capacity to differentiate into specialized tissue types. These cell populations can be isolated from many different tissues; however, they are most commonly isolated from embryonic or adult tissue. Stem cells derived from embryonic tissue have the potential to form all of the specialized cell types of the body, a feature known as *pluripotency*. However, their use is highly controversial and there are many concerns that their unlimited potential could lead to the formation of unwanted tissue types (i.e., tumors). Adult stem cells are more desirable as they can be obtained from many tissues in the body, including bone marrow, fat, and skin. These cells were originally thought to have more limited potential for tissue regeneration due to their more differentiated state. However, recent studies have demonstrated that differentiated cell populations, such as fibroblasts (from skin) and adipocytes (from fat), can be reprogrammed to an embryonic-like stem cell by transient expression of four early developmental transcription factors [2]. This finding opens up numerous therapeutic applications, allowing researchers and doctors to harness the regenerative potential of more readily available cells types and to generate patient specific stem cells for autologous transplantation.

While stem cell therapies appear promising given the potential for muscle regeneration, their clinical application has, to date, been met with limited success [3]. Stem cell-mediated muscle regeneration for Duchenne muscular dystrophy (DMD) has been the most thoroughly investigated and is one of the few pathologies for which stem cell therapy has been translated to clinical trials [3]. DMD is a progressive muscle wasting disorder caused by a loss of the protein dystrophin, resulting in the loss of functional muscle by early teenage years. Early clinical trials of myoblast injections for DMD demonstrated the safety of intramuscular injections and the ability of transplanted cells to contribute to new myoblast formation and muscle regeneration [4–11]. Unfortunately, many studies found that the newly formed fibers did not provide any meaningful functional benefit, as no improvements in muscle strength were observed [5, 8, 9, 12]. In addition, poor donor cell engraftment was often noted [13–15]. Ultimately, the limited clinical success of these early interventions was attributed to rapid cell death, poor migration, and immune rejection of the implanted cells [13–15].

The limited success of stem cell therapies has highlighted a number of barriers to translation. The method of delivery, commonly via direct injection into the target tissue of interest, often leads to formation of a bolus of cells at the injection site [16]. Cells in the center of the injection site are therefore not able to get the nutrients or signals they need to thrive and differentiate, often leading to massive cell death. In addition, cells must be able to migrate away from the injection site in order to effectively integrate into the area of interest, which is not possible with a bolus of cells. On the other hand, there is also concern as to methods to maintain the cells in the location of interest. Without some form of "anchoring" within the target tissue, there is potential that the injected cells may migrate away from the area of interest, negating any regenerative benefit of the implanted cells.

If the cells do remain viable following transplantation, another concern is the ability of stem cells to differentiate into the target tissue. As growth factors are partially responsible for guiding cells to differentiate into one of many tissue types, there is some concern regarding the growth factors to which donor stem cells are exposed in an injured/diseased environment. If the cells are implanted into an inhospitable microenvironment, such as is the case in either "diseased" or acutely injured tissue, cells may be exposed to growth factors that promote further pathogenesis. If cells are exposed to such deleterious factors, there may be a risk for terminal differentiation toward an unwanted phenotype, such as fibrosis [17]. Studies are needed to investigate the optimal time at which stem cells should be introduced into an environment, especially if the introduction of cells occurs following an acute injury.

In an attempt to address many of the known limitations of stem cell therapies, novel techniques to improve donor cell incorporation into muscle tissue are being investigated. Improvements in the isolation and manipulation of stem cells have the potential to improve engraftment potential and encourage functional muscle regeneration. For example, applications utilizing stem cell populations typically involve isolation and expansion of the cell population on tissue culture plates prior to implantation. However, it was later shown that even short-term culture of muscle stem or satellite cells results in a myoblast population with a greatly diminished regenerative potential [18–20]. To avoid the deleterious effects of cell culture, advances in cell sorting using flow cytometry have enabled the improved identification and isolation of fresh muscle satellite cells [19]. Recent studies have demonstrated the robust engraftment efficiency of such freshly isolated satellite cells, as demonstrated by a significant increase in muscle force production, as compared to cultured populations [21]. These studies represent an important step toward clinical translation of these therapies through improved methodology for cell isolation.

8.2.2   Biologic and Engineered Scaffolds

The use of scaffold materials for the replacement of injured or diseased tissue has gained considerable interest in recent years. This process involves the use of either naturally occurring or engineered/synthetic materials, sometimes combined with bioactive molecules, to reconstruct or restore living tissues. These scaffolds not only provide structural support for infiltrating progenitor cells, but, additionally, they facilitate tissue formation by enabling cell attachment, migration, proliferation, and differentiation [22]. To be effective, the scaffold must be able to bridge any tissue defect, interact with the surrounding tissue, and encourage new, functional tissue formation. Arguably, one of the most important properties of any scaffold is that it must not elicit an immune response. Both biologic and engineered scaffolds offer many of these advantages; however, there are specific benefits to each.

8.2.2.1   Biologic Scaffolds

Biologic scaffolds are derived through the decellularization of various source tissues and organs using detergents and/or enzymes, leaving behind only the extracellular matrix (ECM). The ECM is the secreted product of the resident cells of every tissue and organ in the body and is composed of various structural and cell adhesion proteins and glycans. Structural proteins within the ECM, such as collagen and elastin, provide structure and resilience to the tissue, while cell adhesion proteins such as fibronectin and laminin provide integrin-binding sites. These integrin-binding sites activate intercellular signaling pathways that are important for regulating expression of ECM proteins. Glycans are an especially important component of the ECM as they provide a reservoir for signaling molecules and growth factors, which help to direct cell differentiation upon surgical implantation.

In the past 20 years, the FDA has approved many biologic scaffolds for use in soft tissue repair, including the reinforcement of tendon repairs and soft tissue grafts [23]. Commercially available biologic scaffolds have been derived from a variety of tissues, including the small intestine, dermis, urinary bladder, pericardium, and heart values. More recently, these scaffold materials have been repurposed for the treatment of severe muscle injuries. Studies have suggested that surgical implantation of biologic scaffolds encourages the site-appropriate, functional remodeling of muscle tissue in individuals with volumetric muscle loss [24–26]. Upon implantation into a muscle, the ECM is infiltrated by mononuclear cells and is gradually degraded (Fig. 8.1). The resulting degradation products, including bioactive peptides, growth factors, and cytokines, are released from the ECM to influence and direct multipotent stem/progenitor cell recruitment, proliferation, and differentiation,

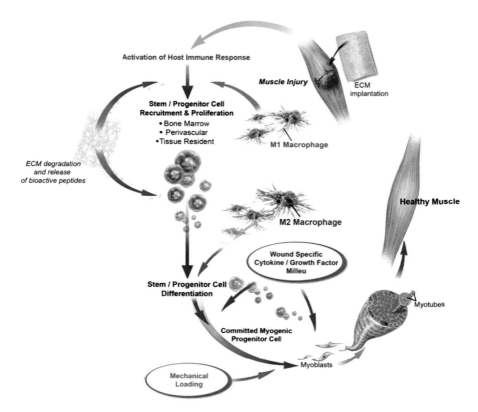

Fig. 8.1 Schematic displaying the cascade of events leading to the formation of healthy skeletal muscle following ECM implantation (adapted from Wolf et al. [22])

all of which contribute to the formation of site-appropriate tissue. In addition, degradation products have been suggested to modulate the innate immune response and encourage tissue remodeling. Preclinical studies have demonstrated that ECM implantation into areas of large volumetric muscle loss promotes an anti-inflammatory, remodeling macrophage phenotype (M2), rather than the default pro-inflammatory macrophage phenotype (M1) [27–29]. Previous studies have demonstrated that the macrophage phenotype (M1 vs. M2) is a major determining factor in the host tissue response, with increased scar tissue formation and poorer functional outcomes observed in individuals presenting with an M1 macrophage phenotype [28–30].

Scaffold degradation is a critical component of constructive tissue remodeling. If scaffold degradation is prohibited, muscle tissue formation will not occur. Badylak et al. demonstrated that chemical cross-linking of scaffold materials inhibits degradation and the release of biologic factors, leading to impaired tissue remodeling [27, 31]. Previously, biologic scaffolds were chemically cross-linked to strengthen the scaffold, allowing it to withstand the large tensile forces generated in vivo. However, as degradation of the scaffold appears to be a critical step in the process of muscle remodeling following ECM implantation, this practice would appear detrimental to muscle regeneration therapies.

8.2.2.2 Biologic Scaffold Implantation for Skeletal Muscle Repair

The successful formation of functional skeletal muscle following ECM implantation has been demonstrated in preclinical models of volumetric muscle loss in the abdominal wall, quadriceps, and gastrocnemius/Achilles tendon complex [26, 31–33]. These studies have utilized many scaffold source materials (e.g., small intestinal submucosa and urinary bladder matrix), different animal model species (e.g., mouse, rat, rabbit, canine), and various defect sizes (approximately 15–75 % of the affected muscle).

This regenerative medicine approach has been investigated in small case studies, with encouraging results reported. Mase et al. evaluated the implantation of ECM into the quadriceps muscle of a former military service member who had sustained a traumatic skeletal muscle injury to the quadriceps muscle 3 years previously [25]. In this proof-of-principle case report, dramatic improvements in knee extensor torque, power, and work were observed 16 weeks following ECM implantation. In addition, the participant reported that his cycling and walking endurance had improved and he was able to walk up and down stairs more easily and with greater stability.

More recently, Sicari et al. evaluated muscle regeneration and patient function following ECM implantation in five individuals with volumetric muscle loss [26]. All patients had a reported 58–90 % loss of muscle volume, as compared to their unaffected extremity. Prior to surgery, subjects underwent a personalized preoperative physical therapy program targeting specific strength and functional deficits until they reached a plateau in performance, defined as a period of 2 weeks with no appreciable improvements (<2 %) in strength or function. The purpose of this preoperative physical therapy program was to maximize strength and function so that any improvements observed following surgery could be attributed to the surgical intervention and not to rehabilitation alone. Following plateau, subjects underwent surgery, which included excision of local scar tissue and ECM implantation into the defect area. After surgery, patients underwent 6 months of physical therapy and then returned for muscle biopsies, imaging, and assessment of muscle strength and function. Histological evaluation revealed perivascular stem cell mobilization, angiogenesis, and de novo formation of skeletal muscle. Imaging results indicated the formation of dense tissue, consistent with the appearance of skeletal muscle, in the region of ECM implantation. In addition, increased force production and/or improvements in activities of daily living were observed in four of the five patients 6 months after ECM implantation.

While the current clinical studies utilizing biologic scaffolds provide promising findings, there are a number of limitations that must be considered when interpreting the results. First, these studies include a heterogeneous sample of patients and do not include control subjects for comparison. In addition, investigators were not blinded as to the surgical limb/location, which could influence the results. It is also important to note that although participants in clinical trials demonstrated improvements in strength and function, there was not a total recovery as compared to control limbs. Subjects in these trials were only followed up to 6 months, and it's possible that muscle regeneration may continue well beyond that. Future studies should include later time points to determine if additional gains in strength and function occur. Finally, there is a possibility that the scar tissue debridement performed during the surgery may play a role in the improvements observed following surgery. However, this is unlikely due to the fact that the majority of these patients have previously undergone such surgeries without any appreciable improvement.

Although biologic scaffolds are advantageous due to their native complex structure and the availability of bioactive molecules within the matrix [34, 35], there are additional factors that need to be considered with their use. Given that the ECM is the secreted product of resident cells, each scaffold will have variations in architecture and biochemical composition. This may be problematic for studying muscle regeneration in patients as different scaffolds may affect the remodeling of muscle tissue in different patients. In addition, while it has been observed that ECM transplantation promotes site-specific tissue remodeling, the underlying mechanism by which this occurs has yet to be fully elucidated. Therefore, it is difficult to determine which factor has the most influence on the process of muscle remodeling.

8.2.2.3 Engineered Scaffolds

Engineered scaffolds have been used for over 50 years for skeletal muscle repair and reconstruction [22]. Their use is desirable for a number of reasons. First, engineered scaffolds are readily available and can be manufactured as needed, unlike biologic scaffolds that may have more limited availability. In addition, engineered scaffolds can be manufactured in a highly reproducible manner, which may allow for more controlled delivery of the product to the patient. Engineered scaffolds are typically made of polypropylene, poly(lactic-co-glycolic acid) (PLGA), poly(ε-caprolactone) (PCL), and polyurethanes and can be made in many different configurations (e.g., meshes, foams, hydrogels, and electrospun scaffolds) [22]. Polypropylene was one of the earliest materials used for muscle repair and was desirable due to its high mechanical strength, durability, and low cost to manufacture. However, as a nonbiodegradable material, it elicits a cascade of immunological events resulting in fibrotic tissue deposition and thus has limited application for muscle regeneration [36, 37]. The use of PLGA has been investigated more extensively for tissue engineering and is most commonly used in biodegradable sutures [38–40]. As a scaffold for muscle tissue regeneration, PLGA is desirable as it is biodegradable and its degradation products are nontoxic [39, 40]. PLGA scaffolds have been shown to promote cell adherence, proliferation, and formation of new three-dimensional tissues, and porous PLGA scaffolds have been shown to promote vascularization and cell infiltration upon implantation [41–43]. Both PCL and polyurethanes are biodegradable. However, their degradation rates are typically slower than biological scaffolds and, thus, they are often used in combination with other components, such as bioactive molecules or growth factors, or are chemically modified [22, 44–46].

While engineered scaffolds have commonly been used for reinforcement and repair of muscle tissue, their application for muscle regeneration continues to be challenging. Engineered scaffolds lack the bioactive molecules found in biologic tissue that facilitate progenitor cell recruitment upon scaffold remodeling. The addition of specific growth factors, such as hepatocyte growth factor, insulin-like growth factor-1, and fibroblast growth factor, may overcome some of the limitations of engineered scaffolds and recreate some of the "niche" properties vital for site-specific tissue remodeling. In addition, synthetic scaffolds tend to elicit a pro-inflammatory foreign body reaction upon implantation, leading to scar tissue formation both within and around the implanted scaffold [37, 47]. Given this response, considerable research has been focused on the development of hybrid scaffolds, including the addition of bioactive coatings and biologically derived materials. Scaffolds are now being developed that are capable of providing the timed release of specific factors necessary for different stages of tissue repair, providing both the spatial and temporal cues necessary to support the normal regenerative process in skeletal muscle.

8.2.3 Biologic and Engineered Scaffolds Combined with Cells

While the implantation of biologic or engineered scaffolds is appealing for use in individuals with volumetric muscle loss, it is unknown if the same procedure can be effective for muscle remodeling in the presence of muscle pathology, such as is observed in muscular dystrophy or age-related sarcopenia.

In the case of a "diseased" muscle, it is possible there may not be an adequate supply of healthy progenitor cells to infiltrate the implanted scaffold. In addition,

ECM implantation has thus far been explored only for the replacement of an area of volumetric muscle loss, and not for entire muscle groups. For larger scale replacement of muscle tissue, there may not be enough progenitor cells available to populate the implanted scaffold. To address some of these issues, more recent studies are investigating the combined use of stem cells and biologic/engineered scaffolds as a potential technique to facilitate improved muscle regeneration. This approach provides localized delivery of various cell populations and growth factors to areas of diseased or missing skeletal muscle, providing both healthy progenitor cells and the appropriate biophysical and biochemical cues to encourage site-appropriate skeletal muscle formation.

Many different cell populations, including mesenchymal stem cells, skeletal muscle satellite cells, and myoblasts, have been used to prepare cell-seeded constructs. Following selection of the desired cell type, cells are placed on the scaffold and subsequently cultured in a bioreactor. The bioreactor is an apparatus that allows for the maintenance of a sterile environment while approximating in vivo conditions, including temperature, pH, oxygen levels, nutrients, metabolites, and regulatory molecules. In addition, physiologically relevant signals can be applied (i.e., interstitial fluid flow, shear, pressure, compression, and stretch), allowing for recreation of the in vivo physical environment. Scaffolds are maintained in the bioreactor until ready for transplantation, the duration of which may vary depending on the bioreactor conditions and cell type used.

There are many factors that need to be considered in developing the optimal cell/scaffold combination for skeletal muscle regeneration. Cell adhesion is critical for survival; therefore, any engineered construct must provide the appropriate biophysical cues to permit adhesion to the ECM. Previous studies attempting stem cell injections have failed partially due to the inability of the injected cells to attach to the host ECM. The addition of the cell adhesion ligand Arg-Gly-Asp (RGD) (cell binding domain for fibronectin) to both biologic and synthetic matrices allows stem cells to interact with the ECM and improves cell viability [48, 49]. In vitro studies have demonstrated that a minimum RGD ligand density (36 nm spacing) is required for myoblast growth on alginate gels [49, 50], and in vitro models have demonstrated that RGD-coupled alginate gels seeded with cells enhance cell viability following transplantation in mice [48]. In addition, the inclusion of ECM proteins, such as collagen, laminin, and fibronectin, along with recreation of the appropriate architecture and material stiffness, is important to recapitulate the mechanical properties of the cellular environment. Collagen VI, an important component of the ECM, has been shown to improve maintenance and survival of muscle satellite cells in vitro [51].

Along with specific ECM composition, studies have demonstrated that substrate biophysical characteristics are potent regulators of stem cell responses. Engler et al. demonstrated that mesenchymal stem cells seeded on matrices mimicking the stiffness of young, healthy skeletal muscle differentiated into myoblasts, while those seeded onto stiffer matrices differentiated toward a fibrogenic lineage [52]. Along these lines, architectural properties, such as porosity and topography, are also important considerations in the creation of synthetic environments, as these characteristics play a role in the exchange of oxygen and nutrients crucial for cell survival.

Studies have demonstrated that cells can tolerate macropore sizes ranging from 100 to 500 μm (average myofiber size ~100 μm); however, they demonstrate reduced viability when the pore size falls below 10–20 μm [53, 54]. Finally, scaffolds mimicking the collagen fibril alignment of native skeletal muscle ECM have also been found to promote regeneration of skeletal muscle in partial thickness muscle defects [55].

Although the use of cell-seeded scaffolds have not yet reached clinical trials, preclinical investigations have demonstrated promising results. Nseir et al. demonstrated that a synthetic scaffold, combined with a coculture of mouse myoblasts and either human embryonic endothelial cells or umbilical vein endothelial cells, demonstrated formation of endothelial networks both in between and around differentiating skeletal muscle fibers [56]. Shandalov et al. additionally demonstrated the fabrication of an engineered scaffold to act as a substitute for an autologous muscle flap for transplantation into a large soft tissue defect [57]. A biodegradable polymer scaffold was utilized and embedded with endothelial cells, fibroblasts, and/or myoblasts, which was then implanted into a full-thickness abdominal wall defect. After 1 week, the scaffold was shown to be highly vascularized, well integrated into the surrounding musculature, and had sufficient mechanical strength to support the abdominal viscera.

8.3 The Role of Mechanical and Electrical Stimulation in Tissue Healing and Remodeling

Skeletal muscle is a mechanosensitive tissue. That is, it responds to physical cues not only from the external environment, but also from its local microenvironment, including the ECM and surrounding cells. Both electrical and mechanical stimulation provide such physical cues to skeletal muscle and may therefore be valuable modalities to promote improved tissue healing and muscle remodeling following regenerative medicine therapies.

8.3.1 Mechanical Stimulation

Mechanical stimulation of skeletal muscle influences muscle growth, morphology, and cellular differentiation. Tensile stain, compressive loads, and hydrostatic pressure all cause structural alterations in the ECM and increase force transmission both across and between the ECM and neighboring cells. Cells respond to this stimuli through the activation of intercellular signaling pathways that regulate a multitude of cellular functions that are essential for tissue development, homeostasis, and recovery from injury. The importance of mechanical stimulation on tissue healing and regeneration has been elegantly described in murine hind-limb unloading studies, which demonstrate an inhibition of the regenerative potential of skeletal muscle following injury under conditions of unloading [58, 59].

The process by which mechanical stimuli are converted into a cellular response is called "mechanotransduction." Mechanotransduction consists of three distinct phases: (1) signal transduction at the level of the receptors, (2) signal propagation, and (3) cellular response (Fig. 8.2). Briefly, during the signal transduction phase, mechanical stimuli are transmitted to mechanosensors that reside in the ECM and both within and outside the cell. A mechanosensor is a receptor that responds to changes in mechanical force. Mechanosensors include stretch activated ion channels in the plasma membrane, focal adhesion complexes (including integrins) that bridge the cytoskeleton and ECM, and basement membrane proteins in the ECM that unfold/activate in response to increased force. These sensors deform in response to the application of force, and this change triggers a cascade of biochemical signals that will ultimately influence cellular function. During the next phase, signal propagation, biochemical conversion and propagation of the transmitted mechanical signal occurs through cell signaling pathways that can either enhance or diminish the intracellular spread of the converted biochemical signal. These signals will reach a final downstream target that then modulates cell function. In the final phase, cellular response, the cell responds to the received signal. This response can occur immediately or may be delayed. In the case of an immediate response, there is as an increase or decrease in intracellular tension, changes in adhesive properties, cytoskeletal reorganization, or cellular priming for migration. Delayed responses include changes in gene expression and the synthesis of proteins that influence cell proliferation, differentiation, structural properties, and viability.

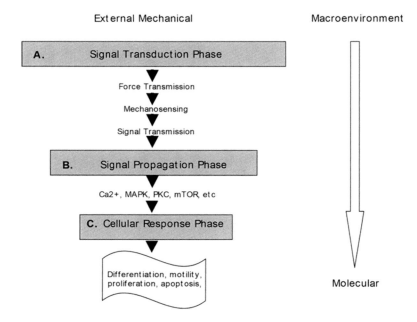

Fig. 8.2 Schematic of the phases of mechanotransduction (adapted from [60])

The importance of mechanical stimulation on tissue healing and regeneration has been elegantly described in murine hind-limb unloading studies, which demonstrate an inhibition of the regenerative potential of skeletal muscle following injury under conditions of unloading [58, 59].

While the in vivo mechanical stimulation occurs through muscle contraction or stretching, in vitro models have been developed to study of the direct influence of mechanical stimulation on cellular function [61]. There are various methods that have been implemented to apply mechanical stimulation to cells, including stretching and compressing cells [61, 62]. These methods can allow for the investigation of the underlying mechanisms by which cells directly respond to mechanical stimuli. One method to apply a mechanical stretch to cells involves culturing cells on a membrane or gel. Stretching of the membrane can then be generated in two ways, either through (1) multiaxial strain or (2) uniaxial strain (Fig. 8.3) [61]. For multiaxial strain, the membrane is stretched around a rigid frame or is deformed by applying a vacuum to the membrane, thus applying strain along multiple axes. For uniaxial strain, a stepper motor is used to increase uniaxial tension on cells seeded within either a 3D collagen gel or on a membrane. For either mode, stretching can be applied in either a cyclic or static mode to mimic different physiologic conditions, such as muscle contractions or prolonged stretching. While these methods can be applied to many different types of cells, using different stretching parameters, studies have specifically looked at the application of these methods for studying the effects on muscle cells. Specifically, studies utilizing such methods on muscle cells have demonstrated an increase in myofiber length and diameter, protein expression, and contractility when compared to static controls [61, 63–67].

8.3.2 Electrical Stimulation

Electrical stimulation is another mechanism used to modulate the tissue microenvironment. Electrical stimulation can be applied in two ways: directly to the area of interest (direct current), or indirectly, through stimulation of the nerve innervating the muscle [neuromuscular electrical stimulation (NMES)]. Direct currents have traditionally been utilized in wound healing to encourage infiltration of cells into an area of tissue damage. Following an injury, endogenous electrical currents are generated in the damaged tissue, which promotes and directs migration of cells into the area for wound healing. Efforts to enhance wound healing have therefore utilized this property to encourage cell migration through the application of external electric fields. Given that currents have a direction, the application of electrical stimulation can therefore be used to promote cell migration and alignment [68]. The use of electrical currents to influence cell alignment is of particular interest in the case of skeletal muscle where orientation of muscle fibers (aligned in parallel) is required for proper tissue functioning. While the use of electrical currents in rehabilitation is common for wound healing, there is much yet to be understood about their application in concert with regenerative medicine therapies for skeletal muscle

A Multiaxial stretch

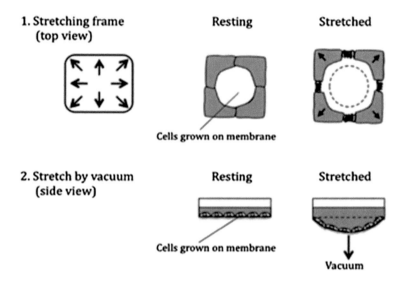

1. Stretching frame (top view)

Resting   Stretched

Cells grown on membrane

2. Stretch by vacuum (side view)

Resting   Stretched

Cells grown on membrane

Vacuum

Uniaxial stretch

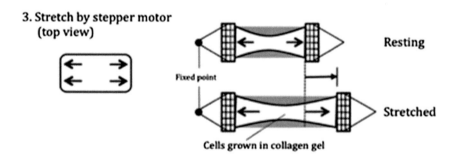

3. Stretch by stepper motor (top view)

Resting

Fixed point

Stretched

Cells grown in collagen gel

Fig. 8.3 In vitro methods to apply mechanical stimulation to cells (adapted from Passey et al. [61])

regeneration. Future studies are needed to investigate the ability of both direct and alternating currents as a method to encourage donor stem cell infiltration into an area of injury.

Electrical stimulation of the motor unit to elicit a muscle contraction may be achieved through neuromuscular electrical stimulation (NMES). NMES is a rehabilitation modality that can be used to mimic the physiologic action of neurons and recreates the mechanical environment experienced by resident muscle cells through the induced contraction of innervated muscle fibers. The transmission of electrical signals via nerve innervation is well known to play a major role in directing the process of terminal differentiation of skeletal muscle cells [69]. The application of

NMES to muscle can be used to stimulate increased cellular proliferation and survival rates, desired differentiation, and improved functionality [69]. Prior studies utilizing electrical currents to study the effects on muscle cell behavior have demonstrated increased satellite cell activation, improved differentiation, and enhanced muscle force output [70–76].

8.4 The Synergy of Rehabilitation with Regenerative Medicine Therapies to Enhance Muscle Remodeling

As described above, mechanical and electrical stimulation are powerful methods to trigger mechanotransductive responses of resident and infiltrating cells. Given the importance of mechanical and electrical stimulation on endogenous cell function and muscle healing, the prescription of targeted exercise or NMES as part of a rehabilitation protocol is a logical adjunct therapy to the application of regenerative medicine therapies. Preclinical models have provided the strongest evidence to support the use of such modalities to promote improved functional muscle regeneration. Although still in the early stages of investigation, results from recent studies are providing exciting evidence to support the use of rehabilitation approaches applied in synergy with regenerative medicine therapies to facilitate skeletal muscle regenerative potential.

8.4.1 Preclinical Models

Preclinical have demonstrated improved force production, both in vitro using cell-seeded scaffolds and in vivo, following the application of mechanical and electrical stimulation in different models of muscle regeneration. Ito et al. applied electrical stimulation (bidirectional, continuous pulses; 24 % of peak force initially, up to 50–60 % of peak force as the tissue developed) to tissue-engineered constructs seeded with C2C12 myoblasts in vitro [73]. Following stimulation, muscle constructs were fixed with two pins, one attached to a force transducer, and one to the bottom of the culture plate well. Constructs were stimulated and force production was measured. Ito et al. determined that the application of pulsed electrical stimulation resulted in increased force production in vitro, as compared to constructs that did not receive stimulation [73]. Increased force production was hypothesized to occur due to improved sarcomere organization and increased expression of myosin heavy chain, the motor protein of muscle thick filaments. Similarly, Machingal et al. demonstrated that mechanical preconditioning of muscle precursor cells seeded on biologic scaffolds prior to implantation in murine models of volumetric muscle loss demonstrated 44 % greater force production as compared to animals receiving unstimulated cell constructs [77]. Distefano et al. evaluated the ability of NMES to improve stem cell engraftment in a murine model of muscular dystrophy (*mdx* mouse) [71]. Muscle-derived stem cells were isolated from wild-type mice and used for the experiments. The *mdx* mice were randomized into one of four treatment groups: saline injection, NMES alone, muscle-derived stem cell injection, or NMES plus muscle-derived stem cell injection. Animals in the NMES groups were stimulated 5×/week over the course of 4 weeks. The mice treated with NMES following stem cell transplantation demonstrated a twofold increase in the number of dystrophin-positive myofibers, an increased vascularity, and an accelerated recovery from a fatigue protocol, when compared to control animals that did not receive electrical stimulation [71]. In addition, several other studies have similarly demonstrated enhanced stem cell transplantation efficiency when stem cell transplantation is followed by mechanical loading, elicited via various methods such compensatory overloading [78, 79], swimming [80], or treadmill running [79, 81].

8.4.2   Clinical Applications

There are very few studies investigating the use of mechanical stimulation, or rehabilitation, for muscle regeneration in clinical trials. However, recent studies provide evidence in further support of the importance of mechanical stimulation in the success of regenerative medicine therapies. As reviewed above, Mase et al. described a case study of a military service member who received biologic ECM implantation following volumetric muscle loss of his quadriceps [25]. Given preclinical findings demonstrating the importance of mechanical stimulation for effective functional remodeling following scaffold implantation [77], investigators initiated a rehabilitation protocol 4 weeks after surgery. Rehabilitation then continued for 12 weeks. As described above, 16 weeks postsurgery, marked gains in isokinetic performance were observed, and CT scans revealed the formation of new tissue at the implantation site. More recently, as additionally described above, five subjects underwent implantation of a porcine bladder-derived ECM into a region of volumetric muscle loss [26]. Once again, investigators included a postoperative rehabilitation protocol with the goal of maximizing functional incorporation of the implanted ECM. Within 24 h of the ECM implantation, subjects began a rehabilitation program that continued for the next 6 months postoperatively. At the end of the program, performance on different tests of strength and function were quantified and compared to preoperative baseline measures. As a part of the same ECM transplantation trial, Gentile et al. provided an in-depth description of the rehabilitation protocol implemented in the case of a military veteran who received surgical ECM implantation for volumetric muscle loss to his quadriceps muscle [24]. Following surgery, the subject underwent a targeted physical therapy program, including site-specific range of motion and strengthening exercises, progressing to more dynamic functional activities over 6 months. At the end of the program, the subject demonstrated gains in all of the strength and functional outcome variable quantified, with the most marked gains observed during more dynamic tasks such as the single leg hop for distance, in which he improved 1820 % compared to presurgical values.

8.4.2.1 Limitations of Current Studies

While clinical studies demonstrate promising findings with respect to improvements in strength and/or function, along with evidence of new muscle and blood vessel formation, following ECM implantation for VML, the direct role of rehab is difficult to discern. None of these studies included control subjects receiving surgery alone without rehabilitation. Without these controls, it is difficult to directly assess the benefit of rehabilitation and the contribution to the improvements observed. In addition, each rehabilitation program included different components so it's additionally difficult to determine which aspect of the rehabilitation program may be most beneficial. Preclinical studies using both electrical and mechanical stimulation provide greater direct evidence that these modalities may play a critical role in the translation of cellular therapeutics for the treatment of muscle injuries or diseases. However, much has yet to be understood regarding the selection of the most appropriate interventions, the optimal loading frequencies and intensities, as well as the best time to initiate such interventions. For example, the application of an electrical stimulation protocol that is too aggressive has been associated with altered cellular metabolism, impaired cell viability, and even cell death [82]. Studies evaluating the utility of mechanical and electrical stimulation to enhance muscle regeneration following regenerative medicine technologies will need to take these parameters into careful consideration.

8.5 Future Directions

The synergy of physical rehabilitation and regenerative medicine therapies to optimize muscle regeneration is an exciting field of study. Although muscle regeneration is the main focus of this chapter, the use of rehabilitation to facilitate tissue remodeling following regenerative medicine therapies has been suggested for a number of different applications, including cardiac regeneration, gene therapies for muscle pathologies, and stroke [83–85]. The early evidence presented above suggests that mechanical and electrical stimulation, as applied during rehabilitation, may be powerful tools to encourage functional muscle formation following regenerative medicine therapies. Physical therapy and rehabilitation play a key role in the recovery of strength and function in individuals suffering from musculoskeletal injuries and pathologies. As such, new protocols are needed to accomplish these same goals in individuals receiving regenerative medicine therapies for muscle regeneration [86–88]. As such, there is a need for rehabilitation professionals to work closely with basic scientists toward the goal of developing clinically relevant protocols with an eye on maximized functional outcomes [87]. There is a need for rehabilitation professionals to understand the most recent advances in regenerative medicine therapies so as to design effective programs to optimize tissue regeneration at every stage. Future studies should include well-designed clinical trials with blinded investigators and placebo control groups. The collaboration of basic scientists and clinicians promises to yield exciting advances in the treatment of a multitude of muscle injuries and pathologies in the coming years.

References

1. Ramalho-Santos M, Willenbring H. On the origin of the term "stem cell". Cell Stem Cell. 2007;1(1):35–8.
2. Takahashi K, Yamanaka S. Induction of pluripotent stem cells from mouse embryonic and adult fibroblast cultures by defined factors. Cell. 2006;126(4):663–76.
3. Cezar CA, Mooney DJ. Biomaterial-based delivery for skeletal muscle repair. Adv Drug Deliv Rev. 2015;84:188–97.
4. Tedesco FS, Cossu G. Stem cell therapies for muscle disorders. Curr Opin Neurol. 2012;25(5):597–603.
5. Karpati G, et al. Myoblast transfer in Duchenne muscular dystrophy. Ann Neurol. 1993;34(1):8–17.
6. Law PK, et al. Feasibility, safety, and efficacy of myoblast transfer therapy on Duchenne muscular dystrophy boys. Cell Transplant. 1992;1(2–3):235–44.
7. Law PK, et al. Myoblast transfer therapy for Duchenne muscular dystrophy. Acta Paediatr Jpn. 1991;33(2):206–15.
8. Mendell JR, et al. Myoblast transfer in the treatment of Duchenne's muscular dystrophy. N Engl J Med. 1995;333(13):832–8.
9. Morandi L, et al. Lack of mRNA and dystrophin expression in DMD patients three months after myoblast transfer. Neuromuscul Disord. 1995;5(4):291–5.
10. Skuk D, et al. Dystrophin expression in muscles of Duchenne muscular dystrophy patients after high-density injections of normal myogenic cells. J Neuropathol Exp Neurol. 2006;65(4):371–86.
11. Skuk D, et al. First test of a "high-density injection" protocol for myogenic cell transplantation throughout large volumes of muscles in a Duchenne muscular dystrophy patient: eighteen months follow-up. Neuromuscul Disord. 2007;17(1):38–46.
12. Tremblay JP, et al. Results of a triple blind clinical study of myoblast transplantations without immunosuppressive treatment in young boys with Duchenne muscular dystrophy. Cell Transplant. 1993;2(2):99–112.
13. Fan Y, Maley M, Beilharz M, Grounds M. Rapid death of injected myoblasts in myoblast transfer therapy. Muscle Nerve. 1996;19(7):853–60.
14. Qu Z, et al. Development of approaches to improve cell survival in myoblast transfer therapy. J Cell Biol. 1998;142(5):1257–67.
15. Skuk D, et al. Resetting the problem of cell death following muscle-derived cell transplantation: detection, dynamics and mechanisms. J Neuropathol Exp Neurol. 2003;62(9):951–67.
16. Vilquin JT, Catelain C, Vauchez K. Cell therapy for muscular dystrophies: advances and challenges. Curr Opin Organ Transplant. 2011;16(6):640–9.
17. Shiras A, et al. Spontaneous transformation of human adult nontumorigenic stem cells to cancer stem cells is driven by genomic instability in a human model of glioblastoma. Stem Cells. 2007;25(6):1478–89.
18. Gilbert PM, et al. Substrate elasticity regulates skeletal muscle stem cell self-renewal in culture. Science. 2010;329(5995):1078–81.

19. Montarras D, et al. Direct isolation of satellite cells for skeletal muscle regeneration. Science. 2005;309(5743):2064–7.
20. Sacco A, et al. Self-renewal and expansion of single transplanted muscle stem cells. Nature. 2008;456(7221):502–6.
21. Cerletti M, et al. Highly efficient, functional engraftment of skeletal muscle stem cells in dystrophic muscles. Cell. 2008;134(1):37–47.
22. Wolf MT, et al. Naturally derived and synthetic scaffolds for skeletal muscle reconstruction. Adv Drug Deliv Rev. 2015;84:208–21.
23. Chen J, Xu J, Wang A, Zheng M. Scaffolds for tendon and ligament repair: review of the efficacy of commercial products. Expert Rev Med Devices. 2009;6(1):61–73.
24. Gentile NE, et al. Targeted rehabilitation after extracellular matrix scaffold transplantation for the treatment of volumetric muscle loss. Am J Phys Med Rehabil. 2014;93(11 Suppl 3):S79–87.
25. Mase Jr VJ, et al. Clinical application of an acellular biologic scaffold for surgical repair of a large, traumatic quadriceps femoris muscle defect. Orthopedics. 2010;33(7):511.
26. Sicari BM, et al. An acellular biologic scaffold promotes skeletal muscle formation in mice and humans with volumetric muscle loss. Sci Transl Med. 2014;6(234):234ra58.
27. Badylak SF, et al. Macrophage phenotype as a determinant of biologic scaffold remodeling. Tissue Eng Part A. 2008;14(11):1835–42.
28. Brown BN, et al. Macrophage phenotype as a predictor of constructive remodeling following the implantation of biologically derived surgical mesh materials. Acta Biomater. 2012;8(3):978–87.
29. Brown BN, et al. Macrophage polarization: an opportunity for improved outcomes in biomaterials and regenerative medicine. Biomaterials. 2012;33(15):3792–802.
30. Brown BN, et al. Macrophage phenotype and remodeling outcomes in response to biologic scaffolds with and without a cellular component. Biomaterials. 2009;30(8):1482–91.
31. Valentin JE, Turner NJ, Gilbert TW, Badylak SF. Functional skeletal muscle formation with a biologic scaffold. Biomaterials. 2010;31(29):7475–84.
32. Sicari BM, et al. A murine model of volumetric muscle loss and a regenerative medicine approach for tissue replacement. Tissue Eng Part A. 2012;18(19–20):1941–8.
33. Turner NJ, Badylak JS, Weber DJ, Badylak SF. Biologic scaffold remodeling in a dog model of complex musculoskeletal injury. J Surg Res. 2012;176(2):490–502.
34. Keane TJ, et al. Preparation and characterization of a biologic scaffold and hydrogel derived from colonic mucosa. J Biomed Mater Res B Appl Biomater. 2015 Oct 27. [epub ahead of print]
35. Keane TJ, Swinehart IT, Badylak SF. Methods of tissue decellularization used for preparation of biologic scaffolds and in vivo relevance. Methods. 2015;84:25–34.
36. Anderson JM, Rodriguez A, Chang DT. Foreign body reaction to biomaterials. Semin Immunol. 2008;20(2):86–100.
37. Klinge U, Klosterhalfen B, Muller M, Schumpelick V. Foreign body reaction to meshes used for the repair of abdominal wall hernias. Eur J Surg. 1999;165(7):665–73.
38. Athanasiou KA, Niederauer GG, Agrawal CM. Sterilization, toxicity, biocompatibility and clinical applications of polylactic acid/polyglycolic acid copolymers. Biomaterials. 1996;17(2):93–102.
39. Grizzi I, Garreau H, Li S, Vert M. Hydrolytic degradation of devices based on poly(DL-lactic acid) size-dependence. Biomaterials. 1995;16(4):305–11.
40. Li S. Hydrolytic degradation characteristics of aliphatic polyesters derived from lactic and glycolic acids. J Biomed Mater Res. 1999;48(3):342–53.
41. Harris LD, Kim BS, Mooney DJ. Open pore biodegradable matrices formed with gas foaming. J Biomed Mater Res. 1998;42(3):396–402.
42. Peters MC, Polverini PJ, Mooney DJ. Engineering vascular networks in porous polymer matrices. J Biomed Mater Res. 2002;60(4):668–78.
43. Smith MK, et al. Locally enhanced angiogenesis promotes transplanted cell survival. Tissue Eng. 2004;10(1-2):63–71.

44. Lee JH, Ju YM, Kim DM. Platelet adhesion onto segmented polyurethane film surfaces modified by addition and crosslinking of PEO-containing block copolymers. Biomaterials. 2000;21(7):683–91.
45. Li D, Chen H, Glenn McClung W, Brash JL. Lysine-PEG-modified polyurethane as a fibrinolytic surface: effect of PEG chain length on protein interactions, platelet interactions and clot lysis. Acta Biomater. 2009;5(6):1864–71.
46. Guelcher SA. Biodegradable polyurethanes: synthesis and applications in regenerative medicine. Tissue Eng Part B Rev. 2008;14(1):3–17.
47. Leber GE, Garb JL, Alexander AI, Reed WP. Long-term complications associated with prosthetic repair of incisional hernias. Arch Surg. 1998;133(4):378–82.
48. Borselli C, et al. The role of multifunctional delivery scaffold in the ability of cultured myoblasts to promote muscle regeneration. Biomaterials. 2011;32(34):8905–14.
49. Rowley JA, Mooney DJ. Alginate type and RGD density control myoblast phenotype. J Biomed Mater Res. 2002;60(2):217–23.
50. Boontheekul T, et al. Quantifying the relation between bond number and myoblast proliferation. Faraday Discuss. 2008;139:53–70. discussion 105–28, 419–20.
51. Urciuolo A, et al. Collagen VI regulates satellite cell self-renewal and muscle regeneration. Nat Commun. 2013;4:1964.
52. Engler AJ, Sen S, Sweeney HL, Discher DE. Matrix elasticity directs stem cell lineage specification. Cell. 2006;126(4):677–89.
53. Hill E, Boontheekul T, Mooney DJ. Designing scaffolds to enhance transplanted myoblast survival and migration. Tissue Eng. 2006;12(5):1295–304.
54. Ikada Y. Challenges in tissue engineering. J R Soc Interface. 2006;3(10):589–601.
55. Page RL, et al. Restoration of skeletal muscle defects with adult human cells delivered on fibrin microthreads. Tissue Eng Part A. 2011;17(21–22):2629–40.
56. Nseir N, et al. Biodegradable scaffold fabricated of electrospun albumin fibers: mechanical and biological characterization. Tissue Eng Part C Methods. 2013;19(4):257–64.
57. Shandalov Y, et al. An engineered muscle flap for reconstruction of large soft tissue defects. Proc Natl Acad Sci U S A. 2014;111(16):6010–5.
58. Kohno S, et al. Unloading stress disturbs muscle regeneration through perturbed recruitment and function of macrophages. J Appl Physiol (1985). 2012;112(10):1773–82.
59. Mozdziak PE, Truong Q, Macius A, Schultz E. Hindlimb suspension reduces muscle regeneration. Eur J Appl Physiol Occup Physiol. 1998;78(2):136–40.
60. Garay E, et al. Mechanotransduction as a Tool to Influence Musculoskeletal Tissue Biology. Hughes C, ed. ISC 23.2, Applications of Regenerative Medicine to Orthopaedic Physical Therapy. La Crosse, WI: Orthopaedic Section APTA; 2014.
61. Passey S, Martin N, Player D, Lewis MP. Stretching skeletal muscle in vitro: does it replicate in vivo physiology? Biotechnol Lett. 2011;33(8):1513–21.
62. Baraniak PR, et al. Stiffening of human mesenchymal stem cell spheroid microenvironments induced by incorporation of gelatin microparticles. J Mech Behav Biomed Mater. 2012;11:63–71.
63. Candiani G, et al. Cyclic mechanical stimulation favors myosin heavy chain accumulation in engineered skeletal muscle constructs. J Appl Biomater Biomech. 2010;8(2):68–75.
64. du Moon G, et al. Cyclic mechanical preconditioning improves engineered muscle contraction. Tissue Eng Part A. 2008;14(4):473–82.
65. Powell CA, Smiley BL, Mills J, Vandenburgh HH. Mechanical stimulation improves tissue-engineered human skeletal muscle. Am J Physiol Cell Physiol. 2002;283(5):C1557–65.
66. Sasai N, et al. Involvement of PI3K/Akt/TOR pathway in stretch-induced hypertrophy of myotubes. Muscle Nerve. 2010;41(1):100–6.
67. Vandenburgh HH, Karlisch P. Longitudinal growth of skeletal myotubes in vitro in a new horizontal mechanical cell stimulator. In Vitro Cell Dev Biol. 1989;25(7):607–16.
68. Zhao M, et al. Electrical signals control wound healing through phosphatidylinositol-3-OH kinase-gamma and PTEN. Nature. 2006;442(7101):457–60.

69. Handschin C, Mortezavi A, Plock J, Eberli D. External physical and biochemical stimulation to enhance skeletal muscle bioengineering. Adv Drug Deliv Rev. 2015;82–83:168–75.
70. Guo BS, et al. Electrical stimulation influences satellite cell proliferation and apoptosis in unloading-induced muscle atrophy in mice. PLoS One. 2012;7(1), e30348.
71. Distefano G, et al. Neuromuscular electrical stimulation as a method to maximize the beneficial effects of muscle stem cells transplanted into dystrophic skeletal muscle. PLoS One. 2013;8(3), e54922.
72. Fujita H, Nedachi T, Kanzaki M. Accelerated de novo sarcomere assembly by electric pulse stimulation in C2C12 myotubes. Exp Cell Res. 2007;313(9):1853–65.
73. Ito A, et al. Induction of functional tissue-engineered skeletal muscle constructs by defined electrical stimulation. Sci Rep. 2014;4:4781.
74. Langelaan ML, et al. Advanced maturation by electrical stimulation: differences in response between C2C12 and primary muscle progenitor cells. J Tissue Eng Regen Med. 2011;5(7):529–39.
75. Pedrotty DM, et al. Engineering skeletal myoblasts: roles of three-dimensional culture and electrical stimulation. Am J Physiol Heart Circ Physiol. 2005;288(4):H1620–6.
76. Serena E, et al. Electrophysiologic stimulation improves myogenic potential of muscle precursor cells grown in a 3D collagen scaffold. Neurol Res. 2008;30(2):207–14.
77. Machingal MA, et al. A tissue-engineered muscle repair construct for functional restoration of an irrecoverable muscle injury in a murine model. Tissue Eng Part A. 2011;17(17–18):2291–303.
78. Ambrosio F, et al. Functional overloading of dystrophic mice enhances muscle-derived stem cell contribution to muscle contractile capacity. Arch Phys Med Rehabil. 2009;90(1):66–73.
79. Palermo AT, et al. Bone marrow contribution to skeletal muscle: a physiological response to stress. Dev Biol. 2005;279(2):336–44.
80. Bouchentouf M, Benabdallah BF, Mills P, Tremblay JP. Exercise improves the success of myoblast transplantation in mdx mice. Neuromuscul Disord. 2006;16(8):518–29.
81. Ambrosio F, et al. The synergistic effect of treadmill running on stem-cell transplantation to heal injured skeletal muscle. Tissue Eng Part A. 2010;16(3):839–49.
82. Thelen MH, Simonides WS, van Hardeveld C. Electrical stimulation of C2C12 myotubes induces contractions and represses thyroid-hormone-dependent transcription of the fast-type sarcoplasmic-reticulum Ca2+-ATPase gene. Biochem J. 1997;321(Pt 3):845–8.
83. Behfar A, Terzic A, Perez-Terzic CM. Regenerative principles enrich cardiac rehabilitation practice. Am J Phys Med Rehabil. 2014;93(11 Suppl 3):S169–75.
84. Boninger ML, Wechsler LR, Stein J. Robotics, stem cells, and brain-computer interfaces in rehabilitation and recovery from stroke: updates and advances. Am J Phys Med Rehabil. 2014;93(11 Suppl 3):S145–54.
85. Braun R, Wang Z, Mack DL, Childers MK. Gene therapy for inherited muscle diseases: where genetics meets rehabilitation medicine. Am J Phys Med Rehabil. 2014;93(11 Suppl 3):S97–107.
86. Ambrosio F, et al. Guest editorial: emergent themes from second annual symposium on regenerative rehabilitation, Pittsburgh, Pennsylvania. J Rehabil Res Dev. 2013;50(3):vii–xiv.
87. Ambrosio F, et al. The emerging relationship between regenerative medicine and physical therapeutics. Phys Ther. 2010;90(12):1807–14.
88. Perez-Terzic C, Childers MK. Regenerative rehabilitation: a new future? Am J Phys Med Rehabil. 2014;93(11 Suppl 3):S73–8.

Chapter 9
Practical Nutrition Guidelines for Individuals with Duchenne Muscular Dystrophy

Zoe E. Davidson, Greg Rodden, Davi A.G. Mázala, Cynthia Moore, Carol Papillon, Angela J. Hasemann, Helen Truby, and Robert W. Grange

Abbreviations

BMI   Body mass index
DRI   Daily recommended intake
DXA   Dual-energy X-ray absorptiometry

Z.E. Davidson (✉) • H. Truby
School of Clinical Sciences, Monash University,
Level 1, 264 Ferntree Gully Road, Melbourne, VIC 3168, Australia
e-mail: zoe.davidson@monash.edu; Helen.truby@monash.edu

G. Rodden • R.W. Grange
Human Nutrition, Foods, and Exercise, Virginia Tech,
ILSB, 1981 Kraft Dr., Blacksburg, VA 24060, USA
e-mail: grodd91@vt.edu; rgrange@vt.edu

D.A.G. Mázala
Department of Kinesiology, School of Public Health, University of Maryland College Park,
215 Valley Drive, College Park, MD 20742, USA
e-mail: dmazala@umd.edu

C. Moore
Nutrition Counselling Center, University of Virginia,
2955 Ivy Road, Suite 1500, Charlottesville, VA 22908, USA
e-mail: clp6g@virginia.edu

C. Papillon
Human Nutrition, Foods and Exercise, Virginia Polytechnic Institute and State University,
338 Wallace Hall, 295 West Campus Drive, Blacksburg, VA 24061, USA
e-mail: cpapillo@vt.edu

A.J. Hasemann
School of Clinical Sciences, University of Virginia Children's Hospital,
800673, Charlottesville, VA 22908-0673, USA
e-mail: ajh5j@hscmail.mcc.virginia.edu

© Springer Science+Business Media New York 2016   225
M.K. Childers (ed.), *Regenerative Medicine for Degenerative Muscle Diseases*,
Stem Cell Biology and Regenerative Medicine,
DOI 10.1007/978-1-4939-3228-3_9

METs	Metabolic equivalents
NOAEL	No observed adverse effect level
OSL	Observed safety limit
REE	Resting energy expenditure
RD	Registered dietitian
ROS	Reactive oxygen species
UL	Upper limit

9.1 Introduction

Fundamental to life is the ingestion of nutrients. Nutrients provide energy and the basic molecules to maintain and repair tissues in the body to maintain their function. Water, an essential nutrient to cell survival, comprises ~65 % of the body. A neuro-muscular disease such as DMD imposes a unique set of nutritional considerations because several of the primary nutrient consumers in the body (cardiac, skeletal, and smooth muscle) are compromised. In addition, DMD individuals have an increased proportion of adipose tissue and are physically inactive, especially if wheelchair-dependent. Furthermore, the ability to digest food is also compromised. Remarkably, even though nutrition is fundamental to life and there are numerous nutrition-related issues associated with the condition, there have been few dietary studies conducted in DMD boys, and at present there are few practical nutrition guidelines specific to DMD that physicians, parents, and caregivers can follow. The primary purpose of this chapter is to provide practical nutritional assessment and dietary interventions for individuals with DMD.

9.2 Brief Overview of DMD and Current Research on Nutritional Issues

DMD is one of the most severe and fatal of the muscular dystrophies. It is caused by the absence of the protein dystrophin [1] in several tissues due to genetic mutations in the dystrophin gene [2]. The dystrophin gene is found on the X chromosome (i.e., an X-linked disease), and so boys are primarily afflicted. Mechanisms responsible for the disease have not yet been clearly defined, though there are many possibilities (see review by Markert et al. [3]). There are several isoforms of dystrophin, but a full-length form is found in skeletal, cardiac, and smooth muscles [4–6]. Striated muscles such as skeletal and cardiac have been well-studied, particularly in the dystrophic mdx mouse model. Less well-studied are the effects of absent dystrophin in smooth muscle [6–8]. Both striated and smooth muscle activations are required for digestion, and digestion is essential to process nutrients. In the absence of dystrophin, digestion is likely to be compromised impacting nutritional status even in the presence of a nutritious diet. Therefore, complications that arise from dysfunctional muscle function from the mouth to the gastrointestinal tract must be considered as part of the nutritional assessment.

Glucocorticosteroids are now accepted as best practice for the medical management for DMD usually given either as prednisolone or deflazacort [9, 10]. Steroids have clear benefits in the maintenance of muscle strength and functional ability. Side effects of treatment such as excessive weight gain, cushingoid features, osteoporosis, impaired growth, hirsutism, glucose intolerance, cataracts, and behavioral changes can be of sufficient concern to cause their withdrawal. Multidisciplinary care is essential and works in conjunction with steroid therapy.

Nutritional considerations for patients with various muscular dystrophies (e.g., muscular dystrophy type 1, facioscapulohumeral, limb-girdle, Becker) vary greatly across the lifespan. Energy excess is characteristic of childhood, while adolescents and young adults may be nutrient-deficient because of mobility limitations and oropharyngeal weakness [11]. Many older patients with these muscular dystrophies demonstrate inadequate nutrient intake of protein, energy, vitamins (e.g., E), and minerals (calcium, selenium, and magnesium). Though a similar detailed nutritional assessment has not been reported for DMD individuals, it is likely that similar deficiencies would be evident. Unfortunately, a review of available literature concluded that there was a dearth of studies that focused specifically on nutritional care for DMD [12]. It is our primary intent with this chapter to provide some initial practical nutritional assessment and dietary guidelines for DMD. In addition, we hope to promote (1) continuing discussion of best practices among physicians, RDs, and other caregivers in the field and (2) initiation of directed controlled research studies that will define the nutritional needs of DMD individuals across their lifespans.

Boys with DMD can move from over- to undernutrition within their shortened life span. In 1993, Willig first reported nutritional issues associated with a French cohort of 252 boys [13]. By 13 years of age, 54 % of the cohort was obese (weight >90th percentile). Conversely, undernutrition (weight <10th percentile) affected 54 % of boys by the age of 18 years. In 1995, in an American cohort, 44 % of boys aged 9–13 years were >90th percentile in weight, whereas after the age of 17 years, 65 % weighed <10th percentile [14]. Data from a French cohort described in 2011 also suggests an increasing and maintained high prevalence of obesity [15]. By 13 years of age, 73 % of these participants were obese, whereas at 18 years of age, 47 % were obese and 34 % were underweight. All of these cited studies were conducted in steroid-naive cohorts. In addition, short stature is also prominent in boys with DMD, although its precise etiology is unclear [13, 14, 16, 17].

More recently, growth outcomes in a group of steroid-treated DMD boys have been documented using BMI Z-scores. A Z-score represents the number of standard deviations an individual score is from the mean. A positive Z-score above the mean indicates a greater BMI (e.g., overweight); a negative Z-score below the mean indicates a smaller BMI (e.g., underweight). In relation to BMI, Z-scores correspond directly with BMI percentiles and are often used in research exploring growth

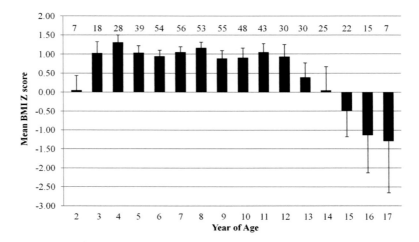

Fig. 9.1 BMI Z-score in DMD boys. Solid bars represent the mean BMI Z-score for each age group, while the *number above each bar* represents the number of observations. The *thin vertical lines* represent the standard error of the mean [19]

(Centers for Disease Control and [18]). In a cohort of DMD boys, the highest prevalence of obesity based on the BMI Z-score was 50 % at the age of 10 years. Longitudinally, BMI Z-scores from the age of 2–12 years plot approximately one standard deviation above the mean, after which there is a marked and progressive decline [19]. These data suggest that DMD boys are initially overweight and then become progressively underweight after age 12 years.

Although weight changes have been observed to have deleterious effects on functional ability in the clinical environment [20], there is a lack of empirical research documenting the relationship between weight and muscle strength and function. At present, little is understood about the relationship between weight status and clinical outcomes. Because of the many complications associated with DMD, dietary recommendations must derive from a detailed clinical nutritional assessment (Fig. 9.1).

9.3 Nutrition Care

Essential to the nutritional care is the team approach including the physician, registered dietitian, individual with DMD, and caregivers. Included in all nutritional assessments are a client/family interview, anthropometric and biochemical data, food-drug interaction, and nutrition-focused physical findings. There are pivotal times in which a nutrition assessment is indicated (Table 9.1). These include:

- At or soon after diagnosis
- At commencement of steroids or other drugs having nutrition implications
- If growth is deviating
- Yearly nutritional assessment

Table 9.1 Timeframes and key items to be included in a nutrition assessment, problem identification, and interventions for key clinic visits

	Key clinic visits		
	Initial diagnosis	Commencing steroids	Yearly review or with weight changes
Timeframe	Within 6 months of diagnosis	At clinic visit when steroids begun	Yearly or when weight charts demonstrate a change in classification (underweight, overweight, obese)
Assessment/nutritional diagnosis	• Client/family interview – Dietary intake – Supplement intake – Gastrointestinal symptoms – Feeding issues (parent role, child's responsibility) – Activity level – Food and nutrition knowledge/skills – Resources and support • Anthropometric (growth chart review) • Biochemical data and tests • Food-drug interaction review • Nutrition-focused physical findings	Same as initial with focus on weight status and bone health	Same as initial with other treatment/changes impacting nutrition status
Intervention	• Introduce the dietitian and role in the neuromuscular clinic • Develop rapport and counsel client/caregiver on: – Basic nutrition goals – Identifying potential nutrition issues (calcium and vitamin D intake for bone health) – Weight progression digestive issues related to inactivity	• Counseling/education of client/caregiver on effect of steroids on: – Nutritional status and potential increase in appetite – Strategies for preventing weight gain through food intake and physical activity – Bone health • Consider calcium and Vitamin D supplementation	• Counseling/education of client/caregiver on effect of steroids on nutritional status: – Strategies for preventing weight gain through food intake and physical activity – Bone health – Supplement use • Consider calcium and Vitamin D supplementation • Weight management strategies for over or under weight (including enteral feedings)

9.4 Growth Assessment

Anthropometric measures should be obtained at every clinic visit to provide serial measures of both weight and height which are essential when assessing growth or weight status. Height and weight can be measured by any trained member of the clinical team, but the RD or physician should assess the measures. For children and adolescents (<18 years), this involves plotting serial measures on growth charts. The best assessment of growth will be made with clinical judgment based on an individual's serial anthropometric measures. Standard growth charts such as "CDC 2000" [21] or national percentile charts are appropriate [9, 10]. Percentiles can be converted to Z-scores to facilitate longitudinal growth monitoring.

9.4.1 Height

If the child is able to stand, height should be measured using a wall-mounted stadiometer with shoes and socks removed. Heels should be hip-width apart and touching the back mounting of the stadiometer, ensuring the Frankfurt plane is horizontal. If the child is unable to stand or an accurate height measure cannot be obtained (child unable to place heels flat on the floor; child has significant lumbar lordosis), height can be estimated from ulnar length using the equation by Gauld et al. [22]:

$$\text{Height}\,(cm) = 4.605(u\ln ar\,length^*) + 1.308(age^\dagger) + 28.003$$
$$(^*centimetres\,to\,the\,nearest\,millimetre,^\dagger\,decimal\,years)$$

9.4.2 Weight

Weight should be measured with either seated scales or wheelchair scales. If the wheelchair weight is unknown or if it has had recent modifications, a new wheelchair weight should be obtained with the use of a hoist.

9.4.3 Body Mass Index

Body Mass Index (BMI) should be calculated using the following equation: BMI $(kg/m^2) = mass(kg)/height(m)^2$. BMI is a common measure to classify an individual as normal weight, overweight, or obese. A convenient calculator for either adult or child and teen can be found at http://www.cdc.gov/healthyweight/assessing/bmi/Index.html. Interpretation of the BMI value is also provided at this website.

However, BMI may underestimate obesity [23]. A measure of body composition should be considered in conjunction with basic anthropometry. Skinfolds are inadequate to assess body composition in DMD because they cannot determine the fat within muscles. Dual-energy x-ray absorptiometry (DXA) is the gold standard measure of body composition and is used to monitor bone density annually for those boys receiving steroids. Alternatively, if DXA is not available, bioelectrical impedance can be used at the bedside to measure body composition changes [24].

9.4.4 Pubertal Status

Pubertal status should be assessed in accordance with Tanner stages. Corticosteroid treatment can have considerable impact on growth through delayed pubertal development. For example, a boy who has not commenced pubertal development will have considerably slower height growth (possibly even static height growth) compared to his developing peers, and this will be reflected in serial growth monitoring.

9.5 Pediatric Nutrition and Feeding Considerations

Upon diagnosis, DMD individuals would benefit from similar food intake guidance provided for well children. The MyPlate recommendations published by the US Department of Agriculture based on the *2010 Dietary Guidelines for Americans* (www.choosemyplate.gov) provide extensive resources. The dietary guidelines provide recommendations for dietary allowances of nutrients essential to healthy living. MyPlate simplifies these recommendations into five food groups: fruits, vegetables, grains, protein, and dairy which are represented by a mealtime visual, the place setting icon. There are a number of resources at this website including the 2010 Dietary Guidelines Brochure, 10 Tips Nutrition Education Series, Sample Menus for a week, and Food Group-Based Recipes. Even after the DMD individual is assessed to have unique nutritional problems, these standards can be adapted as gastrointestinal changes occur (volume, texture, consistency).

There is no evidence to support specific changes in nutrient requirements beyond recommendations for Dietary Reference Intakes (Table 9.2). It is particularly important for the physician or RD to ensure that boys are consuming an Adequate Intake of protein, calcium, vitamin D, fiber, and fluid. Individual fluid requirements vary based on activity, age, and weather conditions, but two liters of fluid intake daily is a good starting point. While water is an optimal choice, dairy-based beverages can assist in meeting calcium requirements. However, dairy and sugar-sweetened beverages can provide excess calories if consumed in large amounts. Consideration should also be given to achieving adequate fiber intake from both soluble and insoluble sources.

Essential to pediatric nutritional care is establishing a positive feeding relationship between the child and parent. Feeding interactions are an extension of good parental attachment as early as birth and should be maintained after diagnosis. It is important that the child and parent partner on food choices resulting in a positive feeding environment. Behavioral considerations should also be considered so that food is not offered as reward or punishment. The parents' roles are to provide healthy foods, and the child's role is to determine what and how much is consumed (Table 9.3).

Table 9.2 Dietary reference intakes for macronutrients for males [25]

Age range (years)	Recommended dietary Allowance (g/day)[a]		Acceptable macronutrient distribution range as % of energy intake[b]			Adequate intake (g/day)[c]
	Carbohydrate	Protein	Carbohydrate	Fat	Protein	Fiber
1–3	130	13	45–65	30–40	5–20	19
4–8	130	19	45–65	25–35	10–30	25
9–13	130	34	45–65	25–35	10–30	31
14–18	130	52	45–65	25–35	10–30	38
19–30	130	56	45–65	20–35	10–35	38
31–50	130	56	45–65	20–35	10–35	38

[a]Recommended Dietary Allowance (RDA) for fat not determined. The RDA is the average daily dietary intake level sufficient to meet the nutrient requirements of nearly all healthy individuals in a group
[b]Is the range of intake for a particular energy source that is associated with reduced risk of chronic disease while providing intakes of essential nutrients
[c]An Adequate Intake is set when the is insufficient evidence to specify a Recommended Dietary Allowance

This table presents Dietary Reference Intakes (DRIs) with the exception of sodium in which case the Adequate Intake (AI) value is presented. The DRI is the average daily dietary intake level sufficient to meet the nutrient requirements of nearly all healthy individuals in a group. If there is insufficient scientific evidence to establish a DRI, an AI is usually developed. DRI reports include DRI for Calcium, Phosphorus, Magnesium, Vitamin D, and Fluoride [26]; DRI for Thiamin, Riboflavin, Niacin, Vitamin B6, Folate, Vitamin B12, Pantothenic Acid, Biotin, and Choline [27]; DRI for Vitamin C, Vitamin E, Selenium, and Carotenoids [28]; DRI for Vitamin A, Vitamin K, Arsenic, Boron, Chromium, Copper, Iodine, Iron, Manganese, Molybdenum, Nickel, Silicon, Vanadium, and Zinc [29]; DRI for Water, Potassium, Sodium, Chloride, and Sulfate [30]; and Dietary Reference Intakes for Calcium and Vitamin D [31]. These reports may be accessed viawww.nap.edu.

9.6 Considerations for Weight Management

Weight management in its simplest form is a balance of energy intake as food compared to energy expenditure. Energy expenditure, measured in kcals, is the sum of resting energy expenditure, the thermic effect of food, energy associated with activities of daily living, and exercise/activity energy cost [32]. When energy is in balance, weight and appropriate growth are maintained. Weight is gained when energy intake exceeds energy output. Growth may be stunted, and weight loss occurs when energy expenditure is greater than energy intake. In DMD, often the excess weight gained is adipose tissue, further complicated by the increased replacement of functional muscle tissue by fatty and connective tissue associated with progression of DMD.

Table 9.3 Dietary reference intakes for micronutrients for males

Age range (years)	Calcium (mg/day)	Iron (mg/day)	Magnesium (mg/day)	Zinc (mg/day)	Sodium (g/day)	Vitamin D (µg/day)	Vitamin A (µg/day)	Vitamin C (mg/day)	Folate (µg/day)	Vitamin B_{12} (µg/day)
1–3	700	7	80	3	1.0	15	300	15	150	0.9
4–8	1000	10	130	5	1.2	15	400	25	200	1.2
9–13	1300	8	240	8	1.5	15	600	45	300	1.8
14–18	1300	11	410	11	1.5	15	900	75	400	2.4
19–30	1000	8	400	11	1.5	15	900	90	400	2.4
31–50	1000	8	420	11	1.5	15	900	90	400	2.4
Key food sources	Dairy (milk, yogurt, cheese)	Meat products (beef, lamb, chicken, pork), eggs, legumes	Nuts and seeds, legumes (haricot, lima, red kidney and soya beans)	Meat products (beef, lamb, chicken, pork), cheddar cheese	Packaged food products	Dairy (milk, yogurt, cheese)	Fruits and vegetables, animal livers, cheese	Fruits and vegetables	Fruits and vegetables (especially dark green leafy vegetables), fruits, nuts, legumes, fortified cereal products	Meat products (beef, lamb, chicken, pork), eggs, dairy

In reality, this balance and weight change is more complex. For example, Davidson and Truby [12] report that delayed growth, short stature, muscle wasting, and fat mass in DMD boys all impact nutritional and energy status. Further, fuel utilization and preferences of fuel used have been shown to be impacted by various other factors including sleep deprivation, body composition, and fed vs. starved state [33].

9.7 Energy Intake and Assessing Nutrient Intake

A detailed diet history can be conducted at each nutritional assessment. Alternatively, food diaries recorded by the family over 3–5 days can provide helpful information on dietary intake [34]. Important to recognize are inherent errors of recalling what was eaten and the impact of record keeping on actual food intake. The information obtained will allow the practitioner to assess the quality of the diet (total calories, nutrients, fiber, and fluid). Also included in the nutritional assessment interview is a review of supplement intake, feeding issues, and knowledge and skills. From this information, suggested changes in dietary habits can be recommended as needed. Optimally, an RD would conduct the dietary assessment and make recommendations in consultation with the physician. Documentation of this assessment will allow the practitioners to compare intake over time. Valuable information can be obtained by asking the client and caregiver to review any changes in dietary intake since the previous clinic visit.

The dietitian or physician should also collect information on supplement use, including type and dosage. This information will inform recommendations and assist the family to monitor the safety and efficacy of food or supplement regimes. Specific supplements that are commonly used are outlined later in this chapter. Various resources exist that allow for the RD and medical team members to stay current with effectiveness and safety evidence of varied nutritional constituents (e.g., animal vs. human studies, source of constituent, appropriateness, and dosage for use in children). One such resource is the Natural Standard Database (www. naturalstandard.com).

Gastrointestinal symptoms including swallowing difficulties, constipation, diarrhea, gastroesophageal reflux, abdominal pain, and bloating should be assessed at each clinic visit.

9.8 Energy Expenditure: Resting Energy Expenditure, Physical Activity, and the Thermic Effect of Food

Resting energy expenditure (REE) in DMD can be reasonably estimated using the Schofield Weight Equation [35]. An activity factor of 1.4 for ambulatory boys would be an appropriate starting point for activity level. The activity factor should be adjusted based on assessment of physical activity levels and the weight of the patient at each clinic visit.

Physical activities, which include activities of daily living as well as structured physical activity, are essential considerations in the estimation of energy needs.

Schofield weight equation for males

0–3 years	$0.249 \times wt - 0.127$
3–10 years	$0.095 \times wt + 2.110$
10–18 years	$0.074 \times wt + 2.754$

REE resting energy expenditure in MJ, *Wt* weight
in kilograms (kg) [36]

Boys with DMD are a heterogeneous population with considerable variation in activity with disease progression. As a result, it is important to assess physical activity level regularly over time. Energy prescription should therefore reflect changes in physical capability. Physical activity intake can be estimated based on an exercise history conducted at the same time as the diet history.

Alternatively, a physical activity diary completed by parents is also an inexpensive instrument to provide a quantitative assessment [34]. Energy output can be estimated by the use of the American College of Sports Medicine (ACSM) tables that report various physical activities in Metabolic Equivalents (METs). Over the course of the day, summation of the time spent at each activity multiplied by the MET cost of that activity provides a reasonable estimate of daily energy cost. A MET is considered a metabolic equivalent of the oxygen consumption of an individual at rest (3.5 ml O_2/kg/min); activities that require greater than resting oxygen consumption can be expressed as METs.

The thermic effect of food is the energy required to digest the food/drink consumed and is estimated as 10 % of the total daily energy expenditure [37].

9.9   Nutritional Implications of Steroid Therapy

One of the most commonly prescribed glucocorticoid for DMD boys, prednisone [9, 10], has the unfortunate side effect of increased appetite. Increased energy consumed coupled with a decrease in energy expended particularly if the individual is wheelchair-bound will lead to weight gain as adipose tissue. If the individual becomes overweight or obese because of additional adipose deposition, then there are potential consequences with comorbidities such as diabetes and heart disease. How these longer-term complex comorbidities will fully impact and interact with muscular dystrophy are not yet understood. Given the difficulties with achieving weight loss for boys with DMD at any age, prevention of obesity is key. This can be facilitated through regular growth monitoring and nutrition assessment to ensure growth deviations are identified and managed early.

The physician or RD should note the presence of steroids in the treatment plan, the type of steroids, and the dosing regimen. Possible side effects are then monitored which include delayed puberty (in particular, lack of appropriate height), weight gain secondary to increased appetite, decreased bone mineral density, and other endocrine disturbances such as increased blood glucose levels or insulin resistance.

Because of the predisposition to low bone mineral density exacerbated by corticosteroid treatment and reduced bone impact associated with inactivity, supplements of vitamin D and calcium are required [12].

9.10 Managing Overweight

When energy input and expenditure have been estimated, imbalances can be noted along with additional clinical nutrition assessment details. From this information, strategies to manage weight can be implemented in partnership with the DMD individual and caretaker. Factors the practitioner must consider include the family patterns with food (prepackaged, frequency of eating out, or home prepared), attention to food as a reward or struggle, and socioeconomic factors influencing the child's nutrition.

The primary goal for management of overweight/obesity in boys with DMD should be determined on an individual basis with clinical discretion. However, management goals in most instances will fit into one of three categories:

- Weight stabilization so that weight tracked on growth chart returns to expected growth
- Weight loss with consideration given to benefits of weight loss versus the burden of additional weight given that lean tissue may be lost

Management strategies should focus on achieving energy balance.

- Energy intake can be reduced by:
  - Establishing client-centered goals with parents and child
  - Encouraging core foods (i.e., fruit, vegetables, protein foods, dairy, and grains) with focus on protein-rich or high nutrient density foods to manage increased appetite resulting from steroids
  - Limiting empty calorie foods and beverages. Encouraging water and other calorie-free beverages instead of sweet tea, soft drinks, and juice
  - Encouraging low energy snacks.
  - Behavioral strategies during meals and snacks can be tailored to the family (e.g., use of snack decision tree that gives children choice of several of the healthiest options and only one or two of the less healthy snack options)
  - More frequent phone or in-person coaching/nutrition visits with dietitian/RD if specific goals are set at a clinic visit
- Physical activity can be encouraged through consultation with the treating physical therapist to ensure that appropriate activities are recommended.

If dietary and physical activity interventions don't achieve weight stabilization, the following options may be considered:

- Consultation with a neurologist regarding alterations to steroid regime. If the patient is treated with prednisolone, a change to deflazacourt may be considered.

Alternatively, a trial of a reduced dose or alternative dosing schedule (such as higher dose on weekends) may be considered.
• Consultation with an endocrinologist with additional goals established.

9.11   Managing Underweight

As boys with DMD age, the likelihood that they will become underweight is increased. Significant muscle wasting will result in weight loss. This is accentuated by eating difficulties experienced at meal times usually leading to a reduced food intake. Specific recommendations to manage undernutrition include:

• Dietary strategies to ensure high protein and high energy intake while limiting food volume. Oral nutrition supplements may be used.
• Monitor fatigue considering feeding with frequent small meals.
• Speech-language pathologist consultation for recommendations regarding texture modification and to ensure safety for oral feeding.
• Introduction of gastrostomy feeding: The nutritional benefits of enteral feeding have now been demonstrated by several authors. Collectively, gastrostomy feeding improved nutritional status across all cohorts and was well tolerated [38–40].

9.12   Digestion Dysfunction

In the absence of dystrophin, striated and smooth muscle contractions are compromised, including those muscles necessary for consumption of foods and liquids. The influence of absent dystrophin can impact all the stages of digestion from mastication to swallowing to gastric emptying and motility.

9.13   Swallowing Dysfunction

Digestion begins in the mouth, in response to specific enzymes in the saliva. Mastication requires contraction of the striated muscles of the jaw such as the masseter. The tongue contributes to food movements in the mouth to position it for chewing and swallowing. Swallowing involves several coordinated movements between the tongue, cheeks, and palate as well as closing of the glottis so that food can pass first through the pharynx and then down into the esophagus to avoid food deposition into the larynx and trachea [41]. Thus, food consumption requires a tight coordination between eating, swallowing, and breathing. Swallowing is dominant to respiration, which results in an exhale-swallow-exhale rhythm during eating [41]. In DMD individuals, both dysphagia and aspiration are likely.

Dysphagia, or abnormal swallowing, results because there is a progressive loss of oral muscle control. The loss of control is often observed between the ambulatory to the late nonambulatory stage [41–44]. Though not specific to muscular dystrophy, dysphagia can lead to dehydration, malnutrition, pneumonia, or airway obstruction.

Aspiration is the passage of material through the vocal cords which can occur before, during, or after swallowing. There are a number of mechanisms that may be responsible for aspiration, so it is critical that clinicians diagnose the mechanism(s) and suggest appropriate corrective actions. For example, swallowing studies and services of a speech pathologist may be required. The normal response to aspiration is a strong cough reflex or throat clearing; however, in some individuals with severe dysphagia, aspiration may not be noted. Severe uncorrected aspiration can lead to aspiration pneumonia or airway obstruction [41]. Once swallowing problems are identified, two simple recommendations include:

(1) Decrease the solid food in a given feeding and implement a dysphagia level I–III eating plan. In addition, recommendations from a speech-language pathologist assessment can be implemented to adjust food consistency.
(2) Drink water after meals to clear the oropharyngeal area [43].

9.14 Esophageal and Gastric Dysfunction

Once the food/drink bolus has been swallowed, additional striated muscles contribute to the movement of the food/drink bolus into the esophagus. A peristaltic wave moves the bolus to the stomach under autonomic nervous control. Gravity aids peristalsis if the individual is in an upright position [41]. Digestion continues in the stomach with additional enzyme reactions. Normally, movement of food or drinks within and then from the stomach into the small intestine is dependent on smooth muscle contractions of the stomach walls. In the DMD individual, gastroparesis is an outcome of poor gastric motility.

Gastroparesis describes poor gastric muscle contractions likely due to altered function of gastric smooth muscle cells. These poor contractions disrupt motility and slow emptying time [45, 46].

9.15 Intestinal Dysfunction

Nutrient absorption occurs in the small intestine, while the large intestine absorbs water and vitamins and reduces acidity, among other functions, including stool formation. Movements of the stomach contents through the intestines occur by peristalsis or rhythmical contractions of the smooth muscles in the intestinal walls [7]. Abnormalities in the gastrointestinal tract can include edema, atrophy, loss of smooth muscle cells, and fibrosis. In addition, pseudo-obstruction can occur.

Pseudo-obstruction is defined as obstructive symptoms unrelated to a mechanical cause, but secondary to disrupted intestinal motility. In one case study, pseudo-obstruction presented as nausea, vomiting, and abdominal distention over many years. In this individual, smooth muscle fibrosis was present in the entire gastrointestinal tract and was markedly evident in the esophagus and stomach [47].

These physiological complications of digestion should be assessed and taken into account when making dietary recommendations. For example, what can the individual with DMD tolerate as regards quantity, consistency, and taste/flavor so that meals and snacks are both nutritious, enjoyable, and beneficial?

Case Studies In summary for this chapter section, two case studies are presented to illustrate application of the nutrition assessment and recommendations described above (Tables 9.4 and 9.5).

9.16   Potential Nutrient Constituents/Supplements

Many of the nutrient constituents/supplements thought to be beneficial in the treatment of DMD are consumed in foods, though in much smaller proportions than are used in animal or human research (Table 9.6). Most of these purified components have been tested in animals, some by injection and some orally. Consumption or use of various nutrient constituents/supplements in humans (adults or children) and animals is reported when they are available. Please note that for the animal studies, the *mdx* mouse is most often used, as it is considered a good but not perfect model of DMD [62, 63].

Summarized below are studies conducted in DMD boys (Table 9.6) and mdx mice (Table 9.7). These tables are followed by a brief review for each of the nutrient constituents/supplements listed in the tables. Despite the studies conducted in mdx mice and because of the limited studies conducted in DMD boys for many of the nutrient constituents/supplements, recommendations for their use are difficult. Among the studies conducted in DMD boys, creatine and vitamin D are two nutrient constituents/supplements that have evidence to support their use. Among the studies conducted in mdx mice, green tea extract has shown positive benefits. Additional nutrient constituent/supplements to consider are arginine-butyrate, taurine, CoQ10, and idebenone. The reader is encouraged to read the brief reviews for each nutrient constituent/supplement carefully to make an informed choice about their use.

9.17   Review of Nutrient Constituents/Supplements Reported in Tables 9.6 and 9.7

9.17.1   Amino Acids

Protein metabolism studies in DMD suggest that afflicted patients are hypercatabolic, with accelerated protein turnover [84, 85]. For example, fractional protein synthesis and degradation rates (estimated from urinary 3-methylhistidine and

Table 9.4 A. Case study 1

Sean is a 5-year and 1-month-old boy who was diagnosed with DMD approximately 12 months ago. This is his second visit to the neuromuscular clinic. Since last clinic visit, his neurologist and physiotherapist have noticed his muscle strength and function have declined. The neurologist has recommended that Sean commence corticosteroids
As the clinic dietitian, you have been asked to see this family today before they start steroids. You meet with Sean and his mother and father for the first time. His parents seem defensive at first as they are unsure why they are seeing a dietitian today. Immediately, you observe that Sean is snacking on potato chips as the family enters the room
Sean's weight and height in the clinic today were 18.6 kg (50th percentile) and 102.9 cm (10th percentile)
Clinical approach
First, anthropometric measures should be assessed. Sean's BMI is 17.6 kg/m². This BMI plots on the 92nd percentile and would classify Sean as overweight. Ideally, Sean's current weight status should be contextualized using serial measures of height and weight
The nutrition assessment with Sean and his family would consider: dietary intake, diet history or food diary; activity level, activity history or diary; supplement use; and gastrointestinal symptoms
Considering Sean's current weight status together with the planned initiation of corticosteroid treatment, priorities for the dietetic consult would include:
(a)   Identifying dietary and activity strategies for weight management and managing increasing appetite related to the commencement of steroid therapy
(b)   Education regarding appropriate calcium and vitamin D requirements for bone health
(c)   Because the dietitian has not met this family before and given the defensive nature of the parents, it would be valuable to spend some time introducing the role of the dietitian in clinic so that the family do not feel threatened by the consultation. By establishing rapport and trust, successful clinical outcomes can be achieved
After commencing steroids, Sean's BMI remained stable. Sean has seen the dietitian on numerous occasions. Although his BMI remains around the 90th percentile, the dietitian is pleased with his progress because his BMI is tracking steadily
Sean is now 10 years and 4 months old. He is having increasing difficulty walking and can no longer climb stairs or get up off the floor himself. His physiotherapist is in the process of ordering an electronic wheelchair for him. In the meantime, Sean uses his motorized scooter for all movements outside of the home. In clinic today, his BMI has increased to the 99th percentile and is now classified as obese
Sean and his parents are visibly surprised and upset by his recent weight gain because his dietary intake has remained stable. The family are keen for strategies to stop further weight increases. Given Sean's recent declines in strength, the family is also very keen to know what supplements they could try at this stage to assist with his muscles. They have not tried any supplements before this
Clinical approach
Sean's weight gain is likely due to a decline in physical activity. He is using his motorized scooter for most daily activities which would limit energy expended in activity. At this juncture, a physical activity diary would be useful to quantify Sean's physical activity level and guide energy prescription
To assist Sean with weight management, the RD needs to help Sean shift back into a state of energy balance (energy in=energy out). Appropriate steps to establishing energy balance would be:

(continued)

Table 9.4 (continued)

(a)	Establishing Sean's energy needs by calculating his energy requirements using the Schofield Weight Equation for males together with a physical activity level determined from an activity history or diary
(b)	Recommending dietary changes to reduce energy intake, such as limiting empty calories and encouraging core foods and encouraging low-energy snacks
(c)	Working with Sean's physiotherapist to recommend appropriate activities that increase energy output

There are many supplements that are being investigated as potential adjunct therapies for treatment of DMD; however, most investigations are yet to reach clinical trials in humans. The most promising supplement is creatine monohydrate as there is evidence from several clinical trials that have shown positive effects on increased muscle mass and strength. This would be an appropriate supplement for Sean to trial

Table 9.5 B. Case study 2

Chris is 16 years old and has DMD. Chris lost the ability to ambulate independently not long after his 12th birthday and now uses an electronic wheelchair full time. He was treated with steroids for only a very short period (6 months); the treatment was stopped because his parents reported that his behavior was unmanageable on steroids

As a dietitian, you have been reviewing Chris annually for the last 4 years, in which time his body mass index (BMI) has steadily tracked the 75th percentile (healthy weight range). Today, Chris' BMI has dropped to the 50th percentile. Chris is surprised as he has not been trying to lose weight. However, he does mention that he probably has been eating less because he has been coughing at meal times, which has been slowing meals down even more

You only have 15 min with Chris in clinic today because both the respiratory physician and the occupational therapist have requested additional time with him. His lung function has declined since his last clinic visit, and he is losing strength in his upper body, making activities of daily living harder

Clinical approach

Based on Chris' comments, it would be essential to explore the potential feeding and swallowing difficulties that he appears to be experiencing. Because his lung function is declining, potential aspiration risk needs to be assessed and identified. Also, it is likely that feeding issues are impacting on dietary intake and consequently BMI, so a food diary or diet history would be highly relevant. The RD should also consider other gastrointestinal symptoms that could be reducing appetite

At this juncture, it would be timely to engage with other allied health professions. A speech therapist can conduct a thorough swallowing assessment and provide recommendations for appropriate food textures to ensure safe oral feeding. An occupational therapist can provide suggestions to assist with the practicalities of feeding. Secondly, the RD should consider food strategies to optimize energy and protein intake. Oral supplements may be useful especially if Chris is fatiguing during meals. At this stage, oral nutrition strategies should be trialed; however, if weight and clinical status continues to decline (e.g., lung function), then gastrostomy feeding may need to be considered

Table 9.6 Summary of studies using various nutritional supplements in DMD boys

Supplement	Study	Dose	Duration	Results
L-*Leucine*	Mendell et al. [48]	0.2 g/kg/day	12 months	Small ↑ in muscle strength, joint contracture, and pulmonary function
DRI = 34 mg/kg/day				
OSL = 550 mg/kg/day				
α-*keto acid salts*	Stewart et al. [49]	0.45 g/kg/day	4 days	↓ Urinary 3-methylhistidine
DRI = ND				
OSL = 0.45 g/kg/day				
L-*Glutamine*	Mok et al. [50]	500 mg/kg/day	10 day	↓ Protein degradation, but no more than isonitrogenous control
DRI = ND	Mok et al. [51]	500 mg/kg/day	4 months	No effects on protein degradation or ambulatory function
OSL = 14 g/day				
Creatine	Kley et al. [52]	Various: Cochrane systematic review with meta-analysis		↑ in muscle strength and ↑ in lean body mass
DRI = ND	Walter et al. [53]	5 g/day	8 weeks	Modest ↑ in muscle strength and Neuromuscular Symptom Scores
UL = 20 g/day acute	Tarnopolsky et al. [54]	100 mg/kg/day	4 months	↓ Bone breakdown
5 g/day chronic				↑ Grip strength and fat free mass
	Louis et al. [55]	3 g/day	3 months	↑ Maximal voluntary strength, resistance to fatigue, and bone mineral density
	Banerjee et al. [56]	5 g/day	8 weeks	Modest ↑ in muscle strength, no change in functional scale
	Escolar et al. [57]	5 g/day	6 months	No change is strength or function
Vitamin D	Bianchi et al. [20]	25-OHD (0.8 ug/kg/day) with calcium intake above 1 g/day	24 months	↓ Bone resorption
DRI = 15 ug/day				↑ Lumbar spine and whole-body BMD
UL = 100 ug/day				No change in gross motor function

L-*Carnitine*	Paulson et al. [58]	50 mg/kg	Single injection	DMD boys exhibit greater ketogenic response to carnitine
DRI=ND	Escobar-Cecillo et al. [59]	50 mg/kg/day; twice daily	12 months	No effect on muscle function
OSL=3 g/day				
Coenzyme-Q10	Folkers & Simonsen [60]	100 mg/day	3 months	↑ cardiac function and ↑ physical function[a]
ADI=12 mg/kg/day	Spurney et al. [160]	>90 mg/day[a]	6 months	↑ muscle strength
NOAEL=1200 mg/kg/day				
Idebenone	Buyse et al. [61]	450 mg/day (3×150 mg per day)	12 months	↑ respiratory strength and trend for improvement of peak systolic radial strain in the left ventricle lateral wall
DRI=ND				
UL=75 mg/kg/day				

Note: Information on UL was reported when available, but if not unavailable, OSL and NOAEL information from human studies were reported instead

DRI Daily Recommended Intake, *OSL* Observed Safety Limit, *UL* Upper Limit, *ND* Not Determined, *ADI* Acceptable Daily Intake, *NOAEL* No Observed Adverse Effect Level

[a]Initial doses at 90 mg/day (increased doses until serum levels reached 2.5 µg/ml); dosing in patients ranged from 90 to 510 mg/day

Table 9.7 Summary of studies using various nutritional supplements in *mdx* mice

Supplement	Study	Experimental dose	Duration	Results
L-Arginine DRI=ND OSL=20 g/day	Voisin et al. [64]	200 mg/kg/day (on 5 days/week)	6 weeks	↓ SM necrosis, EBD uptake, and serum CK ↑ Diaphragm specific force
	Hnia et al. [65]	200 mg/kg/day	2 weeks	↓ % non-muscle area and inflammatory markers in SM ↑ % CNF
	Wehling-Henricks et al. [66]	5 mg/mL in drinking water	17 months	↑ Fibrosis in heart and muscle
Arginine-butyrate salt DRI=ND OSL=2 g/kg/day	Vianello et al. [67]	5–300 mg/kg/day (on 5 days/week)	6 weeks	100 mg/kg/day dose ↓ markers of muscle damage 100 mg/kg/day dose ↑ utrophin levels in the heart, brain, and SM and soleus isometric force
Arginine-butyrate DRI=ND UL=ND	Vianello et al. [68]	50–100 mg/kg/day (on 4 days/week)	7 weeks (treated every other wk)	↓ Necrotic fibers in the heart and serum CK Normalized cardiac lipid composition ↑ Grip strength and fatigue resistance while hanging
L-Glutamine DRI=ND OSL=14 g/day	Mok et al. [69]	500 mg/kg/day	3 days	↑ Antioxidative capacity and GSSG:GSH in SM
Taurine DRI=ND OSL=3 g/day	De Luca et al. [70]	10 % of daily chow	4–8 weeks	↑ Forelimb SM strength Improved markers of regenerative cycling and membrane excitability
	Cozzoli et al. [71]	1 g/kg/day	4–8 weeks	↓ Plasma LDH ↑ Forelimb SM strength
Green Tea Extract DRI=ND OSL=800 mg/day[a]	Buetler et al. [72]	0.01–0.05 % of daily chow	4 weeks	↓ SM necrosis and regenerative cycling
	[73]	0.50 % of daily chow	3 weeks	↓ Serum CK ↑ Voluntary running performance
	Evans et al. [74]	0.25–0.50 % of daily chow	6 weeks	↓ % Regenerating fibers
Omega-3's	Machado et al. [125]	EPA only (300 mg/kg/day)	16 days	↓ %CNF, inflammatory biomarkers, and EBD uptake

	Reference	Dose	Duration	Outcomes
EPA only DRI=ND	Mauricio et al. [75]	EPA and DHA (300 and 150 mg/kg/day, respectively)	16 days	↓ Serum CK, %CNF, EBD uptake, and inflammatory biomarkers
EPA+DHA DRI=250 mg/day EPA only OSL=1.8 g/day EPA+DHA OSL=5 g/day	de Carvalho et al. [76]	EPA only (300 mg/kg/day)	16 days	Promoted anti-inflammatory M1 to M2 macrophage shift ↓ IFN-γ Modestly ↑ IL-10
L-Carnitine DRI=ND UL=3 g/day	Oh et al. [77]	75 mg/kg/day	6 weeks	↓ Serum CK and EBD uptake ↑ Exercise tolerance
Resveratrol DRI=ND OSL=5 g/day	Gordon et al. [78]	100 mg/kg/day	8 weeks	↓ %CNF and oxidative stress in SM No change in inflammatory biomarkers in SM ↑ SM strength
	Kostek et al. [79]	100 mg/kg/day	10 day or 8 weeks	10-day treatment ↑ SIRT1 and IL-6 expression 8-week treatment ↑ SM strength but not fatigue resistance
	Hori et al. [80]	4 g/kg/day	32 weeks	↓ SM ROS and fibrosis ↓ Maintained muscle mass
N-acetylcysteine DRI=ND OSL=400 mg/day	Whitehead et al. [74]	1 % of drinking water (60 mM)	6 weeks	Improved various markers of cardiac pathology
	Whitehead et al. [81]	1 % of drinking water (60 mM)	6 weeks	↓ EBD uptake, %CNF, and NFkB in SM ↑ β-DG and utrophin expression in SM
	Pinto et al. [82]	150 mg/kg/day	2 weeks	↓ EBD uptake, inflammatory area, and TNF-α in diaphragm
Idebenone DRI=ND UL=75 mg/kg/day	Buyse et al. [83]	200 mg/kg/day	9 months	↓ cardiac fibrosis, ↓ inflammation, ↑ voluntary running performance, and prevention of cardiac dysfunction

DRI Daily Recommended Intake, *OSL* Observed Safety Limit, *UL* Upper Limit, *ND* Not Determined, *ADI* Acceptable Daily Intake, *NOAEL* No Observed Adverse Effect Level, *SM* skeletal muscle, *EBD* Evans Blue Dye, *CK* creatine kinase, *CNF* centrally nucleated fibers, *LDH* lactate dehydrogenase, *ROS* reactive oxygen species: IL interleukin, *NFkB* nuclear factor kappa-B, *GSSG:GSH* oxidized glutathione to reduced glutathione ratio

Note: Information on UL was reported when available, but if not unavailable, OSL and NOAEL information from human studies were reported instead

creatinine) were markedly higher in DMD boys on a meat-free diet (i.e., no exogenous muscle consumed), compared to age-matched controls ($n = 20$ DMD boys; ages 4–20) [84]. Additionally, Hankard et al. [85] showed that protein metabolism of DMD boys was hypercatabolic, compared to height- and mass-matched boys, in the postabsorptive state [85]. Negative protein balance with accelerated protein turnover was also observed in young *mdx* mice [86]. To correct negative protein balance and to preserve muscle mass and function in DMD, many research groups have investigated the therapeutic potential of supplemental amino acids (AA).

9.17.2 Branched-Chain Amino Acids

Branched-chain amino acids (BCAA), which include leucine, isoleucine, and valine, are known to potently stimulate protein synthesis [87, 88]. In 1982, Stewart and colleagues administered metabolites of BCAAs as a mixture of ornithine salts of α-keto acids (α-ketoisocaproate, α-ketoisovalerate, and α-keto-β-methylvalerate; 4:1:1 ratio; 0.45 g/kg/day) to DMD boys ($n = 9$) for 4 days after a 7-day run-in period. The α-keto acid mixture reduced excretion of 3-methylhistidine with no adverse effects, indicating that administration of BCAA metabolites reduces systemic protein degradation in DMD boys [49].

Leucine has limited study in DMD. In a pilot study with young mdx mice, 4 weeks of leucine supplementation improved running endurance and extensor digitorum longus (EDL) muscle strength, but EDL muscle mass, cross-sectional area (CSA), and central nucleation were unchanged [89]. However, in a 12-month study of leucine supplementation (0.2 g/kg/day) in DMD boys ($n = 47$ leucine, 44 placebo), BCAA treatment elicited only marginal effects on a variety of clinical and functional outcomes (e.g., muscle strength, joint contracture, pulmonary function, etc.) [48].

Side Effects: Decreased appetite, anorexia, nausea, stomach discomfort [48].

Conclusion: Leucine's role in promoting positive protein balance in DMD is still unclear; further study is warranted.

9.17.3 L-Arginine

L-arginine is a conditionally essential amino acid naturally found in animal proteins such as meat, fish, poultry, and dairy foods. L-arginine has been studied for its therapeutic potential in animal models of DMD because of its potential roles in nitric oxide synthase (NOS) metabolism and utrophin expression in skeletal muscle [90]. Many research groups are interested in regulating utrophin expression in DMD to compensate for dystrophin deficiency because of its structural/functional similarity to dystrophin [91–94].

Chaubourt et al. [90] showed that *mdx* mice injected intraperitoneally (IP) with L-arginine (200 mg/kg/day) for 3 weeks increased sarcolemmal utrophin expression and nitric oxide (NO) synthase (NOS) activity. These results in *mdx* mice suggest that the compensatory expression of sarcolemmal utrophin may have resulted from upregulated NO signaling. However, a weakness of Chaubourt et al.'s study was their lack of data on the functional responses to L-arginine treatment (i.e., measures of muscle strength, ambulatory capacity, etc). Another research group tested the effects of 6 weeks of L-arginine IP injections (200 mg/kg/day, 5 days/week) and found that *mdx* mice treated with L-arginine exhibited 35 % less necrosis in the lower limb, 57 % less serum creatine kinase (CK) activity, reduced Evans Blue dye (EBD) uptake in gastrocnemius and diaphragm, modestly stronger diaphragm specific force, and greater utrophin expression in various tissues (e.g., the heart, diaphragm, soleus, extensor digitorum longus, gastrocnemius, tibialis anterior, and semitendinosus), compared to untreated *mdx* mice. Their results also suggested that increased NOS activity mediates some of the benefits experienced by L-ARGININE-SUPPLEMENTED *mdx* mice [64].

Hnia et al. [65] also evaluated the effects of L-arginine treatment in *mdx* mice. Five-week-old *mdx* mice were treated for 2 weeks with IP injections of 200 mg/kg of L-arginine [65]. L-arginine treatment decreased the percentage of non-muscle area in *mdx* diaphragm muscles; yet, centrally nucleated fibers were increased [65]. L-arginine treatment inhibited interleukin (IL)-1β, IL-6, and tumor necrosis factor-alpha (TNF-α) secretion by macrophages as well as other infiltrating inflammatory cells. L-arginine treatment also decreased nuclear factor kappa-B (NFκB) as well as MMP-2 and MMP-9, suggesting that L-arginine can have anti-inflammatory effects in dystrophic muscle. Moreover, levels of full-length β-dystroglycan and utrophin were higher in L-arginine-treated mdx mice compared to non-treated mice [65]. Although several studies have shown the benefits of L-arginine supplementation in *mdx* mice, such studies only evaluated the effects of a short-term treatment [64, 65, 67, 90].

Not all studies have found positive effects of L-arginine in DMD models. Wehling-Henricks et al. [66] evaluated the effects of a long-term treatment (from 3 weeks of age until 18 months) of L-arginine (added to drinking water; 5 mg/mL) in *mdx* mice, and results demonstrated that, unlike short-term treatment, long-term treatment with L-arginine exacerbated fibrosis in dystrophic heart and muscle. Findings from this study suggest that there might be differences in short- vs. long-term treatment of L-arginine in dystrophic muscles in dystrophic animals. However, there are no human studies using L-arginine for treatment of DMD either short term or long term.

Side Effects: High doses of oral L-arginine (up to 20 g/day) in humans do not seem to elicit major adverse reactions; however, complaints include nausea and diarrhea. Due to L-arginine's vasodilatory properties, hypotension may occur [95].

Conclusion: The effects of L-arginine on dystropathophysiology are unclear, and further study is needed to better evaluate its potential benefits or detriments.

9.17.4 Arginine-Butyrate

Arginine-butyrate, a combination of L-arginine with *n*-butyric acid (either as a salt or covalently linked), has also been evaluated in *mdx* mice as a potential treatment for DMD. Vianello et al. [67] evaluated the effects of continuous (five injections per week for 6 weeks; dose range 5–300 mg/kg/day) or intermittent (four daily injections every 2 weeks for 6 weeks; dose range 100–1000 mg/kg/day) arginine-butyrate salt complex injections in young (2–3 days) and old (8 weeks) *mdx* mice. These researchers found that continuous IP injections (100 mg/kg/day) of arginine-butyrate in *mdx* mice (1) increased utrophin protein levels by twofold in the skeletal muscle, heart, and brain, (2) decreased markers of muscle damage, and (3) improved soleus muscle isometric force. Meanwhile, intermittent IP injections (800 mg/kg/day) increased utrophin protein levels by twofold, decreased markers of muscle damage, and improved grip strength.

In a follow-up study, Vianello et al. [68] evaluated the effects of treating *mdx* mice with different isoforms of covalently linked arginine-butyrate (3-hydroxybutyrate or N-butyryl arginine), rather than using a salt complex of the two constituents. Covalently linking L-arginine and *n*-butyric acid increases the stability and bioavailability of arginine-butyrate [68]. Every 2 weeks, over a span of 7 weeks, Vianello et al. [68] administered the arginine-butyrate compounds (most effective at 50, 80, and 100 mg/kg/day) for 4 consecutive days starting at 6 weeks of age. The *mdx* mice treated with low doses (50–100 mg/kg/day) had ~60 % improvement in grip strength force production and 3.5-fold increase in resistance to fatigue during inverted grid tests [68]. Moreover, treated *mdx* mice exhibited normalized nerve excitability, ~90 % decrease in cardiac necrotic fibers, and increased cardiac utrophin expression. Furthermore, the *mdx* mice on arginine-butyrate appeared to have cardiac lipid composition different from untreated *mdx* mice, which resembled wild-type (C57BL/10) mice; these results coincided with decreased serum CK levels (by >50 %) and ~75 % reductions in EBD uptake. This study also showed that arginine-butyrate reduced spontaneous Ca^{2+} spikes by ~50 % and markedly increased utrophin staining in cultured DMD myotubes [68]. Overall, the results reported by Vianello et al. [68] show that arginine-butyrate can improve membrane integrity, calcium regulation, and functional outcomes in DMD models. The authors posit that arginine-butyrate's pharmacological effects may be mediated by enhanced histone deacetylase activity and subsequently increased utrophin expression. Increased utrophin expression has previously been shown to improve membrane integrity and functional outcomes in mouse models of DMD [68, 96, 97].

Side Effects: In a clinical trial of patients afflicted by Epstein-Barr virus, administration of arginine-butyrate salts with ganciclovir was associated with nausea, vomiting, headaches, and elevated liver enzymes [98]. Other reported adverse effects at high doses (2 g/kg/day) were lethargy, stupor, confusion, acoustic hallucinations, dyspnea, and hypokalemia [98].

Conclusion: To further evaluate the effects of arginine-butyrate, Vianello and colleagues indicated that they will conduct a clinical trial in ambulant DMD patients, which will use a protocol similar to their 2014 study in *mdx* mice.

9.17.5 Glutamine

Glutamine has also been studied for its effects on DMD pathophysiology in both mice and humans. Glutamine has been proposed as a "conditionally essential" amino acid in DMD patients due to its decreased bioavailability in the post-absorptive state [85]. In 1998, Hankard and others showed that acute oral L-glutamine administration (800 µmol/kg/h for 5 h) in DMD patients decreased leucine appearance and oxidation rates (markers of protein degradation) but had no effect on non-oxidative leucine disposal [99]. Their results suggest that glutamine may help spare nitrogen precursors and whole-body protein in DMD boys. However, in a slightly longer-term trial, Mok et al. [50] showed that 10 days of glutamine supplementation (500 mg/kg/day) in DMD boys ($n = 13$ DMD boys on glutamine, 13 DMD boys on control) did not suppress leucine appearance rates any more than an isonitrogenous control (0.8 g/kg/day) mixture. In a double-blinded, randomized crossover trial, 30 DMD patients were either treated with glutamine (500 mg/kg/day) or placebo for 4 months each (1 month washout period between treatments) [51]. In these patients, glutamine supplementation did not alter muscle function, muscle mass, or markers of protein breakdown. However, there was no deterioration in functional measures after completion of the study (~9 months), and glutamine supplementation (500 mg/kg/day) was well tolerated. Interestingly, a subgroup analysis of the corticosteroid-treated boys ($n = 5$) revealed that glutamine may have preserved walking speed; after washout and placebo, walking speed significantly declined in corticosteroid-treated boys [51].

In *mdx* mice, Mok et al. [69] assessed the effects of a 3-day glutamine (500 mg/kg/day) IP administration on markers of antioxidative capacity in skeletal muscle. Their results showed that glutamine lowered the oxidized glutathione to reduced glutathione (GSSG:GSH) ratio in gastrocnemius, indicating that high-dose glutamine supplementation may reduce oxidative stress in dystrophic skeletal muscle. Findings also suggested that the anti-oxidative properties of exogenous glutamine could also have antiproteolytic effects [69].

Side Effects: A suggested UL for glutamine is ~16 g/day in adults [95]. Temporarily elevated liver enzymes have been observed at high doses. In children, no effects of glutamine supplementation were observed at very high doses (0.6 g/kg) [100].

Conclusion: Mixed results of studies in DMD boys and *mdx* mice support the need for further study of L-glutamine's therapeutic applicability in DMD; special attention should be paid to dose, short- vs. long-term effects, and patient sub-populations (e.g., age and corticosteroid status).

9.17.6 Creatine

Creatine is an amino acid popularized by athletes who take purified versions as a supplement (e.g., creatine monohydrate) to enhance performance. In metabolism, creatine plays an important role in the phosphocreatine metabolic system, which helps skeletal muscle maintain adequate ATP status during short bursts of physical activity (~0–10 s). Creatine may also play other metabolic roles, such as maintaining optimal sarcoplasmic/endoplasmic reticulum calcium ATPase (SERCA) function [101].

Most [53–56, 102] but not all [57] studies of creatine supplementation have demonstrated positive effects in DMD boys. Creatine supplementation (5 g/day for 8 weeks) has been shown to beneficially alter creatine-phosphate metabolites in DMD patients ($n = 13$ creatine, 14 placebo) [56]; these changes coincided with small improvements in average performance of 34 different muscles/groups in manual muscle testing (MMT). Despite enhanced MMTs in these patients, creatine supplementation failed to improve Vignos functional grades (i.e., functional mobility was unchanged in these patients). Additionally, supplying creatine (5 g/day) to dystrophic patients [$n = 32$ total (8 DMD)] for 8 weeks has been shown to modestly improve muscle strength (+3 %) and Neuromuscular Symptom Scores (+10 %) [53]. In a case study of one 9-year-old boy with DMD, exercise performance improved while the patient was on either intermittent creatine (4×3 g/day) or continuous creatine (2×3 g/day) supplement, and the patient's performance declined while off creatine [102]. Further, this patient regained the ability to climb stairs while on the supplement.

Tarnopolsky and colleagues [54] performed a double-blind, randomized, crossover trial with 4 months of creatine (100 mg/kg/day) or placebo (6-week washout between conditions) in DMD patients. Results demonstrated that treated patients had increased grip strength, fat free mass, and decreased bone breakdown [54]. In another randomized double-blind cross-over study, Louis et al. [55] treated DMD patients with creatine (3 g/day) or placebo for 3 months (washout period of 2 months), and results demonstrated that treated patients increased their maximal voluntary contractile strength and resistance to fatigue. Moreover, bone mineral density was increased in ambulant patients supplemented with creatine [55]. In a Cochrane Review, Kley et al. [52] discuss creatine supplementation in several muscular diseases, including DMD. The report indicates that short-term creatine supplementation offers modest functional benefits to DMD patients, but they highlight that the long-term effects of creatine need to be assessed with standardized, quantitative measures of muscle strength and function [52].

Finally, Payne et al. [103] carried out an 8-week nutritional therapy study of creatine (2 % food mass), conjugated linoleic acid (1 % food mass), α-lipoic acid (1 % food mass +1 % soybean oil), β-hydroxy-β-methylbutyrate (HMB; a leucine metabolite; 0.5 % food mass + 1 % soybean oil), and a combination of all four, with and without methylprednisolone (14 mg/kg/week given in two injections) in *mdx* mice, beginning at 4 weeks of age. Their results demonstrated that combination therapy with methylprednisolone provided the best functional outcomes overall (i.e., best outcomes for grip strength, fatigue resistance, rotarod performance, serum CK, and centrally nucleated muscle fibers in gluteus) [103]. Combined therapy without prednisone exhibited similar effects but had no effect on serum CK. Creatine supplementation alone improved grip strength fatigue resistance, retroperitoneal fat, and centralized nuclei in gluteus muscle, while HMB treatment alone improved serum CK activity, fatigue resistance, and centralized nuclei.

Side Effects: Generally, creatine supplements are well tolerated. Several known side effects of creatine are weight gain, gastrointestinal distress, renal dysfunction, and potential cytotoxicity [104]. Interstitial nephritis has been observed a previously healthy 20-year-old male patient taking 20 g/day of creatine for 4 weeks [105]. In DMD, creatine has been well tolerated.

Conclusion: Creatine is the most studied of the constituents/supplements with high level evidence. Based on the data available in humans and *mdx* mice, supplementing with creatine may be a helpful adjunct therapy for DMD patients to improve functional strength outcomes.

9.17.7 Taurine

Taurine is a ubiquitous free amino acid (i.e., not involved in protein synthesis), which plays various metabolic and immunomodulatory roles [106, 107]. Taurine may even assist in maintaining sarcolemmal integrity [106, 107]. Taurine is abundantly found in red meats, salmon, and mackerel. McIntosh et al. [109] reported that taurine levels in skeletal muscle of *mdx* mice seemed to track with disease progression. In young *mdx* mouse muscle (~3 weeks), when degenerative cycles are initiated, taurine concentrations were lower than wild-type controls, and then taurine levels climbed and exceeded normal levels during ages 3–6 weeks [109]. However, taurine levels remain elevated at 6–12 months in *mdx* mice, but disease progression still worsens, indicating that taurine plays a secondary role in amelioration of dystrophic pathophysiology [110]. With these interesting findings in mind, several research groups have studied the effects of taurine supplementation in dystrophic mice.

De Luca and others [70] studied the effects of supplementing 3–4-week-old *mdx* mice with taurine (10 % of chow mass) and other compounds (e.g., creatine and IGF-1) for 4 or 8 weeks while the animals were on a treadmill running routine (30 min; 12 m/min; 2× per week). Treadmill running exacerbated the disease phenotype in untreated *mdx* mice, whereas *mdx* mice treated with taurine displayed greater forelimb muscle strength, similar chloride membrane conductance to wild-type control mice in EDL and diaphragm (this conductance is a marker of regenerative cycling in skeletal muscle [111]), and similar EDL strength-duration curves to wild-type control mice (a measure of baseline membrane potential) [70].

In a follow-up study, Cozzoli et al. [71] assessed the solitary and combined effects of taurine (1 g/kg/day; via oral administration) and α-methylprednisolone (PDN; 1 mg/kg/day via IP injection) for 4 or 8 weeks in *mdx* mice (beginning at age 4–5 weeks) while the mice were on the same treadmill running routine as the De Luca et al. [70] study. While taurine, PDN, and taurine+PDN had no effect on plasma CK, taurine treatment alone decreased plasma lactate dehydrogenase. In vivo, their results showed that taurine and PDN each improved forelimb strength alone, and when combined there was a marked synergistic improvement. Taurine and taurine+PDN also restored resting membrane potential and rheobase potential to wild-type control levels in EDL muscle.

Side Effects: Taurine is a cytochrome P450 3A4 agonist, indicating that it could accelerate drug metabolism and thereby reduce effects of other drugs [112].

Conclusion: The above preclinical studies involving taurine supplementation in *mdx* mice indicate that taurine supplementation could serve as a helpful secondary treatment in DMD, possibly by improving calcium homeostasis [113, 114]. Clinical studies are needed to further investigate the effect of taurine for boys with DMD.

9.17.8   *L-Carnitine*

Carnitine, a dipeptide of the amino acids lysine and methionine, is known to play a prominent role in fat metabolism, ketogenesis, and thermogenesis [115]. While carnitine is made endogenously, it can also be found in large quantities in meat, poultry, fish, and dairy. Natural sources of carnitine are absorbed more efficiently than supplemental carnitine [116]. In humans, muscles concentrate the majority of the L-carnitine (~95 %), and alterations in L-carnitine homeostasis affect muscle function [117]. Skeletal muscle concentrations of carnitine are lower in dystrophic patients [118], and Berthillier et al. [119] observed abnormal carnitine metabolism in young DMD boys [119]. In *mdx* mice, Oh et al. [77] demonstrated that mdx mice supplemented with L-carnitine (75 mg/kg/day) for 6 weeks improved exercise tolerance and reduced serum CK levels compared to control [77].

Le Borgne et al. [117] showed that muscle cells isolated from DMD patients have lower levels of L-carnitine and L-carnitine uptake. These lower levels were associated with increased rigidity of the DMD cell membrane (decreased fluidity) compared to control cells. However, when DMD cells were exposed to L-carnitine (500 µmol), membrane fatty acid profile was improved and membrane fluidity was restored [117]. These results suggest that carnitine supplementation could improve muscle membrane stability in DMD. Escobar-Ceclillo and colleagues [59] performed a double-blinded, controlled trial in DMD patients assigned to L-carnitine (50 mg/kg, twice daily) or placebo treatment for 1 year. Overall, L-carnitine was well tolerated but did not improve muscle function in steroid-naïve boys.

Side Effects: At present, none are reported.

Conclusion: Considering the limited evidence base for the effects of carnitine supplementation in DMD, very little can be said about its therapeutic potential.

9.18   Other Nutrient Constituents/Supplements

9.18.1   Green Tea Extract

Oxidative stress has been implicated in the pathogenesis of DMD [120–123]. Green tea extract (GTE) is derived from *Camellia sinensis* leaves, which contain polyphenol catechins [primarily epigallocatechin gallate (EGCG)] that have been shown to counteract oxidative stress and inflammation [124–126]. Several studies have been conducted in *mdx* mice to investigate the effects of GTE on dystrophic muscle. Buetler et al. [72] observed dose-dependent reductions in EDL necrosis and regeneration when *mdx* mice were given GTE-supplemented chow (0.01 and 0.05 % of total food mass, respectively) for 4 weeks. However, the human equivalent doses used in their study would be ~7 to 35 cups of green tea per day; even these doses exceed current safety guidelines. Evans et al. [74] showed that *mdx* mice on even higher doses of GTE for 6 weeks (0.25–0.50 % of total food mass) exhibited ~15 % fewer regenerating fibers in tibialis anterior (TA) and experienced reductions in NFκB staining in regenerating fibers of TA cross sections, relative to untreated *mdx* controls. Furthermore, Call et al. [73] demonstrated that GTE (0.5 % of total food mass) modestly improves voluntary running performance and serum CK activity by ~50 % in *mdx* mice. Finally, Dorchies et al. [127] showed that 5 weeks of GTE diet (0.05–0.25 % of total food mass) improved muscle function (leftward shift in tension-frequency profile) and resistance to fatigue in plantar flexors, compared to untreated *mdx* mice. Together, these studies suggest that dystrophic muscle can benefit from high-dose GTE supplementation. However, the doses used in these studies exceed current safety recommendations.

Side Effects: A review by Sarma et al. [128], on the safety of green tea consumption, generally regards the product as safe (assuming correct formulation by the manufacturer) [128]. The most concerning adverse events described by Sarma et al. [128] are of 34 reports (out of 216) of possible liver damage, resulting from green tea intake [128]. The presence of caffeine in GTE may be the cause of high-dose side effects like liver problems, headache, nervousness, sleep problems, vomiting, diarrhea, irritability, irregular heartbeat, tremor, heartburn, dizziness, ringing in the ears, convulsions, and confusion. Additionally, since caffeine is a diuretic, DMD patients taking GTE should consider the risk of urinary calcium loss and possible risks to bone health. The cardiovascular effects of caffeine may be undesirable for some DMD patients; caffeine-free green tea products may be advised in these cases. A study of high-dose EGCG (up to 800 mg/day for 4 weeks) reported several adverse effects, such as excess gas, upset stomach, nausea, heartburn, abdominal/stomach pain, dizziness, headache, and muscle pain, but the subjects rated these adverse events as mild [129]. This study's authors reported that 8–16 cups of green tea per day is generally safe [129].

Conclusion: At present, we are unaware of any published studies of GTE in human patients with DMD, but based on the preclinical studies in *mdx* mice, further research is warranted to determine the effects of GTEs on oxidative stress and inflammation status in DMD. Drs. Paul Friedemann and Arpad von Moers (Charite University, Berlin, Germany) are currently recruiting ambulatory DMD patients ages 5–10 for a clinical trial of EGCG. Clinical Trial ID: NCT01183767.

9.18.2   Omega-3 Fatty Acids

Omega-3 fatty acids (ω3FA) are polyunsaturated fats (PUFA) that are abundant in various plant and animal sources (e.g., avocado and salmon). The primary ω3FAs are alpha-linoleic acid (ALA), eicosapentaenoic acid (EPA), and docosahexaenoic acid (DHA). ω3FAs are generally known for their anti-inflammatory effects [130–132], which also extend to skeletal muscle [133]. Since heightened inflammatory tone in skeletal muscle has been implicated in the pathogenesis of DMD, several research groups have investigated the effects of ω3FAs on the dystrophic phenotype. Machado et al. [134] showed that *mdx* mice supplemented with EPA (300 mg/kg/day) by oral gavage for 16 days greatly reduced the presence of centralized nuclei and immune cell infiltration in sternomastoid muscle, but not biceps brachii or diaphragm. Additionally, EPA potently decreased EBD uptake and TNF-α protein in sternomastoid, biceps brachii, and diaphragm muscle in *mdx* mice. Similarly, Fogagnolo et al. [135] showed that treatment with combined EPA and DHA (300 and 150 mg/kg/day, respectively) by oral gavage for 16 days improved various markers of skeletal muscle health in *mdx* mice. Specifically, combined EPA and DHA reduced the following measures: serum CK activity (by ~36 %); centrally nucleated fibers in sternomastoid (by ~50 %); EBD uptake by ~50 % (in biceps brachii and sternomastoid); inflammatory area in diaphragm, biceps brachii, and sternomastoid; and TNF-α protein in diaphragm and biceps. Although these results suggest an anti-inflammatory role for combined EPA and DHA, their doses exceeded the recommended upper limit in humans (3 g/day), but the current upper limit may underestimate the maximal safe dose [136].

Carvalho et al. [76] investigated the effects of 16 days of EPA (300 mg/kg/day) by oral gavage in mdx mice (starting at 2 weeks of age) and compared to the effects of deflazacort (1.2 mg/kg/day). These researchers found that EPA was more effective at promoting M1 to M2 macrophage shift and that EPA modestly reduces serum IFN-γ and modestly increases serum IL-10 protein, suggesting that EPA improves the inflammatory environment in *mdx* mice during peak when the rate of degeneration/regeneration cycles are at their highest. Since corticosteroids can have negative side effects (e.g., weight gain and reduced bone mineral density), further study is warranted to assess whether high-dose ω3FAs could serve as an adjunct or alternative therapy to corticosteroids in DMD.

Interestingly, recent evidence suggests that supplementing mdx mice with monounsaturated fatty acids (MUFA), such as oleate, may be a better option to improve the dystrophic phenotype. For example, *mdx* mice on a high-MUFA diet (primarily oleate) exhibited lower levels of serum CK after 8 weeks, compared to mdx mice on a high-PUFA diet (primarily ALA). However, skeletal muscle histopathology of *mdx* mice on the two supplemented diets was not significantly different [137].

Side Effects: A few side effects have been reported in patients supplementing with ω3FAs. Some of these include fishy breath, gas, bloating, belching, upset stomach, diarrhea, and nausea (reviewed by [138]). There is potential for ω3FAs to interrupt normal blood clotting (reviewed by [139]). It is also important to consider potential contaminants/toxins that come from sources of ω3FA (e.g., presence of mercury in fish and fish oils) (reviewed by [140]). Hence, physicians should help patients ensure that they purchase ω3FAs supplements from reputable manufacturers who ensure their supplements are free of heavy metals and other potential contaminants.

Conclusion: Further studies are needed to determine the potential impact of ω3FA and MUFA supplementation in DMD. The American Heart Association encourages eating fatty fish 2 times/week or more often for general heart health. This level of fish consumption is thought to be equivalent of 0.5 g of fish oil daily. Encouraging dietary consumption of several fish meals/week and the naturally occurring ALA-containing plants (e.g., avocado, walnuts, flax) for individuals with DMD is consistent with public health guidelines for healthful nutrition and heart disease prevention [141].

9.18.3   Vitamin D

Vitamin D is essential to, among its other functions, the maintenance of bone health. Excellent sources of vitamin D are fortified dairy products, fortified orange juice, fortified cereal, salmon, sardines, and vitamin D3 (a precursor of active vitamin D) which is endogenously made in the skin with the assistance of sunlight (UVB) exposure. Low serum 25(OH)D has been well documented in boys with DMD [142, 143]. Individuals with DMD have also been shown to have lower bone mineral density (BMD), and these effects are exacerbated by corticosteroid treatment [142, 143]. Reductions in BMD can have deleterious effects on weight-bearing and ambulatory capacity. Bianchi et al. [142] showed that DMD patients on corticosteroids ($n=22$) had lower BMD than steroid-naïve boys ($n=12$). Furthermore, BMD positively correlated with muscle strength scores, indicating that corticosteroid usage may have negatively affected muscle strength in these patients. Söderpalm and colleagues [143] showed similar results when comparing DMD patients ($n=24$; 2.3–19.7 years old) to age-matched healthy controls ($n=24$), but unexpectedly there was no difference in the fracture rate between DMD boys and controls. When studying the matter further, Bianchi et al. [20] gave 25-OH vitamin D supplements (0.8 µg/kg/day) to 33 steroid-dependent DMD boys for 24 months while simultaneously ensuring calcium intake above 1 g/day. After 24 months, the supplementation regime helped reduce bone resorption [measured by serum osteocalcin, urinary N-terminal telopeptide of procollagen type I (NTx), and serum C-terminal telopeptide of procollagen type I (CTx)], helped increase lumbar spine BMD, and improved whole-body BMD. However, the boys on 25-OH vitamin D population did not affect gross motor function. Despite these findings, the appropriate dose of vitamin D supplementation for boys with DMD is unclear. Unpublished findings described in a workshop report detail the variability of responses to vitamin D supplementation in boys with DMD [144, 145]. In boys receiving 1200–1300 IU vitamin D daily, 50 % showed no response to therapy, while the mean increase in 25(OH)D in the remaining children was 13 ng/ml.

Side Effects: Vitamin D is generally considered safe when taken in appropriate amounts, but can cause side effects when taken chronically in high doses (i.e., above the current UL of 4000 IU per day). Side effects of vitamin D include weakness, fatigue, dizziness, nausea, vomiting, headache, anorexia, and others. Excess vitamin D is also associated with symptoms related to hypercalcemia, such as hypertension, anorexia, nausea, and possible kidney damage (http://www.nlm.nih.gov/medlineplus/druginfo/natural/929.html).

Conclusion: Annual monitoring of serum 25(OH)D as well as supplement compliance should be conducted to adjust supplementation dose as required.

9.18.4 Resveratrol

Resveratrol is a polyphenol that is commonly found in the skin of red grapes, cocoa powder, and peanuts. When consumed in its purified/supplemental form, resveratrol has been associated with a variety of health benefits in skeletal muscle, such as increased oxidative capacity [146], decreased protein degradation [147], and decreased NFκB activation and inflammation [147, 148].

Several animal studies have demonstrated that acute and chronic resveratrol may have therapeutic potential in DMD. Kostek et al. [79] administered resveratrol (100 mg/kg/day) to *mdx* mice both acutely (10 days) and chronically (8 weeks). In the acute phase, treated *mdx* mice increased gene expression of SIRT1 and IL-6, but did not change TNF-α expression in gastrocnemius. In the chronic phase, triceps surae muscles of treated mdx mice were stronger but no more resistant to fatigue than untreated *mdx* mice. Hori et al. [80] tested the effects of 32 weeks of resveratrol supplementation (4 g/kg/day of rodent diet) and found that the treatment helped *mdx* mice maintain muscle mass, decrease skeletal muscle ROS levels, and attenuate fibrosis infiltration but not infiltration of inflammatory cells. The same research group found that resveratrol (4 g/kg/day for 32 weeks) also reduces cardiomyopathy and fibrosis in *mdx* mice. Their results in animals suggest that SIRT1 activity is directly involved in resveratrol's cardioprotective effects [149].

Gordon et al. [150] showed that 10 days of resveratrol (range 10–500 mg/kg/day; optimal dose at 100 mg/kg/day) markedly reduced several markers of young *mdx* skeletal muscle inflammation (e.g., inflammatory cell infiltrate, CD45, and F4/80 expression). In 2014, the same research group treated *mdx* mice for 8 weeks (from age 4 to 12 weeks) with 100 mg/kg/day of resveratrol (gavage every other day) and reported there were no changes in oxidative capacity or total inflammatory cell infiltrate between treated and non-treated *mdx* mice [78]. However, treated mice improved in their rotarod performance, in situ peak tension of triceps surae, and decreased central nucleation and oxidative stress [78].

Side Effects: High doses of resveratrol appear to be well tolerated in humans, even in doses up to 5 g/day [151]. However, in vitro studies [152, 153] show that resveratrol inhibits the activity of cytochrome P450 3A4, which is important for the metabolism of calcium channel antagonists (felodipine, nicardipine, etc), immunosuppressants (cyclosporine, tacrolimus, etc), antihistamines (terfenadine), and sildenafil, among other drugs [154].

Conclusion: Further study is needed to determine the effects of resveratrol in dystrophic humans, but these studies suggest that resveratrol could serve as an adjunct therapy in DMD to reduce muscular inflammation and fibrosis.

9.18.5 N-Acetyl Cysteine

Because muscles in dystrophic individuals may produce more ROS and are thought to be more susceptible to oxidative damage [122, 155], antioxidant supplementation may serve as an adjunct therapy that can be easily incorporated into treatment plans. N-acetyl cysteine (NAC) is a potent antioxidant and has been studied as a potential DMD treatment in *mdx* mice.

Williams and Allen [156] gave *mdx* mice NAC (1 % in drinking water) for 6 weeks (ages 3–9 weeks) to study its effects on heart pathology. NAC returned several markers of cardiac health to wild-type (C57BL/10ScSn) ranges (e.g., fractional shortening, calcium concentrations, calcium transit, CD68 staining, and tissue fibrosis). These results showed that the antioxidative properties of NAC may benefit cardiac health in dystrophic models. The following year, the same research group showed that NAC may offer similar benefits to skeletal muscle [81].

Whitehead et al. [81] studied the effects of NAC in *mdx* mice in vitro and in vivo. Their results showed that acute treatment of EDL muscles from 8 to10-week-old *mdx* mice in vitro (20 mM) improved EDL resistance to stretch-induced force reductions and decreased ROS presence in EDL cross sections. Additionally, in vivo supplementation of NAC for 6 weeks [ages 3–9 weeks; 1 % of drinking water (60 mM)] reduced EBD uptake and centralized nuclei in EDL cross sections, increased β-dystroglycan and utrophin expression in EDL, and decreased NFκB and caveolin protein in TA. Further, Pinto and others [82] gave IP injections of NAC (150 mg/kg/day) to 2-week-old *mdx* mice for 2 weeks. Their results showed that NAC treatment, compared to untreated *mdx* mice, reduced EBD uptake and inflammatory area in diaphragm cross sections and reduced TNF-α and 4-HNE protein in diaphragm homogenates [82].

Terrill et al. [157] studied the effects of short-term (1 week) and long-term (6 weeks) NAC treatment (1 % and 4 % in drinking water, respectively) in *mdx* mice. The researchers reported that mice on the short-term higher dose consumed ~2 g/kg/day of NAC. Both NAC treatments reduced myofiber necrosis and serum CK following downhill treadmill exercise, but only the high dose decreased the oxidation of protein thiols (following exercise) and glutathione [157]. The following year, the same research group [158] experimented with a compound called OTC in *mdx* mice, which acts as a precursor to NAC. They found that OTC's effects on skeletal muscle were very similar to NAC (e.g., reduced necrosis, enhance antioxidant status, improved muscle function), but they concluded that OTC may work by increasing taurine concentrations and reducing thiol oxidation of contractile proteins in skeletal muscle [158].

Side Effects: Side effects of NAC include nausea, vomiting, diarrhea, constipation, fatigue, rashes, fever, headache, drowsiness, hypotension, and possible liver stress [159–161].

Conclusion: Further studies are needed to determine whether NAC's antioxidative and anti-inflammatory effects, when taken as a dietary supplement, can be useful for DMD boys.

9.18.6   CoQ10

CoQ10 is a naturally occurring compound that plays a fundamental role in cell bioenergetics as a cofactor in the mitochondrial electron transport chain [162]. Specifically, CoQ10 is a hydrophilic enzyme that is located in the inner mitochondrial membrane that has roles in complexes I and II of mitochondrial respiration. CoQ10 has also been proposed to improve glycemic control in type 2 diabetic patients [163]. CoQ10 is available as a supplement with different concentrations (15–100 mg) [164]. Various doses of CoQ10 have been used in cardiac patients and doses range from 30 to 600 mg/day (see review from [165]). Higher doses have been reported in trials for patients with Huntington's disease (600 mg/day), Parkinson's disease (1200 mg/day), and amyotrophic lateral sclerosis (3000 mg/day) [166–168]. Although such doses might seem very high, CoQ10 has shown excellent safety and tolerability [164]. Several studies have tested the effects of CoQ10 in DMD. Folkers and Simonsen [60] demonstrated that one patient with DMD receiving CoQ10 for 3 months (100 mg/day) had improved cardiac function as well as overall physical function (note, this finding was observational and not actually evaluated). Moreover, Spurney and colleagues [169] performed an open-label, "add-on" pilot study for DMD patients between 5 and 10 years of age, and results demonstrated that treated patients had an increase in muscle strength of 8.5 %, but this increase did not correlate with an increase in function. An interesting comment from the authors was related to the variability between subjects to achieve the target concentration of CoQ10 [169]. Children required doses ranging between 90 and 510 mg/day to achieve the same serum concentration of CoQ10 [169].

9.18.7 Idebenone

Idebenone is a synthetic analogue of CoQ10 and an organic compound of the quinone family [170]. Overall, the benefits from idebenone are related to its ability to protect against mitochondrial chain dysfunction and to decrease oxidative stress [83, 171]. Studies have evaluated the benefits of idebenone treatment in both *mdx* and DMD patients. Buyse et al. [83] demonstrated that *mdx* mice treated with idebenone (200 mg/kg/day) from 4 weeks until 10 months had decreased cardiac fibrosis and inflammation and improved voluntary running performance compared to vehicle-treated mice. Moreover, idebenone treatment prevented cardiac diastolic dysfunction as well as mortality from cardiac pump failure induced by dobutamine stress in *mdx* mice [83]. Buyse and colleagues [61] also completed a phase IIa double-blinded, randomized, placebo-controlled trial in DMD patients receiving idebenone (450 mg/day given as $150 \text{ mg} \times 3/\text{day}$). DMD patients treated with idebenone demonstrated potential improvement of peak systolic radial strain in the left ventricle lateral wall [61]. Moreover, patients on idebenone showed improvements in respiratory strength, while patients on placebo showed deterioration [61]. The authors also commented that idebenone was safe and well tolerated during treatment [61]. In a subsequent study, Buyse et al. [172] provided additional evidence to support the beneficial effects of idebenone on respiratory function in DMD patients. In this double-blind, randomized, placebo-controlled trial, patients aged 10–18 years who were not on glucocorticoids received 900 mg/day (3×300 mg) idebenone for 52 weeks. After the 1-year idebenone treatment, there was a significant attenuation of the fall in peak expiratory flow as a predicted percentage, suggesting that the typical loss of respiratory function in DMD patients was blunted when treated with idebenone. As with previous studies, treatment was well tolerated with similar adverse events across both groups.

Side Effects: Some subjects consuming high doses might experience mild gastrointestinal symptoms such as nausea and stomach upset [173].

Conclusion: Overall, both CoQ10 and idebenone have shown promising results in dystrophic mouse models as well as DMD patients. However, more studies are needed to fully address the potential of these two compounds in the treatment of DMD (i.e., studies with larger sample sizes, different doses, etc.).

9.19 Drug-Nutrient Interaction

Individuals with DMD will likely be prescribed one or more drugs, so it is important to assess the specific dietary recommendations including supplements for potential drug-nutrient interactions. However, as noted in Table 9.8, to our knowledge, there have been no studies conducted in DMD investigating potential drug-nutrient interactions. The references used to compile Table 9.8 are discussed below.

Table 9.8 Summary of interactions between commonly used drugs in DMD treatment and nutrients/foods

Drug		Interaction with nutrients/foods?	Type of interaction	Suggested approach for interaction	Study evaluating interaction with nutrient/food in DMD?
Glucocorticoids	Deflazacort	Yes	Decreases calcium absorption	Calcium (750 mg/day) and vitamin D (1000 IU/day) supplementation	None
	Prednisone/ prednisolone	Yes	Decreases calcium absorption	Calcium (750 mg/day) and vitamin D (1000 IU/day) supplementation	None
Metformin		Yes	Decreases vitamin B12 absorption	Daily vitamin B12 and calcium supplementation	None
ACE inhibitors	Losartan	Yes	(1) High-sodium intake might decrease losartan effects (2) Low sodium intake might potentiate losartan effects (3) Grapefruit juice decreases the conversion of losartan to its active form	Monitor potassium intake; avoid consumption of grapefruit juice	None
	Captopril	Yes	Decreased absorption if ingested with food/meal	Should be ingested 1 h before meal	None
	Enalapril	No	NA	NA	None
	Lisinopril	No	NA	NA	None
	Cilazapril	Yes	Small delay in action if ingested with food (degree and duration of ACE inhibition are not affected)	It can be ingested with or without food	None
Beta-blockers	Propranolol	Yes	Bioavailability is decreased if ingested with protein-rich meal; decreases activity of CoQ10-NADH-oxidase	Avoid consumption of protein-rich meal with propranolol; no existent recommendation for interaction with CoQ10-NADH-oxidase	None
	Diazoxide	Yes	Decreases activity of Co-Q10-succinoxidase	No existing recommendation for interaction with Co-Q10-succinoxidase	None
Bisphosphonates		Yes	Food/meal decreases bisphosphonate absorption; magnesium inhibits bisphosphonate absorption	Avoid ingestion of food/meal and magnesium with bisphosphonates	None

9.20   Definition and Overview of Drug-Nutrient Interaction

Drug-nutrient interaction is commonly defined as "an alteration of kinetics or dynamics of a drug or a nutritional element, or a compromise in nutritional status as a result of the addition of a drug" [174, 175]. Interactions that are considered clinically significant are known to alter the therapeutic drug response and/or compromise nutrition status of a patient. For example, deleterious interactions might require the drug use be stopped, a decrease in dosage, or supplementation of an essential nutrient due to the drug's action on nutrient availability [176]. Moreover, patients on drugs for long periods might be at a greater risk for drug-nutrient interactions [176]. In general, there are limited studies that evaluate drug-nutrient interactions in patient populations, including DMD.

According to Boullata and Hudson (2012), pharmacokinetic, pharmaceutic, and pharmacodynamic are key terms that can be mechanistically viewed in drug-nutrient interactions. Pharmacokinetic interactions are related to body absorption, distribution, metabolism, and excretion of drugs. Pharmaceutic interactions are physico-chemical reactions that potentially take place in the GI tract lumen or that happen in a type of delivery device [177]. Pharmacodynamic interactions are related to the effects of a drug on the body. Drug-nutrient interaction can be simplified by the model proposed below (Fig. 9.2).

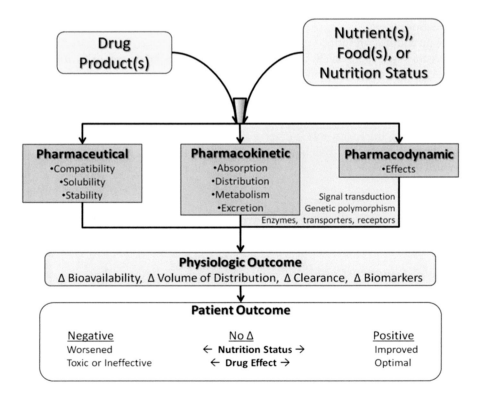

Fig. 9.2   Working model for food-drug interaction (from Boullata et al. [177])

In theory, medications should have specific and similar effects across all patients, never be affected by food or other medications, and not be toxic. However, most medications do not follow such a pattern [178]. Drug-nutrient interactions are as important as drug-drug interactions in patient care but receive much less attention in the clinic. Potential drug interactions can occur with a single nutrient, multiple nutrients, food in general, food components, or nutritional status [177].

Insufficient nutritional status might alter drug metabolism and might predispose individuals for drug-nutrient interactions. The risk for drug-nutrient interactions may be increased in individuals (1) with impaired hepatic, renal, or gastrointestinal function; (2) who are nutritionally compromised due to chronic disease; (3) with significant weight loss; (4) who are receiving multiple and prolonged drug therapy; and (5) who are undergoing major changes in lean body mass, total body fluids, and plasma protein concentration [179]. DMD patients are known to have most, if not all, of the listed complications above, which strongly suggests that they are more susceptible or have a greater risk for drug-nutrient interactions. For instance, DMD patients are known to (1) develop hepatic and gastrointestinal complications, (2) have nutritional deficits, (3) undergo significant weight loss in the late teens and early 20s, and (4) ingest different drugs throughout their lives [e.g., glucocorticoids, beta-blockers, and/or angiotensin-converting enzyme (ACE) inhibitors]. The limited knowledge about the nutritional status of patients with DMD raises serious concerns about potential drug-nutrient interactions that could occur in this patient population.

Typical drugs prescribed to DMD boys include corticosteroids (e.g., deflazacort or prednisone), angiotensin-converting enzyme (ACE) inhibitors (e.g., losartan), beta-blockers (e.g., propranolol), metformin, bisphosphonates, and idebenone. Also, daily supplementation with vitamin D and calcium is recommended. Patients with DMD may also be taking a variety of nutrient constituents/supplements. Currently, there is a lack of studies evaluating drug-nutrient interactions in DMD patients; thus, the following sections will discuss current knowledge. Other potential interactions that have not been yet evaluated are also listed with suggested future studies to determine the presence or absence of interactions.

9.21 Interactions between Commonly Used Drugs in DMD Treatment and Nutrients/Foods

9.21.1 Glucocorticoids

Glucocorticoids are important and commonly recommended as a treatment for DMD. Glucocorticoids provide anti-inflammatory benefits, yet their long-term use leads to adverse side effects which limit their clinical utility. Some of the common adverse effects are immunosuppression, glucose intolerance, osteoporosis, decreased intestinal calcium absorption, and increased renal calcium excretion. The most common glucocorticoids for DMD are deflazacort and prednisone/prednisolone. The recommended dose for deflazacort and prednisone are 0.9 mg/kg/day and 0.75 mg/kg/day, respectively. Prednisone is rapidly metabolized to the active form, prednisolone [180]. Thus, prednisone bioavailability is measured by plasma levels of prednisolone. Known interactions between glucocorticoids and nutrients/foods are described below.

9.21.2   Effects of Glucocorticoids on Calcium and Vitamin D

Sustained glucocorticoid treatment leads to decreases in calcium levels and intestinal calcium absorption [181–184]. As a potential therapy to counter the effects of glucocorticoids on calcium absorption, vitamin D is usually recommended because it can increase calcium absorption and reabsorption from the kidneys [185]. DMD patients undergoing glucocorticoid treatment are usually prescribed calcium and vitamin D supplements because intestinal calcium absorption is decreased. Thus, the recommended daily intake of both calcium and vitamin D is 750 mg and 1000 IU, respectively [12]. Patients are also encouraged to monitor calcium daily intake, serum 25(OH)-D levels, and urine excretion (calciuria) [9, 10].

9.21.3   Effects of Simultaneous Food and Glucocorticoid Intake

Glucocorticoids are usually recommended to be ingested with food (http://www.nlm.nih.gov/medlineplus/druginfo/meds/a601102.html). There are no current studies that have evaluated the effects of concomitant food and glucocorticoid intake in DMD patients. Thus, the summary below describes potential interactions reported from studies in other populations. Studies have evaluated the effects of food ingestion in bioavailability of both deflazacort and prednisone/prednisolone.

Rao et al. [186] evaluated the bioavailability of deflazacort tablets in 12 healthy adult males after consumption of a low- and high-fat meal. Results demonstrated that administration of deflazacort following low- and high-fat meals did not affect the extent of deflazacort absorption compared to administration in a fasting state. However, findings suggested that absorption of deflazacort was slower after a meal compared to administration in a fasting state; nevertheless, slower absorption after a meal was proposed as unlikely of clinical significance. Tembo et al. [187] performed a study to determine whether food decreased, increased, or had no effect on the absorption of prednisone. There were no differences between fasting and nonfasting conditions for time of peak plasma level, area under the curve, or half-life (peak plasma level is the maximal plasma level achieved by the drug; area under the curve (AUC) is the plot of drug concentration over time; half-life is the time it takes for the drug to lose half of its pharmacologic activity). Henderson et al. [188] evaluated the effects of food on prednisolone levels in kidney transplant patients. On day one, patients were administered prednisolone

20 min after ingesting a meal, while on day two patients received prednisolone after a 6 h fast. Peak plasma prednisolone levels were reached within 4 h when patients were fasted, while peak levels under non-fasted conditions were reached between 7 and 10 h. Additionally, when larger doses of prednisolone were administered, there were even greater differences between fasting and non-fasting conditions [188]. [189] performed a 2-way crossover study in six healthy subjects to evaluate the effects of a meal on serum prednisolone concentrations. Results demonstrated that food consumption delayed peak serum prednisolone levels; however, there were no differences in area under the serum concentration curve, peak plasma levels, and serum half-life. Overall, their results demonstrated that meals delay absorption but do not affect the extent of availability. [190] evaluated the differences in prednisolone levels between fasted and non-fasted healthy subjects, and the presence of food had no effect on peak plasma, time to reach peak, and plasma half-life concentration of prednisolone.

Al-Habet and Rogers [191] compared the absorption and bioavailability of enteric-coated and uncoated prednisolone tablets in subjects under fasting and non-fasting conditions. Both absorption and pharmacokinetics of enteric-coated prednisolone tablets were altered by consumption of food. There was also great variability in prednisolone absorption: some subjects had a normal absorption pattern, while others had a delay of 12 h. Hollander et al. [192] performed a study to evaluate the effects of grapefruit juice on bioavailability of prednisolone. Grapefruit juice is known to inhibit cytochrome P450 enzymes and has been reported to play a role in prednisolone metabolism [193]. In contrast, Hollander et al. [192] reported no effects of grapefruit ingestion on prednisolone bioavailability. Overall, studies have shown that concomitant intake of glucocorticoids with food has no major effects on drug bioavailability. However, most of the studies have been performed in healthy not disease state subjects. Therefore, due to the gastrointestinal dysfunction experienced by DMD patients, we suggest the effects of glucocorticoid intake with and without food should be studied

9.21.4   Metformin

Metformin is important in the standard care for type 2 diabetes [194]. Metformin is considered for DMD patients treated with glucocorticoids because of the associated increased weight and risk of developing type 2 diabetes [195]. Although metformin has been shown to be potentially beneficial for DMD patients under glucocorticoid treatment [195], the prolonged use of metformin in type 2 diabetic patients may lead to vitamin B12 deficiency in a dose-dependent manner [196]. To our knowledge, no studies have evaluated changes in vitamin B12 levels in DMD patients chronically treated with metformin. If vitamin B12 levels fall below reference ranges (blood levels below 200 picograms per milliliter), levels can be increased by calcium supplementation as reported for type 2 diabetic patients treated with metformin [197]. Moreover, screening strategies should be implemented to prevent vitamin B12 deficiency in patients treated with metformin [198]. It is suggested that vitamin B12 levels be assessed at intervals of no more than 3–6 months because metformin may depress vitamin B12 levels after 3 months of treatment [199–201]. The ingestion of metformin is usually recommended with food to reduce the likelihood of gastrointestinal events; however, food may decrease the absorption of metformin compared to drug ingestion alone, although the changes are not thought to be clinically relevant (http://packageinserts.bms.com/pi/pi_glucophage.pdf; [202]).

9.21.5   ACE Inhibitors and Beta-Blockers

DMD patients are often prescribed ACE inhibitors and/or beta-blockers due to cardiac complications (e.g., dilated cardiomyopathy). Although the effectiveness of some of these drugs has been evaluated in dystrophic animal models and DMD patients, no studies, to our knowledge, have evaluated their potential interactions with nutrients/foods. Therefore, future studies should evaluate potential effects/interactions. Below is a short summary of studies on other disease populations that have evaluated the effects of nutrient/foods on the pharmacokinetics and pharmacodynamics of ACE inhibitors and beta-blockers.

The ACE inhibitor losartan is used for treatment of hypertension, and it may also delay the progression of diabetic nephropathy. Losartan is commonly prescribed to DMD patients to decrease development of type 2 diabetes. No studies have investigated potential losartan interaction(s) with nutrient/food(s) in the DMD population; furthermore, few studies have evaluated the potential interaction(s) of losartan with nutrients/foods in other populations. High-sodium intake might decrease the effects of losartan [203]. Also, a low sodium diet potentiates the effects of losartan in type 2 diabetic patients, as shown by [203]. This suggests that DMD patients ingesting losartan should have their daily sodium intake monitored to ensure the effectiveness of the drug. Grapefruit juice also affects the bioavailability of losartan. Concomitant consumption of grapefruit juice with losartan leads to a reduction in the conversion of losartan to its active form; thus, this can decrease efficacy [204].

Captopril, an ACE inhibitor, must be ingested 1 h before a meal [176]. If ingested with food, absorption may decrease 30–40 %, which might not allow the drug to reach therapeutic levels [205]. On the other hand, absorption of enalapril, lisinopril, and losartan, also ACE inhibitors, are not affected by the presence of food [176, 206]. Cilazapril, another ACE inhibitor, has a small delayed onset of action when administered with food, yet the degree and duration of ACE inhibition are not affected [207].

The bioavailability of the beta-blocker propranolol is decreased when ingested concomitantly with a protein-rich meal [208–210]. Findings from Liedholm and Melander [208] demonstrated that concomitant intake of propranolol with a protein-rich meal (~19.5 g of protein) increased propranolol bioavailability (there was no effect of concomitant intake of propranolol with a carbohydrate-rich, protein-poor meal). However, the authors reported high interindividual variability, with some subjects demonstrating a decrease while others showed an increase, of as much as 250 %, on propranolol bioavailability [208].

In summary, there are no studies evaluating drug-nutrient interactions in DMD patients receiving ACE inhibitors and/or beta-blockers. The lack of knowledge on any potential interaction suggests that studies are needed to address this topic, since studies in other disease populations have shown interactions that affect the drug bioavailability and, potentially, the effectiveness of a treatment.

9.21.6 Bisphosphonates

Bisphosphonates may be prescribed to DMD patients undergoing glucocorticoid treatment to reduce the risk of bone fractures [211]. Bisphosphonates are known to decrease bone turnover by inhibiting osteoclast function and are commonly used in the prevention or treatment of osteoporosis [62]. Bisphosphonates have very low bioavailability due to poor GI tract absorption, and patients are usually advised to ingest this medication at least 1 h prior to ingesting food [212]. Moreover, bisphosphonates should be ingested with at least a 2 h difference from magnesium ingestion; if ingested concomitantly, magnesium can affect bisphosphonate absorption (http://lpi.oregonstate.edu/infocenter/minerals/magnesium/). To our knowledge, there are no studies that have evaluated potential interaction(s) between bisphosphonates with nutrient(s)/food(s) in DMD patients. Thus, further studies are needed to evaluate any potential interactions.

9.21.7 Effects of Concomitant Food and Drug Intake

DMD patients develop gastrointestinal problems (oropharyngeal, esophageal, and gastric) as the disease progresses [45, 213]. These complications can potentially influence the digestion of nutrients/foods as well as drug absorption. However, there is very limited knowledge about the effects of concomitant food and drug intake in this population. The ingestion of drugs with food might alter the drugs' fate for the following reasons: (1) the drug can be destroyed, due to extensive time in the stomach; (2) the drug is absorbed slowly; and (3) food can potentiate the drug's effects. The simultaneous intake of a drug with food can influence the rate and/or extent of drug absorption; thus, this can alter the physicochemical conditions within the GI tract. According to the FDA, "the nutrient and caloric contents of the meal, the meal volume, and the meal temperature can cause physiological changes in the GI tract in a way that affects drug product transit time, luminal dissolution, drug permeability, and systemic availability" (Food-Effect Bioavailability and Fed Bioequivalence Studies 2002—FDA). Food-effect studies are recommended to evaluate the effects of a meal on the drug bioavailability. The FDA recommends that food-effect studies should use a high-fat meal containing 800–1000 kcal, with 50–65 % from fat, 25–30

% from carbohydrates, and 15–20 % from proteins, since this should provide the greatest effect on gastrointestinal physiology and allows systemic drug availability to be maximally tested [214]. Based on the very limited knowledge about concomitant food and drug intake in DMD patients, we propose that studies are needed to identify potential interactions.

9.21.8 Other Interactions Potentially Relevant for DMD Patients

In addition to the drug-nutrient or drug-food interactions discussed in the previous sections, there are also other relevant interactions that should be addressed for DMD patients. For instance, another interaction that has been reported occurs between beta-blockers and CoQ10 [215]. This interaction might be relevant for DMD patients since CoQ10 has a very important role in cell bioenergetics due to its participation in the mitochondrial electron transport chain and, therefore, energy production [164]. Furthermore, DMD patients are often prescribed heart medications, such as beta-blockers, to counteract some of the heart complications associated with the disease (e.g., dilated cardiomyopathy). Kishi and colleagues [205] demonstrated that the beta-blockers propranolol and diazoxide inhibit the activity of CoQ10-NADH-oxidase and Co-Q10-succinoxidase, respectively. Although not shown in their study [215], this inhibition could potentially alter cardiac function due to the role of CoQ10 in cell energy production. Also, individuals with preexisting CoQ10 deficiency in the myocardium might have greater cardiac complications [215]. To our knowledge, no studies have evaluated the interaction between beta-blockers and CoQ10 in DMD. Therefore, studies should evaluate any potential interaction between beta-blockers and CoQ10 in DMD patients. For now, CoQ10 supplementation should be recommended to DMD patients treated with the beta-blockers propranolol or diazoxide to avoid the potential detriments on cardiac function.

9.22 Conclusion

There is limited published information that describes practical nutrition assessment and dietary recommendations for individuals with DMD. Furthermore, there are limited studies that provide reasonable evidence for the benefits of nutrient constituents/supplements, with only a few exceptions (e.g., vitamin D, creatine). In addition, there are virtually no studies that have explored potential drug-nutrient interactions in DMD. In this chapter, we have provided some practical dietary guidelines and emphasized the need for additional studies to address identified deficiencies in the field. It is our hope that this summary of available literature will provide impetus for additional investigations to more clearly define nutritional care for DMD boys.

References

1. Hoffman EP, Brown Jr RH, Kunkel LM. Dystrophin: the protein product of the Duchenne muscular dystrophy locus. Cell. 1987;51(6):919–28.
2. Blake DJ, Weir A, Newey SE, Davies KE. Function and genetics of dystrophin and dystrophin-related proteins in muscle. Physiol Rev. 2002;82(2):291–329.
3. Markert CD, Ambrosio F, Call JA, Grange RW. Exercise and Duchenne muscular dystrophy: toward evidence-based exercise prescription. Muscle Nerve. 2011;43(4):464–78.
4. Bies RD, Phelps SF, Cortez MD, Roberts R, Caskey CT, Chamberlain JS. Human and murine dystrophin mRNA transcripts are differentially expressed during skeletal muscle, heart, and brain development. Nucleic Acids Res. 1992;20:1725–31.
5. Ferlini A, Neri M, Gualandi F. The medical genetics of dystrophinopathies: Molecular genetic diagnosis and its impact on clinical practice. Neuro Disord. 2013;23:4–14.
6. Sharma P, Tran T, Stelmack GL, McNeill K, Gosens R, Mutawe MM, Unruh H, Gerthoffer WT, Halayko AJ. Expression of the dystrophin-glycoprotein complex is a marker for human airway smooth muscle phenotype maturation. Am J Physiol Lung Cell Mol Physiol. 2008;294:L57–68.
7. Bensen ES, Jaffe KM, Tarr PI. Acute gastric dilatation in Duchenne Muscular Dystrophy: A case report and review of the literature. Arch Phys Med Rehab. 1996;77:512–4.
8. Boland BJ, Silbert PL, Groover RV, Wollan PC, Silverstein MD. Skeletal, Cardiac and smooth muscle failure in Duchenne muscular dystrophy. Ped Neurol. 1996;14(1):7–12.
9. Bushby K, Finkel R, Birnkrant DJ, Case LE, Clemens PR, Cripe L, DMD Care Considerations Working Group, et al. Diagnosis and management of Duchenne muscular dystrophy, part 1: diagnosis, and pharmacological and psychosocial management. Lancet Neurol. 2010;9: 77–93.
10. Bushby K, Finkel R, Birnkrant DJ, Case LE, Clemens PR, Cripe L, et al. Diagnosis and management of Duchenne muscular dystrophy, part 2: implementation of multidisciplinary care. Lancet Neurol. 2010;9(2):177–89.
11. Motlagh B, MacDonald JR, Tarnopolsky MA. Nutritional inadequacy in adults with muscular dystrophy. Muscle Nerve. 2005;31(6):713–8.
12. Davidson ZE, Truby H. A review of nutrition in Duchenne muscular dystrophy. Journal of human nutrition and dietetics. 2009;22(5):383–93.
13. Willig TN, Carlier L, Legrand M, Riviere H, Navarro J. Nutritional assessment in Duchenne muscular dystrophy. Dev Med Child Neurol. 1993;35(12):1074–82.
14. McDonald CM, Abresch RT, Carter GT, Fowler Jr WM, Johnson ER, Kilmer DD, et al. Profiles of neuromuscular diseases. Duchenne muscular dystrophy. Am J Phys Med Rehabil. 1995;74(5 Suppl):S70–92.
15. Martigne L, Salleron J, Mayer M, Cuisset J-M, Carpentier A, Neve V, et al. Natural evolution of weight status in Duchenne muscular dystrophy: a retrospective audit. Br J Nutr. 2011;105:1486–91.
16. Eiholzer U, Boltshauser E, Frey D, Molinari L, Zachmann M. Short stature: a common feature in Duchenne muscular dystrophy. Eur J Pediatr. 1988;147(6):602–5.
17. Nagel BHP, Mortier W, Elmlinger M, Wollmann HA, Schmitt K, Ranke MB. Short stature in Duchenne muscular dystrophy: a study of 34 patients. Acta Paediatr. 1999;88(1):62–5.
18. Centers for Disease Control and Prevention. Percentile data files with LMS values. 2009; Available from: http://www.cdc.gov/growthcharts/percentile_data_files.htm
19. Davidson ZE, Ryan MM, Kornberg AJ, Sinclair K, Cairns A, Walker KZ, et al. Observations of body mass index in Duchenne muscular dystrophy: a longitudinal study. Eur J Clin Nutr. 2014;68(8):892–7.
20. Bianchi ML, Biggar D, Bushby K, Rogol AD, Rutter MM, Tseng B. Endocrine aspects of Duchenne muscular dystrophy. Neuromuscul Disord. 2011;21(4):298–303.
21. Kuczmarski R, Odgen C, Grummer-Strawn L, Flegal K, Guo S, Wei R, et al. CDC growth charts: United States. Advance data from vital and health statistics; No 314. Hyattsville, MD: National Center for Health Statistics; 2000.

22. Gauld LM, Kappers J, Carlin JB, Robertson CF. Height prediction from ulna length. Developmental Medicine & Child Neurology. 2004;46(7):475–80.
23. Pessolano FA, Suarez AA, Monteiro SG, Mesa L, Dubrovsky A, Roncoroni AJ, et al. Nutritional assessment of patients with neuromuscular diseases. Am J Phys Med Rehabil. 2003;82:182–5.
24. Elliott SA, Davidson ZE, Davies PS, Truby H. Accuracy of Parent-Reported Energy Intake and Physical Activity Levels in Boys With Duchenne Muscular Dystrophy. Nutr Clin Pract. 2014. doi:10.1177/0884533614546696.
25. National Research Council. Dietary reference intakes for energy, carbohydrate, fiber, fat, fatty acids, cholesterol, protein, and amino acids (macronutrients). Washington, DC: The National Academies Press; 2005.
26. Institute of Medicine. Dietary Reference Intakes for Calcium, Phosphorus, Magnesium, Vitamin D, and Fluoride. Washington, DC: The National Academies Press; 1997.
27. Institute of Medicine. Dietary Reference Intakes for Thiamin, Riboflavin, Niacin, Vitamin B6, Folate, Vitamin B12, Pantothenic Acid, Biotin, and Choline. Washington, DC: The National Academies Press; 1998.
28. Institute of Medicine. Dietary Reference Intakes for Vitamin C, Vitamin E, Selenium, and Carotenoids. Washington, DC: The National Academies Press; 2000.
29. National Research Council. Dietary reference intakes for vitamin a, vitamin k, arsenic, boron, chromium, copper, iodine, iron, manganese, molybdenum, nickel, silicon, vanadium, and zinc. Washington, DC: The National Academies Press; 2001.
30. National Research Council. Dietary reference intakes for water, potassium, sodium, chloride, and sulfate. Washington, DC: The National Academies Press; 2005.
31. Institute of Medicine. Dietary Reference Intakes for Calcium and Vitamin D. Washington, DC: The National Academies Press; 2011.
32. Levine JA. Measurement of energy expenditure. Pub Hlth Nut. 2005;8:1123–32.
33. St-Onge M. The role of sleep duration in the regulation of energy balance: effects on energy intakes and expenditure. Journal of Clinical Sleep Medicine. 2013;9(1):73–80.
34. Elliott SA, Davidson ZE, Davies PSW, Truby H. A Bedside Measure of Body Composition in Duchenne Muscular Dystrophy. Pediatr Neurol. 2014. doi:10.1016/j.pediatrneurol. 2014.08.008.
35. Elliott SA, Davidson ZE, Davies PS, Truby H. Predicting resting energy expenditure in boys with Duchenne muscular dystrophy. Eur J Paediatr Neurol. 2012;16(6):631–5.
36. Schofield WN. Predicting basal metabolic rate, new standards and review of previous work. Hum Nutr Clin Nutr. 1985;39 Suppl 1:5–41.
37. Whitney E, Rolfes SR, Crowe T, Cameron-Smith D, Walsh A. Understanding Nutrition: Australia and New. Zealandth ed. Melbourne: Cengage Learning; 2014.
38. Diamanti A, Panetta F, Tentolini A. Efficacy and safety of gastrostomy feeding in Duchenne muscular dystrophy. Clin Nutr. 2011;30(2):263.
39. Goldstein M, Meyer S, Freund HR. Effects of overfeeding in children with muscle dystrophies. JPEN J Parenter Enteral Nutr. 1989;13(6):603–7.
40. Martigne L, Seguy D, Pellegrini N, Orlikowski D, Cuisset JM, Carpentier A, et al. Efficacy and tolerance of gastrostomy feeding in Duchenne muscular dystrophy. Clin Nutr. 2010;29(1):60–4.
41. Matsuo K, Palmer JB. Anatomy and Physiology of feeding and swallowing – normal and abnormal. Phys Med REhabil Clin N Amer. 2008;19(4):691–707.
42. Aloysius A, Born P, Kinali M, Davis T, Pane M, Mercuri E. Swallowing difficulties in Duchenne muscular dystrophy: Indications for feeding assessment and outcome of videofluroscopic swallow studies. Eur J of Paed Neurol. 2008;12:239–45.
43. van den Engel-Hoek L, Erasmus CE, Hendriks JCM, Geurts ACH, Klein WM, Sigrid Pillen S, Sie Bert LT, de Swart BJM JM, de Groot IJM. Oral muscles are progressively affected in Duchenne muscular dystrophy: implications for dysphagia treatment. J Neurol. 2013;260:1295–303.
44. van den Engel-Hoek L, Erasmus CE, van Hulst KCM, Arvedson JC, de Groot IJM, de Swart BJM. Children With Central and Peripheral Neurologic Disorders Have Distinguishable

Patterns of Dysphagia on Videofluoroscopic Swallow Study. J Child Neurol doi:. 2013. doi:10.1177/0883073813501871.

45. Barohn RJ, Levine EJ, Olson JO, Mendell JR. Gastric hypomotility in Duchenne's muscular dystrophy. N Engl J Med. 1988;319(1):15–8.
46. Borrelli O, Salvia G, Mancini V, Santoro L, Tagliente F, Romeo EF, et al. Evolution of Gastric Electrical Features and Gastric Emptying in Children with Duchenne and Becker Muscular Dystrophy. Am J Gastroenterol. 2005;100(3):695–702.
47. Leon SH, Schuffler MD, Kettler M, CA R. Chronic intestinal pseudoobstruction as a complication of Duchenne's muscular dystrophy. Gastroent. 1986;90(2):455–9.
48. Mendell JR, Griggs RC, Moxley 3rd RT, Fenichel GM, Brooke MH, Miller JP, Dodson WE. Clinical investigation in Duchenne muscular dystrophy: IV. Double-blind controlled trial of leucine. Muscle Nerve. 1984;7(7):535–41. doi:10.1002/mus.880070704.
49. Stewart PM, Walser M, Drachman DB. Branched-chain ketoacids reduce muscle protein degradation in Duchenne muscular dystrophy. Muscle Nerve. 1982;5(3):197–201. doi:10.1002/mus.880050304.
50. Mok E, Eleouet-Da Violante C, Daubrosse C, Gottrand F, Rigal O, Fontan JE, et al. Oral glutamine and amino acid supplementation inhibit whole-body protein degradation in children with Duchenne muscular dystrophy. Am J Clin Nutr. 2006;83(4):823–8.
51. Mok E, Letellier G, Cuisset JM, Denjean A, Gottrand F, Alberti C, Hankard R. Lack of functional benefit with glutamine versus placebo in Duchenne muscular dystrophy: a randomized crossover trial. PLoS One. 2009;4(5), e5448. doi:10.1371/journal.pone.0005448.
52. Kley RA, Tarnopolsky MA, Vorgerd M. Creatine for treating muscle disorders. Cochrane Database Syst Rev. 2013;6, CD004760. doi:10.1002/14651858.CD004760.pub4.
53. Walter MC, Lochmuller H, Reilich P, Klopstock T, Huber R, Hartard M, et al. Creatine monohydrate in muscular dystrophies: A double-blind, placebo-controlled clinical study. Neurology. 2000;54(9):1848–50.
54. Tarnopolsky MA, Mahoney DJ, Vajsar J, Rodriguez C, Doherty TJ, Roy BD, Biggar D. Creatine monohydrate enhances strength and body composition in Duchenne muscular dystrophy. Neurology. 2004;62(10):1771–7.
55. Louis M, Lebacq J, Poortmans JR, Belpaire-Dethiou MC, Devogelaer JP, Van Hecke P, et al. Beneficial effects of creatine supplementation in dystrophic patients. Muscle Nerve. 2003;27(5):604–10. doi:10.1002/mus.10355.
56. Banerjee B, Sharma U, Balasubramanian K, Kalaivani M, Kalra V, Jagannathan NR. Effect of creatine monohydrate in improving cellular energetics and muscle strength in ambulatory Duchenne muscular dystrophy patients: a randomized, placebo-controlled 31P MRS study. Magn Reson Imaging. 2010;28(5):698–707. doi:10.1016/j.mri.2010.03.008.
57. Escolar DM, Buyse G, Henricson E, Leshner R, Florence J, Mayhew J, CINRG Group, et al. CINRG randomized controlled trial of creatine and glutamine in Duchenne muscular dystrophy. Ann Neurol. 2005;58(1):151–5. doi:10.1002/ana.20523.
58. Paulson DJ, Hoganson GE, Traxler J, Sufit R, Peters H, Shug AL. Ketogenic effects of carnitine in patients with muscular dystrophy and cytochrome oxidase deficiency. Biochem Med Metab Biol. 1988;39(1):40–7.
59. Escobar-Cedillo RE, Tintos-Hernández JA, Martínez-Castro G, de Oca-Sánchez BM, Rodríguez-Jurado R, Miranda-Duarte A, Lona-Pimentel S, et al. L-carnitine suplemmentation in Duchenne muscular dystroph steroidnaive patients: a pilot study. Curr Top Nutraceut Res. 2013;11(3):97.
60. Folkers K, Simonsen R. Two successful double-blind trials with coenzyme Q10 (vitamin Q10) on muscular dystrophies and neurogenic atrophies. Biochim Biophys Acta. 1995;1271:281–6.
61. Buyse GM, Goemans N, van den Hauwe M, Thijs D, de Groot IJ, Schara U, Ceulemans B, Meier T, Mertens L. Idebenone as a novel, therapeutic approach for Duchenne muscular dystrophy: results from a 12 month, double-blind, randomized placebo-controlled trial. Neuromuscul Disord. 2011;21:396–405.

62. Eriksen EF, Díez-Pérez A, Boonen S. Update on long-term treatment with bisphosphonates for postmenopausal osteoporosis: a systematic review. Bone. 2014;58:126–35.
63. Partridge TA. The mdx mouse model as a surrogate for Duchenne muscular dystrophy. FEBS J. 2013;280(17):4177–86.
64. Voisin V, Sebrie C, Matecki S, Yu H, Gillet B, Ramonatxo M, De la Porte S. L-arginine improves dystrophic phenotype in mdx mice. Neurobiol Dis. 2005;20(1):123–30. doi:10.1016/j.nbd.2005.02.010.
65. Hnia K, Gayraud J, Hugon G, Ramonatxo M, De La Porte S, Matecki S, Mornet D. L-arginine decreases inflammation and modulates the nuclear factor-kappaB/matrix metalloproteinase cascade in mdx muscle fibers. Am J Pathol. 2008;172(6):1509–19. doi:10.2353/ajpath.2008.071009.
66. Wehling-Henricks M, Jordan MC, Gotoh T, Grody WW, Roos KP, Tidball JG. Arginine metabolism by macrophages promotes cardiac and muscle fibrosis in mdx muscular dystrophy. PLoS One. 2010;5(5), e10763. doi:10.1371/journal.pone.0010763.
67. Vianello S, Yu H, Voisin V, Haddad H, He X, Foutz AS, de la Porte S. Arginine butyrate: a therapeutic candidate for Duchenne muscular dystrophy. FASEB J. 2013;27(6):2256–69. doi:10.1096/fj.12-215723.
68. Vianello S, Consolaro F, Bich C, Cancela JM, Roulot M, Lanchec E, et al. Low doses of arginine butyrate derivatives improve dystrophic phenotype and restore membrane integrity in DMD models. FASEB J. 2014;28(6):2603–19. doi:10.1096/fj.13-244798.
69. Mok E, Constantin B, Favreau F, Neveux N, Magaud C, Delwail A, Hankard R. l-Glutamine administration reduces oxidized glutathione and MAP kinase signaling in dystrophic muscle of mdx mice. Pediatr Res. 2008;63(3):268–73. doi:10.1203/PDR.0b013e318163a259.
70. De Luca A, Pierno S, Liantonio A, Cetrone M, Camerino C, Fraysse B, et al. Enhanced dystrophic progression in mdx mice by exercise and beneficial effects of taurine and insulin-like growth factor-1. J Pharmacol Exp Ther. 2003;304(1):453–63. doi:10.1124/jpet.102.041343.
71. Cozzoli A, Rolland JF, Capogrosso RF, Sblendorio VT, Longo V, Simonetti S, et al. Evaluation of potential synergistic action of a combined treatment with alpha-methyl-prednisolone and taurine on the mdx mouse model of Duchenne muscular dystrophy. Neuropathol Appl Neurobiol. 2011;37(3):243–56. doi:10.1111/j.1365-2990.2010.01106.x.
72. Buetler TM, Renard M, Offord EA, Schneider H, Ruegg UT. Green tea extract decreases muscle necrosis in mdx mice and protects against reactive oxygen species. Am J Clin Nutr. 2002;75(4):749–53.
73. Call JA, Voelker KA, Wolff AV, McMillan RP, Evans NP, Hulver MW, Grange RW. Endurance capacity in maturing mdx mice is markedly enhanced by combined voluntary wheel running and green tea extract. J Appl Physiol (1985). 2008;105(3):923–32. doi:10.1152/japplphysiol.00028.2008.
74. Evans NP, Call JA, Bassaganya-Riera J, Robertson JL, Grange RW. Green tea extract decreases muscle pathology and NF-kappaB immunostaining in regenerating muscle fibers of mdx mice. Clin Nutr. 2010;29(3):391–8. doi:10.1016/j.clnu.2009.10.001.
75. Mauricio AF, Minatel E, Santo Neto H, Marques MJ. Effects of fish oil containing eicosapentaenoic acid and docosahexaenoic acid on dystrophic mdx mice. Clin Nutr. 2013;32(4):636–42.
76. Carvalho SC, Apolinario LM, Matheus SM, Santo Neto H, Marques MJ. EPA protects against muscle damage in the mdx mouse model of Duchenne muscular dystrophy by promoting a shift from the M1 to M2 macrophage phenotype. J Neuroimmunol. 2013;264(1-2):41–7. doi:10.1016/j.jneuroim.2013.09.007.
77. Oh J, Kang H, Kim HJ, Lee JH, Choi KG, Park KD. The effect of L-carnitine supplementation on the dystrophic muscle and exercise tolerance of muscular dystrophy (mdx) mice. J Korean Neurolog Assoc. 2005;23:519–27.
78. Gordon BS, Delgado-Diaz DC, Carson J, Fayad R, Wilson LB, Kostek MC. Resveratrol improves muscle function but not oxidative capacity in young mdx mice. Can J Physiol Pharmacol. 2014;92(3):243–51. doi:10.1139/cjpp-2013-0350.

79. Kostek MC, Gordon BS, Diaz DCD. Resveratrol affects inflammation and muscle function in mdx mice. FASEB J. 2011;25(1_MeetingAbstracts):lb597.
80. Hori YS, Kuno A, Hosoda R, Tanno M, Miura T, Shimamoto K, Horio Y. Resveratrol ameliorates muscular pathology in the dystrophic mdx mouse, a model for Duchenne muscular dystrophy. J Pharmacol Exp Ther. 2011;338(3):784–94. doi:10.1124/jpet.111.183210.
81. Whitehead NP, Pham C, Gervasio OL, Allen DG. N-Acetylcysteine ameliorates skeletal muscle pathophysiology in mdx mice. J Physiol. 2008;586(7):2003–14. doi:10.1113/jphysiol.2007.148338.
82. de Senzi Moraes Pinto R, Ferretti R, Moraes LH, Neto HS, Marques MJ, Minatel E. N-acetylcysteine treatment reduces TNF-alpha levels and myonecrosis in diaphragm muscle of mdx mice. Clin Nutr. 2013;32(3):472–5. doi:10.1016/j.clnu.2012.06.001.
83. Buyse GM, Van der Mieren G, Erb M, D'hooge J, Herijgers P, Verbeken E, Jara A, Bergh AVD, Mertens L, Courdier-Fruh I, Barzaghi P, Meier T. Long-term blinded placebo-controlled study of SNT-MC17/idebenone in the dystrophin deficient mdx mouse: cardiac protection and improved exercise performance. European Heart Journal. 2009;30(1):116–24.
84. Ballard FJ, Tomas FM, Stern LM. Increased turnover of muscle contractile proteins in Duchenne muscular dystrophy as assessed by 3-methylhistidine and creatinine excretion. Clin Sci (Lond). 1979;56(4):347–52.
85. Hankard R, Mauras N, Hammond D, Haymond M, Darmaun D. Is glutamine a 'conditionally essential' amino acid in Duchenne muscular dystrophy? Clin Nutr. 1999;18(6):365–9. doi:10.1054/clnu.1999.0054.
86. Radley-Crabb HG, Marini JC, Sosa HA, Castillo LI, Grounds MD, Fiorotto ML. Dystropathology increases energy expenditure and protein turnover in the mdx mouse model of duchenne muscular dystrophy. PLoS One. 2014;9(2), e89277. doi:10.1371/journal.pone.0089277.
87. Blomstrand E, Eliasson J, Karlsson HK, Kohnke R. Branched-chain amino acids activate key enzymes in protein synthesis after physical exercise. J Nutr. 2006;136(1 Suppl):269S–73.
88. Kimball SR, Jefferson LS. Signaling pathways and molecular mechanisms through which branched-chain amino acids mediate translational control of protein synthesis. J Nutr. 2006;136(1 Suppl):227S–31.
89. Davoodi J, Markert CD, Voelker KA, Hutson SM, Grange RW. Nutrition strategies to improve physical capabilities in Duchenne muscular dystrophy. Phys Med Rehabil Clin N Am. 2012;23(1):187–99. doi:10.1016/j.pmr.2011.11.010. xii-xiii.
90. Chaubourt E, Fossier P, Baux G, Leprince C, Israel M, De La Porte S. Nitric oxide and l-arginine cause an accumulation of utrophin at the sarcolemma: a possible compensation for dystrophin loss in Duchenne muscular dystrophy. Neurobiol Dis. 1999;6(6):499–507. doi:10.1006/nbdi.1999.0256.
91. Haenggi T, Fritschy JM. Role of dystrophin and utrophin for assembly and function of the dystrophin glycoprotein complex in non-muscle tissue. Cell Mol Life Sci. 2006;63(14):1614–31. doi:10.1007/s00018-005-5461-0.
92. Love DR, Hill DF, Dickson G, Spurr NK, Byth BC, Marsden RF, Davies KE. An autosomal transcript in skeletal muscle with homology to dystrophin. Nature. 1989;339(6219):55–8. doi:10.1038/339055a0.
93. Mizuno Y, Nonaka I, Hirai S, Ozawa E. Reciprocal expression of dystrophin and utrophin in muscles of Duchenne muscular dystrophy patients, female DMD-carriers and control subjects. J Neurol Sci. 1993;119(1):43–52.
94. Rafael JA, Tinsley JM, Potter AC, Deconinck AE, Davies KE. Skeletal muscle-specific expression of a utrophin transgene rescues utrophin-dystrophin deficient mice. Nat Genet. 1998;19(1):79–82. doi:10.1038/ng0598-79.
95. Shao A, Hathcock JN. Risk assessment for the amino acids taurine, L-glutamine and L-arginine. Regul Toxicol Pharmacol. 2008;50(3):376–99. doi:10.1016/j.yrtph.2008.01.004.
96. Mattei E, Corbi N, Di Certo MG, Strimpakos G, Severini C, Onori A, Passananti C. Utrophin up-regulation by an artificial transcription factor in transgenic mice. PLoS One. 2007;2(8), e774. doi:10.1371/journal.pone.0000774.

97. Tinsley JM, Fairclough RJ, Storer R, Wilkes FJ, Potter AC, Squire SE, et al. Daily treatment with SMTC1100, a novel small molecule utrophin upregulator, dramatically reduces the dystrophic symptoms in the mdx mouse. PLoS One. 2011;6(5), e19189. doi:10.1371/journal.pone.0019189.
98. Perrine SP, Hermine O, Small T, Suarez F, O'Reilly R, Boulad F, Faller DV. A phase 1/2 trial of arginine butyrate and ganciclovir in patients with Epstein-Barr virus-associated lymphoid malignancies. Blood. 2007;109(6):2571–8. doi:10.1182/blood-2006-01-024703.
99. Hankard RG, Hammond D, Haymond MW, Darmaun D. Oral glutamine slows down whole body protein breakdown in Duchenne muscular dystrophy. Pediatr Res. 1998;43(2):222–6. doi:10.1203/00006450-199804001-01321.
100. Garlick PJ. Assessment of the safety of glutamine and other amino acids. J Nutr. 2001;131(9 Suppl):2556S–61.
101. Pulido SM, Passaquin AC, Leijendekker WJ, Challet C, Wallimann T, Ruegg UT. Creatine supplementation improves intracellular Ca2+ handling and survival in mdx skeletal muscle cells. FEBS Lett. 1998;439(3):357–62.
102. Felber S, Skladal D, Wyss M, Kremser C, Koller A, Sperl W. Oral creatine supplementation in Duchenne muscular dystrophy: a clinical and 31P magnetic resonance spectroscopy study. Neurol Res. 2000;22(2):145–50.
103. Payne ET, Yasuda N, Bourgeois JM, Devries MC, Rodriguez MC, Yousuf J, Tarnopolsky MA. Nutritional therapy improves function and complements corticosteroid intervention in mdx mice. Muscle Nerve. 2006;33(1):66–77. doi:10.1002/mus.20436.
104. Persky AM, Brazeau GA. Clinical pharmacology of the dietary supplement creatine monohydrate. Pharmacol Rev. 2001;53(2):161–76.
105. Koshy KM, Griswold E, Schneeberger EE. Interstitial nephritis in a patient taking creatine. N Engl J Med. 1999;340(10):814–5. doi:10.1056/NEJM199903113401017.
106. Schaffer SW, Jong CJ, Ramila KC, Azuma J, Azuma J. Physiological roles of taurine in heart and muscle. J Biomed Sci. 2010;17 Suppl 1:S2. doi:10.1186/1423-0127-17-S1-S2.
107. Schuller-Levis GB, Gordon RE, Wang C, Park E. Taurine reduces lung inflammation and fibrosis caused by bleomycin. Adv Exp Med Biol. 2003;526:395–402.
108. Wright CE, Tallan HH, Lin YY, Gaull GE. Taurine: biological update. Annu Rev Biochem. 1986;55:427–53. doi:10.1146/annurev.bi.55.070186.002235.
109. McIntosh L, Granberg KE, Briere KM, Anderson JE. Nuclear magnetic resonance spectroscopy study of muscle growth, mdx dystrophy and glucocorticoid treatments: correlation with repair. NMR Biomed. 1998;11(1):1–10.
110. Griffin JL, Williams HJ, Sang E, Clarke K, Rae C, Nicholson JK. Metabolic profiling of genetic disorders: a multitissue (1)H nuclear magnetic resonance spectroscopic and pattern recognition study into dystrophic tissue. Anal Biochem. 2001;293(1):16–21. doi:10.1006/abio.2001.5096.
111. De Luca A, Pierno S, Camerino DC. Electrical properties of diaphragm and EDL muscles during the life of dystrophic mice. Am J Physiol. 1997;272(1 Pt 1):C333–40.
112. Matsuda H, Kinoshita K, Sumida A, Takahashi K, Fukuen S, Fukuda T, Azuma J. Taurine modulates induction of cytochrome P450 3A4 mRNA by rifampicin in the HepG2 cell line. Biochim Biophys Acta. 2002;1593(1):93–8.
113. Bakker AJ, Berg HM. Effect of taurine on sarcoplasmic reticulum function and force in skinned fast-twitch skeletal muscle fibres of the rat. J Physiol. 2002;538(Pt 1):185–94.
114. Huxtable RJ. Physiological actions of taurine. Physiol Rev. 1992;72(1):101–63.
115. Borum PR. Carnitine. Annu Rev Nutr. 1983;3:233–59. doi:10.1146/annurev.nu.03.070183.001313.
116. Rebouche CJ. Carnitine. In: Shils ME, Ross AC, Caballero B, Cousins RJ, editors. 10 ed. Philadelphia, PA: Lippincott, Williams & Wilkins; 2006.
117. Le Borgne F, Guyot S, Logerot M, Beney L, Gervais P, Demarquoy J. Exploration of lipid metabolism in relation with plasma membrane properties of Duchenne muscular dystrophy cells: influence of L-carnitine. PLoS One. 2012;7(11), e49346. doi:10.1371/journal.pone.0049346.

118. Borum PR, Broquist HP, Roelops RJ. Muscle carnitine levels in neuromuscular disease. J Neurol Sci. 1977;34(2):279–86.

119. Berthillier G, Eichenberger D, Carrier HN, Guibaud P, Got R. Carnitine metabolism in early stages of Duchenne muscular dystrophy. Clin Chim Acta. 1982;122(3):369–75.

120. Ragusa RJ, Chow CK, Porter JD. Oxidative stress as a potential pathogenic mechanism in an animal model of Duchenne muscular dystrophy. Neuromuscul Disord. 1997;7(6-7):379–86.

121. Rando TA. Oxidative stress and the pathogenesis of muscular dystrophies. Am J Phys Med Rehabil. 2002;81(11 Suppl):S175–186. doi:10.1097/01.PHM.0000029774.56528.A6.

122. Tidball JG, Wehling-Henricks M. The role of free radicals in the pathophysiology of muscular dystrophy. J Appl Physiol (1985). 2007;102(4):1677–86. doi:10.1152/japplphysiol.01145.2006.

123. Haycock JW, Mac Neil S, Jones P, Harris JB, Mantle D. Oxidative damage to muscle protein in Duchenne muscular dystrophy. Neuroreport. 1996;8(1):357–61.

124. Chen PC, Wheeler DS, Malhotra V, Odoms K, Denenberg AG, Wong HR. A green tea-derived polyphenol, epigallocatechin-3-gallate, inhibits IkappaB kinase activation and IL-8 gene expression in respiratory epithelium. Inflammation. 2002;26(5):233–41.

125. Dorchies OM, Wagner S, Buetler TM, Ruegg UT. Protection of dystrophic muscle cells with polyphenols from green tea correlates with improved glutathione balance and increased expression of 67LR, a receptor for (-)-epigallocatechin gallate. Biofactors. 2009;35(3):279–94. doi:10.1002/biof.34.

126. Valcic S, Muders A, Jacobsen NE, Liebler DC, Timmermann BN. Antioxidant chemistry of green tea catechins. Identification of products of the reaction of (-)-epigallocatechin gallate with peroxyl radicals. Chem Res Toxicol. 1999;12(4):382–6. doi:10.1021/tx990003t.

127. Dorchies OM, Wagner S, Vuadens O, Waldhauser K, Buetler TM, Kucera P, Ruegg UT. Green tea extract and its major polyphenol (-)-epigallocatechin gallate improve muscle function in a mouse model for Duchenne muscular dystrophy. Am J Physiol Cell Physiol. 2006;290(2):C616–25. doi:10.1152/ajpcell.00425.2005.

128. Sarma DN, Barrett ML, Chavez ML, Gardiner P, Ko R, Mahady GB, Low Dog T. Safety of green tea extracts: a systematic review by the US Pharmacopeia. Drug Saf. 2008;31(6):469–84.

129. Chow HH, Cai Y, Alberts DS, Hakim I, Dorr R, Shahi F, Hara Y. Phase I pharmacokinetic study of tea polyphenols following single-dose administration of epigallocatechin gallate and polyphenon E. Cancer Epidemiol Biomarkers Prev. 2001;10(1):53–8.

130. Babcock T, Helton WS, Espat NJ. Eicosapentaenoic acid (EPA): an antiinflammatory omega-3 fat with potential clinical applications. Nutrition. 2000;16(11-12):1116–8.

131. Babcock TA, Helton WS, Hong D, Espat NJ. Omega-3 fatty acid lipid emulsion reduces LPS-stimulated macrophage TNF-alpha production. Surg Infect (Larchmt). 2002;3(2):145–9. doi:10.1089/109629602760105817.

132. Singer P, Shapiro H, Theilla M, Anbar R, Singer J, Cohen J. Anti-inflammatory properties of omega-3 fatty acids in critical illness: novel mechanisms and an integrative perspective. Intensive Care Med. 2008;34(9):1580–92. doi:10.1007/s00134-008-1142-4.

133. Magee P, Pearson S, Allen J. The omega-3 fatty acid, eicosapentaenoic acid (EPA), prevents the damaging effects of tumour necrosis factor (TNF)-alpha during murine skeletal muscle cell differentiation. Lipids Health Dis. 2008;7:24. doi:10.1186/1476-511X-7-24.

134. Machado RV, Mauricio AF, Taniguti AP, Ferretti R, Neto HS, Marques MJ. Eicosapentaenoic acid decreases TNF-alpha and protects dystrophic muscles of mdx mice from degeneration. J Neuroimmunol. 2011;232(1-2):145–50. doi:10.1016/j.jneuroim.2010.10.032.

135. Fogagnolo Mauricio A, Minatel E, Santo Neto H, Marques MJ. Effects of fish oil containing eicosapentaenoic acid and docosahexaenoic acid on dystrophic mdx mice. Clin Nutr. 2013;32(4):636–42. doi:10.1016/j.clnu.2012.11.013.

136. Agostoni C, Bresson JL, Fairweather-Tait S, Flynn A, Golly I, Korhonen H, Lagiou P, Løvik M, Marchelli R, Moseley MB, Neuhäuser-Berthold M, Przyrembel H, Salminen S, Sanz Y, Strain S, Strobel S, Tetens I, Tomé D, van Loveren H, Verhagen H. Scientific opinion on the tolerable upper intake level of eicosapentaenoic acid (EPA), docosahexaenoic acid (DHA)

and docosapentaenoic acid (DPA). Paper presented at the EFSA Panel on Dietetic Products, Nutrition and Allergies (NDA), Parma, Italy. 2012.

137. Henderson GC, Evans NP, Grange RW, Tuazon MA. Compared with that of MUFA, a high dietary intake of n-3 PUFA does not reduce the degree of pathology in mdx mice. Br J Nutr. 2014;111(10):1791–800. doi:10.1017/S0007114514000129.

138. Riediger ND, Othman RA, Suh M, Moghadasian MH. A systemic review of the roles of n-3 fatty acids in health and disease. J Am Diet Assoc. 2009;109(4):668–79. doi:10.1016/j.jada.2008.12.022.

139. Harris WS. Expert opinion: omega-3 fatty acids and bleeding-cause for concern? Am J Cardiol. 2007;99(6A):44C–6. doi:10.1016/j.amjcard.2006.11.021.

140. Bays HE. Safety considerations with omega-3 fatty acid therapy. Am J Cardiol. 2007;99(6A):35C–43. doi:10.1016/j.amjcard.2006.11.020.

141. Kris-Etherton PM, Harris WS, Appel LJ, American Heart Association. Nutrition Committee. Fish consumption, fish oil, omega-3 fatty acids, and cardiovascular disease. Circulation. 2002;106(21):2747–57.

142. Bianchi ML, Mazzanti A, Galbiati E, Saraifoger S, Dubini A, Cornelio F, Morandi L. Bone mineral density and bone metabolism in Duchenne muscular dystrophy. Osteoporos Int. 2003;14(9):761–7. doi:10.1007/s00198-003-1443-y.

143. Soderpalm AC, Magnusson P, Ahlander AC, Karlsson J, Kroksmark AK, Tulinius M, Swolin-Eide D. Low bone mineral density and decreased bone turnover in Duchenne muscular dystrophy. Neuromuscul Disord. 2007;17(11-12):919–28. doi:10.1016/j.nmd.2007.05.008.

144. Quinlivan R, Shaw N, Bushby K. 170th ENMC International Workshop: bone protection for corticosteroid treated Duchenne muscular dystrophy. 27-29 November 2009, Naarden, The Netherlands. Neuromuscul Disord. 2010;20(11):761–9.

145. Quinlivan R, Shaw N, Bushby K. Bone protection for corticosteroid treated Duchenne muscular dystrophy. Workshop. 170th ENMC International Workshop. Naarden, Netherlands; 2010.

146. Lagouge M, Argmann C, Gerhart-Hines Z, Meziane H, Lerin C, Daussin F, Auwerx J. Resveratrol improves mitochondrial function and protects against metabolic disease by activating SIRT1 and PGC-1alpha. Cell. 2006;127(6):1109–22. doi:10.1016/j.cell.2006.11.013.

147. Wyke SM, Russell ST, Tisdale MJ. Induction of proteasome expression in skeletal muscle is attenuated by inhibitors of NF-kappaB activation. Br J Cancer. 2004;91(9):1742–50. doi:10.1038/sj.bjc.6602165.

148. Wyke SM, Tisdale MJ. Induction of protein degradation in skeletal muscle by a phorbol ester involves upregulation of the ubiquitin-proteasome proteolytic pathway. Life Sci. 2006;78(25):2898–910. doi:10.1016/j.lfs.2005.11.014.

149. Kuno A, Hori YS, Hosoda R, Tanno M, Miura T, Shimamoto K, Horio Y. Resveratrol improves cardiomyopathy in dystrophin-deficient mice through SIRT1 protein-mediated modulation of p300 protein. J Biol Chem. 2013;288(8):5963–72. doi:10.1074/jbc.M112.392050.

150. Gordon BS, Delgado Diaz DC, Kostek MC. Resveratrol decreases inflammation and increases utrophin gene expression in the mdx mouse model of Duchenne muscular dystrophy. Clin Nutr. 2013;32(1):104–11. doi:10.1016/j.clnu.2012.06.003.

151. Boocock DJ, Faust GE, Patel KR, Schinas AM, Brown VA, Ducharme MP, et al. Phase I dose escalation pharmacokinetic study in healthy volunteers of resveratrol, a potential cancer chemopreventive agent. Cancer Epidemiol Biomarkers Prev. 2007;16(6):1246–52. doi:10.1158/1055-9965.EPI-07-0022.

152. Piver B, Berthou F, Dreano Y, Lucas D. Inhibition of CYP3A, CYP1A and CYP2E1 activities by resveratrol and other non volatile red wine components. Toxicol Lett. 2001;125(1-3):83–91.

153. Regev-Shoshani G, Shoseyov O, Kerem Z. Influence of lipophilicity on the interactions of hydroxy stilbenes with cytochrome P450 3A4. Biochem Biophys Res Commun. 2004;323(2):668–73. doi:10.1016/j.bbrc.2004.08.141.

154. Higdon J, Drake V, Steward WP. Resveratrol. Available from Oregon State University: Linus Pauling Institute Micronutrient Information Center; 2005. http://lpi.oregonstate.edu/infocenter/phytochemicals/resveratrol/
155. Lawler JM. Exacerbation of pathology by oxidative stress in respiratory and locomotor muscles with Duchenne muscular dystrophy. J Physiol. 2011;589(Pt 9):2161–70. doi:10.1113/jphysiol.2011.207456.
156. Williams IA, Allen DG. Intracellular calcium handling in ventricular myocytes from mdx mice. Am J Physiol Heart Circ Physiol. 2007;292(2):H846–55. doi:10.1152/ajpheart.00688.2006.
157. Terrill JR, Radley-Crabb HG, Grounds MD, Arthur PG. N-Acetylcysteine treatment of dystrophic mdx mice results in protein thiol modifications and inhibition of exercise induced myofibre necrosis. Neuromuscul Disord. 2012;22(5):427–34. doi:10.1016/j.nmd.2011.11.007.
158. Terrill JR, Boyatzis A, Grounds MD, Arthur PG. Treatment with the cysteine precursor l-2-oxothiazolidine-4-carboxylate (OTC) implicates taurine deficiency in severity of dystropathology in mdx mice. Int J Biochem Cell Biol. 2013;45(9):2097–108. doi:10.1016/j.biocel.2013.07.009.
159. Bobb AJ, Arfsten DP, Jederberg WW. N-acetyl-L-Cysteine as prophylaxis against sulfur mustard. Mil Med. 2005;170(1):52–6.
160. Pendyala L, Creaven PJ. Pharmacokinetic and pharmacodynamic studies of N-acetylcysteine, a potential chemopreventive agent during a phase I trial. Cancer Epidemiol Biomarkers Prev. 1995;4(3):245–51.
161. Pendyala L, Schwartz G, Bolanowska-Higdon W, Hitt S, Zdanowicz J, Murphy M, Creaven PJ. Phase I/pharmacodynamic study of N-acetylcysteine/oltipraz in smokers: early termination due to excessive toxicity. Cancer Epidemiol Biomarkers Prev. 2001;10(3):269–72.
162. Ernster L, Dallner G. Biochemical, physiological and medical aspects of ubiquinone function. Biochim Biophys Acta. 1995;1271:195–204.
163. Hodgson JM, Watts GF, Playford DA, Burke V, Croft KD. Coenzyme Q10 improves blood pressure and glycaemic control: A controlled trial in subjects with type 2 diabetes. Eur J Clin Nutr. 2002;56:1137–42.
164. Bhagavan HN, Chopra RK. Coenzyme Q10: absorption, tissue uptake, metabolism and pharmacokinetics. Free Radical Research. 2006;40(5):445–53.
165. Langsjoen PH, Langsjoen AM. Coenzyme Q10 in cardiovascular disease with emphasis on heart failure and myocardial ischemia. Asia Pac Heart J. 1998;7:160–8.
166. Ferrante KL, Shefner J, Zhang H, Betensky R, O'Brien M, Yu H, Fantasia M, et al. Tolerance of highdose (3,000 mg/day) coenzyme Q10 in ALS. Neurology. 2005;65(11):1834–6.
167. Kieburtz K, The Huntington Study Group. A randomized placebo-controlled trial of coenzyme Q10 and remacemide in Huntington's disease. Neurology. 2001;57:397–404.
168. Shultz CW, Oakes D, Kieburtz K, Beal FL, Haas R, Plumb S, Juncos JL, Nutt J, Shoulson I, Carter J, Kompoliti K, Perlmutter JS, Reich S, Stern M, Watts RL, Kurlan R, Molho E, Harrison M, Lew M, Parkinson Study Group. Effects of coenzyme Q10 in early Parkinson disease. Arch Neurol. 2002;59:1541–50.
169. Spurney CF, Rocha CT, Henricson E, Florence J, Mayhew J, Gorni K, Pasquali L, et al. CINRG pilot trial of coenzyme Q10 in steroid-treated duchenne muscular dystrophy. Muscle & Nerve. 2011;44(2):174–8.
170. Geng J, Dong J, Jiang K, Shen L, Wu T, Ni H, Shi LL, Wang G, Wu H. Idebenone for the treatment of Duchenne muscular dystrophy. Cochrane Database Syst Rev. 2010;8:Art. No. CD008647. doi: 10.1002/14651858.CD008647
171. Gemperli A, Hufschmid M, Courdier-Fruh I, Haefeli R, Erb M, Dallmann R, et al. Restoring mitochondrial function in Duchenne muscular dystrophy by idebenone. Neuromuscular Disorders. 2009;19(8):616–7.
172. Buyse GM, Voit T, Schara U, Straathof CSM, Grazia D'Angelo M, Bernert G, Cuisset J, Finkel RS, Goemans N, McDonald CM, Rummey C, Meier T. Efficacy of idebenone on respiratory function in patients with Duchenne muscular dystrophy not using glucocorticoids (DELOS): a double-blind randomised placebo-controlled phase 3 trial. Lancet. 2015;385(9979):1748–57.

173. Di Prospero NA, Sumner CJ, Penzak SR, Ravina B, Fischbeck KH, Taylor JP. Safety, tolerability, and pharmacokinetics of high-dose idebenone in patients with Friedreich ataxia. Archives of Neurology. 2007;64(6):803–8.
174. Braun L. An introduction to: drug-nutrient interactions. IMER Meet March, Monash University, 1-41; 2012.
175. Ötles S, Ahmet S. Food and drug interactions: a general review. Acta Sci Pol, Technol Aliment. 2014;13(1):89–102.
176. Mycek MJ. Interaction of drugs with foods and nutrients. Nutrition assessment: a comprehensive guide for planning intervention. 1995;135.
177. Boullata JI, Hudson ML. Drug-nutrient interactions: A broad view with implications for practice. Acad Nutr Diet. 2012;112(4):506–17.
178. Bushra R, Aslam N, Khan AY. Food-drug interactions. Oman medical journal. 2011;26(2):77.
179. Zyl VM. The effects of drugs on nutrition. S Afr J Clin Nutr. 2011;24(3):38–41.
180. Gambertoglio JG, Amend Jr WJ, Benet LZ. Pharmacokinetics and bioavailability of prednisone and prednisolone in healthy volunteers and patients: a review. Journal of pharmacokinetics and biopharmaceutics. 1980;8(1):1–52.
181. Burckhardt P. Corticosteroids and bone: a review. Hormone Res. 1984;20:59–64.
182. Gennari C. Differential effect of glucocorticoids on calcium absorption and bone mass. Br J Rheumatol. 1993;32 suppl 2:11–4.
183. Lindholm TS, Sevastikoglou JA, Lingren U. Treatment of patients with senile, postmenopausal and corticosteroid-induced osteoporosis with l-hydroxyvitamin D3 and calcium: short- and long-term effects. Clin Endocrinol. 1977;7:l83s–9.
184. Shult TD, Bollman S, Kumar R. Decreased intestinal calcium absorption in vivo and normal brush border membrane vesicle calcium uptake in cortisol-treated chickens: Evidence for dissociation of calcium absorption from brush border vesicle uptake. Proc Natl Acad Sci USA. 1982;79:3542–6.
185. Borradale D, Kimlin M. Vitamin D in health and disease: an insight into traditional functions and new roles for the 'sunshine vitamin'. Nutr Res Rev. 2009;22:118–36.
186. Rao N, Eller M, Brougham T, Weir S. The effect of food on the relative bioavailability of deflazacort. Eur J Drug Metab Pharmacokinet. 1996;21:241–5.
187. Tembo AV, Sakmar E, Hallmark MR, Weidler DJ, Wagner JG. Effect of food on the bioavailability of prednisone. The Journal of Clinical Pharmacology. 1976;16(11):620–4.
188. Henderson RG, Wheatley T, English J, Chakraborty J, Marks V. Variation in plasma prednisolone concentrations in renal transplant recipients given enteric-coated prednisolone. British medical journal. 1979;1(6177):1534.
189. Uribe M, Schalm SW, Summerskill WHJ, Go VLW. Effect of liquid diet on serum protein binding and prednisolone concentrations after oral prednisone. Gastroenerology. 1976;71: 362–4.
190. Lee DA, Taylor GM, Walker JG, James VH. The effect of food and tablet formulation on plasma prednisolone levels following administration of enteric-coated tablets. British journal of clinical pharmacology. 1979;7(5):523–8.
191. Al-Habet S, Rogers HJ. Pharmacokinetics of intravenous and oral prednisolone. Br J Clin Pharmacol. 1980;10:503–8.
192. Hollander AA, van Rooij J, Lentjes GW, Arbouw F, van Bree JB, Schoemaker RC, van Es LA, van der Woude FJ, Cohen AF. The effect of grapefruit juice on cyclosporine and prednisone metabolism in transplant patients. Clinical Pharmacology & Therapeutics. 1995;57(3): 318–24.
193. McAllister WA, Thompson PJ, Al-Habet SM, Rogers HJ. Rifampicin reduces effectiveness and bioavailability of prednisolone. Br Med J. 1983;286:923–5.
194. Bailey CJ, Turner RC. Metformin. N Engl J Med. 1996;334(9):574.
195. Weatherspoon SE, Collins J, Sucharew H, Wong BL, Rybalsky I, Rose SR, et al. TP 51 Metformin reduces weight and BMI in Duchenne muscular dystrophy patients on long term glucocorticoid therapy. Neuromuscular Disorders. 2012;22(9):866.

196. Kibirige D, Mwebaze R. Vitamin B12 deficiency among patients with diabetes mellitus: is routine screening and supplementation justified. J Diabetes Metab Disord. 2013;12(1):17.
197. Bauman WA, Shaw S, Jayatilleke E, Spungen AM, Herbert V. Increased intake of calcium reverses vitamin B12 malabsorption induced by metformin. Diabetes Care. 2000;23(9): 1227–31.
198. Mazokopakis EE, Starakis IK. Recommendations for diagnosis and management of metformin-induced vitamin B12 (Cbl) deficiency. Diabetes Research and Clinical Practice. 2012;97(3):359–67.
199. Bell DS. Metformin-induced vitamin B12 deficiency presenting as a peripheral neuropathy. Southern Medical Journal. 2010;103(3):265–7.
200. Buvat DR. Use of metformin is a cause of vitamin B12 deficiency. Am Fam Physician. 2004;69:264.
201. Sahin M, Tutuncu NB, Ertugrul D, Tanaci N, Guvener ND. Effects of metformin or rosiglitazone on serum concentrations of homocysteine, folate, and vitamin B$<$sub$>$12$<$/sub$>$in patients with type 2 diabetes mellitus. Journal of Diabetes and Its Complications. 2007; 21(2):118–23.
202. Metzmann K, Schnell D, Jungnik A, Ring A, Theodor R, Hohl K, Meinicke T, Friedrich C. Effect of food and tablet-dissolution characteristics on the bioavailability of linagliptin fixed-dose combination with metformin: evidence from two randomized trials. Int J Clin Pharmacol Ther. 2014;52(7):549–63.
203. Houlihan CA, Allen TJ, Baxter AL, Panangiotopoulos S, Casley DJ, Cooper ME, Jerums G. A lowsodium diet potentiates the effects of losartan in type 2 diabetes. Diab Care. 2002;25(4):663–71.
204. Zaidenstein R, Soback S, Gips M, Avni B, Dishi V, Weissgarten Y, Golik A, Scapa E. Effect of grapefruit juice on the pharmacokinetics of losartan and its active metabolite E3174 in healthy volunteers. Therapeutic drug monitoring. 2001;23(4):369–73.
205. Singhvi SM, McKinstry DN, Shaw JM, Willard DA, Migdalof BH. Effect of food on the bioavailability of captopril in healthy subjects. The Journal of Clinical Pharmacology. 1982;22(2-3):135–40.
206. Patel AM, Majmudar F, Sharma N, Patel BN. Food effect on pharmacokinetic parameters of Losartan & its active metabolite. 2013.
207. Massarella JW, DeFeo TM, Brown AN, Lin A, Wills RJ. The influence of food on the pharmacokinetics and ACE inhibition of cilazapril. British Journal of Clinical Pharmacology. 1989;27(S2):205S–9.
208. Liedholm H, Melander A. Mechanisms and variations in the food effect on propranolol bioavailability. European Journal of Clinical Pharmacology. 1990;38(5):469–75.
209. Ogiso T, Iwaki M, Tanino T, Kawafuchi R, Hata S. Effect of food on propranolol oral clearance and a possible mechanism of this food effect. Biol Pharm Bull. 1994;17:112–6.
210. Semple HA, Xia F. Interaction between propranolol and amino acids in the single-pass isolated, perfused rat liver. Drug metabolism and disposition. 1995;23(8):794–8.
211. Gordon KE, Dooley JM, Sheppard KM, MacSween J, Esser MJ. Impact of bisphosphonates on survival for patients with Duchenne muscular dystrophy. Pediatrics. 2011;127(2): e353–8.
212. Laitinen K, Patronen A, Harju P, Löyttyniemi E, Pylkkänen L, Kleimola T, Perttunen K. Timing of food intake has a marked effect on the bioavailability of clodronate. Bone. 2000;27(2):293–6.
213. Jaffe KM, McDonald CM, Ingman E, Haas J. Symptoms of upper gastrointestinal dysfunction in Duchenne muscular dystrophy: case-control study. Archives of physical medicine and rehabilitation. 1990;71(10):742–4.
214. Food and Drug Administration. Guidance for industry: food-effect bioavailability and fed bioequivalence studies. Rockville, MD: Food and Drug Administration; 2002.
215. Kishi H, Kishi T, Folkers K. Bioenergetics in clinical medicine. III. Inhibition of coenzyme Q10-enzymes by clinically used anti-hypertensive drugs. Res Commun Chem Pathol Pharmacol. 1975;12:533–40.

Reports found at www.nap.edu

Biotin, and Choline; 1998. Dietary Reference Intakes for Vitamin C, Vitamin E, Selenium, and Carotenoids; 2000. Dietary Reference Intakes for Vitamin A, Vitamin K, Arsenic, Boron, Chromium, Copper, Iodine, Iron.

Dietary Reference Intakes for Calcium, Phosphorous, Magnesium, Vitamin D, and Fluoride. Dietary Reference Intakes for Thiamin, Riboflavin, Niacin, Vitamin B6, Folate, Vitamin B12, Pantothenic Acid; 1997.

http://lpi.oregonstate.edu/infocenter/minerals/magnesium/

Manganese, Molybdenum, Nickel, Silicon, Vanadium, and Zinc; 2001. Dietary Reference Intakes for Water, Potassium, Sodium, Chloride, and Sulfate; 2005. and Dietary Reference Intakes for Calcium and Vitamin D; 2011.

Chapter 10
Identifying Therapies for Muscle Disease Using Zebrafish

Elizabeth U. Parker and Lisa Maves

Abbreviations

BM	Bethlem myopathy
CsA	Cyclosporin A
DMD	Duchenne muscular dystrophy
FDA	United States Food and Drug Administration
MO	Morpholino
PDE	Phosphodiesterase
PTP	Mitochondrial permeability transition pore
SSRI	Selective serotonin uptake inhibitor
UCMD	Ullrich congenital muscular dystrophy

10.1 Introduction

Muscular dystrophies comprise a large family of genetic diseases characterized by muscle weakness and muscle degeneration [1, 2]. The zebrafish (*Danio rerio*) has emerged as a significant animal model not only for studying the mechanisms of muscular dystrophy disease but also for investigating drug therapies for muscular

E.U. Parker
Center for Developmental Biology and Regenerative Medicine,
Seattle Children's Research Institute, 1900 Ninth Avenue, Seattle, WA 98101, USA

L. Maves (✉)
Center for Developmental Biology and Regenerative Medicine, Seattle Children's
Research Institute, 1900 Ninth Avenue, Seattle, WA 98101, USA

Department of Pediatrics, University of Washington, Seattle, WA, USA
e-mail: lmaves@u.washington.edu

© Springer Science+Business Media New York 2016
M.K. Childers (ed.), *Regenerative Medicine for Degenerative Muscle Diseases*,
Stem Cell Biology and Regenerative Medicine,
DOI 10.1007/978-1-4939-3228-3_10

dystrophies [3–6]. Here we review recent progress in the use of zebrafish to identify potential drug therapies for two particularly severe forms of muscular dystrophy: Duchenne muscular dystrophy (DMD) and Ullrich congenital muscular dystrophy (UCMD).

DMD, the most common muscular dystrophy, is a severe, X-linked recessive condition affecting about 1 in 3600–6000 male live births [7]. DMD is caused by mutations in the dystrophin gene, which encodes a protein component of the dystrophin-glycoprotein complex that connects the cytoskeleton with the extracellular matrix [8, 9]. DMD affects both skeletal and cardiac muscle. The typical progression is as follows: gait abnormalities are usually noted by 3–4 years of age, loss of ambulation by 13 years if untreated, and, with increasing age, respiratory and cardiac complications progress and usually are lethal by the second or third decade [7, 10]. Glucocorticosteroids are the current standard of care, with treatment ideally beginning before the child begins significant decline, around age 4–6 years [7, 10]. While the glucocorticoids prednisone/prednisolone and deflazacort have proven successful at slowing the progression of the disease, chronic side effects are significant and prevalent [7, 10].

UCMD is a severe congenital muscular dystrophy caused by mutations in each of the three collagen VI genes, *COL6A1*, *COL6A2*, and *COL6A3* [11, 12]. Collagen VI is a critical component of the extracellular matrix surrounding skeletal muscle cells [11]. During infancy, UCMD patients can show skeletal muscle weakness, proximal joint contractures, and distal joint hyperflexibility, and rapid progression of the disease usually leads to respiratory failure [11, 12]. Current care includes respiratory and orthopedic management [12].

There is a current push to identify beneficial pharmacological interventions for both DMD and UCMD [11, 13, 14]. Ongoing research aimed at developing a cure for DMD through gene therapy or stem cell therapy is promising, but these treatments face many challenges and are not immediately forthcoming [13, 15, 16]. For both DMD and UCMD, animal models have been essential for understanding disease pathogenesis as well as for identifying potential pharmacological therapies [11, 17]. In particular, many different classes of drugs have demonstrated remarkable potential for ameliorating the DMD phenotype in animal models, including zebrafish [6, 17, 18]. For both DMD and UCMD, zebrafish provide critical models for testing pharmacological therapies because they reproduce key features of the human dystrophies, as we will review below [3, 6, 19].

10.2 Advantages of Zebrafish as a Model of Muscle Diseases

Several recent reviews have described the advantages of zebrafish as a model for human muscle diseases and for identifying pharmacological therapies for muscle diseases [3, 4, 6, 20]. In particular, we recently reviewed the many approaches available in zebrafish for analyzing skeletal muscle structure and function [6]. In this section of our chapter, we will highlight the general advantages of zebrafish for chemical treatments and chemical screening, and we will summarize the advantages of zebrafish for examining the structural and functional effects of skeletal muscle drug therapies and for generating genetic muscle disease models.

Zebrafish exhibit many general advantages as a model organism and show remarkable potential for novel drug therapy development, due to the ease of embryo manipulation, visualization, and pharmacological rescue [21, 22]. Zebrafish are notable for their high fecundity. Hundreds or even thousands of embryos can be obtained in a single day. Embryos are externally fertilized and transparent at embryonic stages, allowing for extensive observation of development. Zebrafish develop rapidly. Muscle disease phenotypes, such as defective muscle fiber structure or movements, can be observed as early as 1–2 days of development. Zebrafish embryos are raised in petri dishes and can absorb drugs that are added to the embryo medium. Pharmacological rescue experiments, such as those with the zebrafish DMD and UCMD models that we describe below, can be assayed in as few as 2–4 days, allowing extremely rapid assessment of drug treatments. Zebrafish embryos are small and can be raised in 96-well or even 384-well plates, allowing for large-scale chemical screening. Also, zebrafish muscle structural and functional phenotypes can be scored through simple assays (see below). Because of these advantages, zebrafish have become an important model for drug screening for many zebrafish models of human diseases, including muscle diseases [5, 6, 21].

One advantage of zebrafish as a model for muscle disease drug therapies is the ability to rapidly and quantitatively assess the effects of drug treatments on skeletal muscle structure (see also Ref. [6]). Zebrafish skeletal muscle is structurally very similar to human skeletal muscle [4]. The highly organized sarcomere pattern of zebrafish skeletal muscle is visible using polarized light, a property termed birefringence. A simple birefringence assay through a stereomicroscope can reveal muscle structural defects in zebrafish muscle disease models, such as the zebrafish *dmd* model ([3, 23]; Fig. 10.1). Zebrafish with muscle disease show dark patches, or

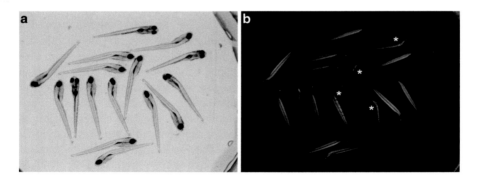

Fig. 10.1 5-day-old zebrafish *dmd* larvae. (**a, b**) Stereomicroscope views, using bright-field (**a**) and polarized light (**b**), of 5-day-old larvae from a cross of *dmd*+/− fish. About 25 % of the larvae from a *dmd*+/− cross are *dmd*−/− and exhibit skeletal muscle lesions, seen as disruptions of the muscle birefringence observed using polarized light (**b**). * in (**b**) label *dmd*−/− larvae with skeletal muscle lesions. See Kawahara et al. [24] and Berger and Currie [3]

muscle lesions, in the bright birefringence pattern due to muscle fiber detachment or degeneration ([3, 4]; Fig. 10.1). Skeletal muscle birefringence assays can be rapidly performed on live or fixed zebrafish larvae and can be used for large-scale drug screening [24–26]. Skeletal muscle can also be easily and rapidly visualized in live zebrafish larvae using transgenic skeletal muscle fluorescent reporters (for examples, see [25, 27]). To quantitate muscle structural defects, and the degree of pharmacological rescue, larvae can simply be scored as affected versus unaffected using a stereomicroscope to view birefringence or muscle transgene expression [24, 25]. Alternatively, more quantitative, but time-consuming, approaches can be employed, such as counting the number of body segments with muscle lesions per larva [25, 28]. Three studies have described highly quantitative approaches to scoring skeletal muscle birefringence in zebrafish larvae [26, 29, 30]. There are thus a range of quantitative approaches for assessing zebrafish muscle structure in muscle disease models and in chemical screens.

Another advantage of zebrafish as a model for muscle disease drug therapies is the ability to quantitatively assess the functional effects of drug treatments (see also [6]). Zebrafish have stereotypical embryonic and larval motor behaviors that are easily observed and measured [31, 32]. Several studies have used assays such as spontaneous embryo coiling, larval touch-evoked escape response, and video monitoring of swimming to measure motor function rescue of zebrafish muscle diseases after drug treatments [4, 19, 28, 33]. More sophisticated muscle function assays employ electrophysiological recordings from muscle in live larvae or from cultured myofibers [34, 35]. Beyond analyzing muscle function in zebrafish embryos and larvae, an important goal of testing pharmacological therapies in zebrafish muscle disease models is assessing the effects on animal survival. Kawahara et al. have shown this is possible in zebrafish by demonstrating improved survival of zebrafish *dmd* mutants following treatment with selected chemicals from their large-scale drug screens, as we discuss further below [24, 36].

The stereotyped muscle structural and movement patterns observed during zebrafish development have allowed for forward genetic screens for zebrafish mutant strains with muscle defects [3, 32, 37]). The characterization of zebrafish muscle mutant strains identified through genetic screens has provided invaluable animal models of human muscle diseases, including the zebrafish *dmd* strain (see below; [3, 4]). Recent reviews have documented known zebrafish models of human muscle diseases [3, 4, 20]. Besides forward genetic screens, other approaches for generating zebrafish models of human muscle diseases have included using antisense morpholinos, transgenic overexpression in zebrafish embryos, reverse genetics through TILLING, or genome editing technology [38–40]. As additional zebrafish mutant strains and muscle disease models are generated, they are documented at the zebrafish model organism database site (ZFIN; http://zfin.org; [41]) or at the Sanger Institute's Zebrafish Mutation Project site (http://www.sanger.ac.uk/sanger/Zebrafish_Zmpbrowse; [40]).

10.3   Using Zebrafish for DMD Drug Screening

Although many pharmacological compounds that show benefits to animal DMD models have already been identified [13, 14, 17], none of these, so far, have been successfully translated into patient therapies. Thus, there continues to be a need to identify pharmacologic treatment for DMD. The zebrafish *dmd* mutant reproduces key features of human DMD [42]. Zebrafish *dmd* mutants begin to exhibit muscle defects by 4 days and show severe, progressive muscle degeneration, fibrosis, inflammation, and motor defects (Granato et al. 1996; [27, 42]). Zebrafish *dmd* mutants die during larval stages, between about 10–30 days postfertilization [24, 42], whereas mouse DMD (*mdx*) mutants are viable and have mild muscle defects relative to human DMD patients [8]. The general advantages of zebrafish as a system for large-scale chemical screens, in addition to the specific advantages of screening muscle defects in live zebrafish *dmd* mutants (see above; [3, 6]), have made zebrafish an outstanding model for large-scale chemical screens for DMD.

Thus far, three large-scale chemical screens have been performed using the zebrafish *dmd* model [24, 36, 43]. The first two screens came from the Kunkel lab, and the third screen came from the Dowling lab. All three screens were designed to identify small molecules capable of suppressing the zebrafish *dystrophin*-null (*dmd*, also known as *sapje*) mutant phenotype. The *dmd* skeletal muscle phenotype can be observed as disruptions, or lesions, in the muscle birefringence pattern (see above; [24, 27, 44]; Fig. 10.1). All three screens took advantage of scoring the *dmd* muscle lesion phenotype through a simple, high-throughput skeletal muscle birefringence assay in 4-day-old larvae [24]. In zebrafish, the *dmd* gene exhibits autosomal recessive inheritance and is not sex-chromosome linked, as in humans and mice [45]. From a cross of heterozygote *dmd*+/− carriers, about 25 % of the larval progeny exhibit muscle lesions ([24, 27, 44]; Fig. 10.1). When larvae from a *dmd*+/− cross are treated with chemicals that ameliorate the skeletal muscle phenotype, significantly less than 25 % of the larvae show the *dmd* muscle lesion phenotype [24]. Following chemical treatments, larvae are then genotyped to confirm the percentages of phenotypically affected versus rescued *dmd*−/− larvae [24].

In their initial large-scale screen, Kawahara et al. [24] tested 1120 small molecules from the Prestwick library of bioreactive compounds approved for human use. More recently, Kawahara et al. [36] have expanded their drug screening on *dmd* zebrafish with an additional 1520 chemicals from the NINDS 2 compound library (1040 chemicals) and the Institute of Chemistry and Cell Biology (ICCB) bioactive molecule library (480 chemicals). The Dowling lab screened a library of 640 FDA-approved drugs from ENZO Biomol [43]. In total, these three screens thus tested 3280 small molecules.

From their initial large-scale screen, Kawahara et al. identified 7 chemicals that ameliorated the zebrafish *dmd* skeletal muscle lesions ([24]; Table 10.1). The chemical that both promoted robust amelioration of the *dmd* skeletal muscle lesion phenotype and, in a secondary long-term drug treatment assay, promoted the highest long-term survival in *dmd* fish was aminophylline, a nonselective phosphodiesterase

Table 10.1 Chemicals that ameliorate skeletal muscle lesions in *dmd* zebrafish

Chemical	Drug class	References
9a,11b-Prostaglandin F2	Vasoconstrictor	[36]
Aminophylline	Nonselective PDE inhibitor	[24, 43]
Androsterone acetate	Steroid	[36]
Ataluren	Nonaminoglycoside nonsense mutation suppressor	[46]
Cerulenin	Antibiotic and antifungal	[36]
Conessine	Anti-allergic agent	[24]
Crassin acetate	Immunosuppressant, cytotoxin	[36]
Dipyridamole	PDE5 inhibitor	[24]
Epirizole	Anti-inflammatory agent	[24]
Equilin	Steroid	[24]
Ergotamine	Monoamine agonist	[43]
Flunarizine	Calcium channel antagonist	[43]
Fluoxetine	Monoamine agonist	[43]
Homochlorcyclizine dihydrochloride	Anti-allergic agent	[24]
Ibudilast	PDE4 inhibitor	[24]
MG123	Proteasomal inhibitor	[30]
Nitromide	Antibacterial	[36]
Pentetic acid	Chelating agent	[24]
Pergolide	Monoamine agonist	[43]
Pomiferin	Flavonoid, antioxidant	[36]
Propantheline bromide	Acetylcholine receptor antagonist	[36]
Proscillaridin A	Cardiotonic agent	[24]
Rolipram	PDE4 inhibitor	[24]
Ropinirole	Monoamine agonist	[43]
Serotonin	Monoamine neurotransmitter	[43]
Sildenafil citrate	PDE5 inhibitor	[24, 36]
Trichostatin A	Histone deacetylase inhibitor	[25]

(PDE) inhibitor [24]. Kawahara et al. further tested a set of six additional PDE inhibitors with different specificities [24]. They found that sildenafil citrate, a PDE5 inhibitor, along with three other PDE4 or PDE5 inhibitors, also could ameliorate the *dmd* skeletal muscle lesion phenotype, similar to aminophylline ([24]; Table 10.1). Because the PDE5 inhibitors sildenafil and tadalafil have been shown to ameliorate the mouse *mdx* model [47–49] and have shown promising effects in DMD patient trials [50], this initial large-scale screen demonstrated that drug screening with *dmd* zebrafish is very relevant to identifying potential DMD therapies.

Additional chemical library screening allowed Kawahara et al. to identify 8 more chemicals that ameliorated the zebrafish *dmd* skeletal muscle lesions, including sildenafil citrate ([36]; Table 10.1). Out of the 18 total drugs identified from both of the Kawahara et al. screens, six compounds target various components of the heme oxygenase signaling pathway [36]. Kawahara et al. showed that four of these drugs, aminophylline, sildenafil citrate, cerulenin, and crassin acetate, indeed act through upregulation of heme oxygenase 1 (Hmox1) by showing that the drugs could not rescue the zebrafish *dmd* phenotype when Hmox1 was knocked down [36]. Overexpression of Hmox1 was sufficient to rescue the zebrafish *dmd* skeletal muscle lesions [36]. Additionally, sildenafil treatments that are beneficial to both *dmd* zebrafish and *mdx* mice increase Hmox1 expression [36]. Thus, these studies reveal Hmox1 as a conserved, novel potential target for DMD therapy.

The Dowling lab recently performed another large-scale chemical screen on *dmd* zebrafish [43]. They screened 640 drugs and identified 6 that improve the zebrafish *dmd* muscle lesions ([43]; Table 10.1), including aminophylline, which was also identified in the Kawahara screen [24]. In addition to confirming the beneficial effects of the PDE inhibitor aminophylline, Waugh et al. also identified a new class of beneficial drugs, monoamine agonists, in particular ergotamine, pergolide, and fluoxetine [43]. They then performed a secondary screen of six additional monoamine agonists and found serotonin could also improve the zebrafish *dmd* skeletal muscle lesion phenotype [43]. The ability of both serotonin and fluoxetine, a selective serotonin uptake inhibitor (SSRI), to improve the zebrafish *dmd* phenotype implicates the serotonin pathway as a therapeutic target for DMD. Waugh et al. went on to perform a transcriptome analysis on fluoxetine-treated versus untreated *dmd* zebrafish [43]. Many of the genes that are differentially expressed between untreated wild type and untreated *dmd* larvae are reversed by fluoxetine treatment, in particular genes associated with calcium homeostasis [43]. This suggests that a possible mechanism of fluoxetine activity is modulating calcium homeostasis [43]. Because SSRIs like fluoxetine are already FDA-approved and are already used in clinical practice, this study in zebrafish points to an important pathway for further therapeutic investigation for DMD.

Together, these large-scale screens highlight the potential of *dmd* zebrafish for identifying new therapeutic compounds and targets for DMD. The mechanistic investigations in these studies also highlight the potential of zebrafish for understanding the molecular mechanisms behind muscle disease and pharmacological treatments. Together, these three chemical screens identified 24 compounds that rescue the zebrafish *dmd* mutant muscle lesion phenotype (Table 10.1). One overlapping finding across these three screens was the identification of PDE inhibitors as beneficial for the *dmd* phenotype. The PDE5 inhibitors sildenafil and tadalafil can ameliorate the mouse *mdx* model [47–49] and have shown promising effects in a clinical trial with young DMD patients [50], although studies in adult DMD and Becker muscular dystrophy patients have not been as promising [51, 52]. Nevertheless, taken together, these studies appear to provide very strong support for PDE inhibitors as possible DMD therapies. Further investigations of PDE-inhibitor treatments on *dmd* zebrafish should allow further insight into the mechanisms of these drugs and perhaps the optimal timing of their administration.

Additional studies have directly tested whether compounds that are known to ameliorate the mouse *mdx* model can also ameliorate the zebrafish *dmd* model. These studies tested the proteosomal inhibitor MG132, the histone deacetylase

(HDAC) inhibitor Trichostatin A (TSA), and ataluren, a drug that can cause read-through of premature stop codons [25, 30, 46]. These drugs all show the ability to rescue the skeletal muscle lesion phenotype in the zebrafish *dmd* mutant ([25, 30, 46]; Table 10.1). Table 10.1 summarizes the list of drugs that have shown beneficial effects in *dmd* zebrafish. The individual drug studies, along with the large-scale screens, further underscore that zebrafish is a suitable model for testing pharmacological therapies for DMD.

10.4 Using Zebrafish to Test Pharmacological Treatments for Collagen VI Myopathies

The zebrafish system has also been useful for directly testing the benefits of specific candidate drug therapies for collagen VI myopathies. Collagen VI is a critical component of the extracellular matrix surrounding skeletal muscle fibers. Collagen VI is made up of subunits of collagen alpha-1(VI), collagen alpha-2(VI), and collagen alpha-3(VI). Mutations in any of the three genes encoding these subunits, *COL6A1*, *COL6A2* and *COL6A3*, can lead to muscle disease, in particular Ullrich congenital muscular dystrophy (UCMD) and the less severe Bethlem myopathy (BM) [11, 12]. Both dominant and recessive mutations in the *COL6A1-3* genes can lead to UCMD and BM [12]. Typically these mutations lead to a reduction in the amount of secreted collagen VI or impaired assembly of functional collagen VI microfibrils [11].

To better understand the contribution of collagen VI to these myopathies, a mouse *Col6a1* knockout model was generated [53]. The *Col6a1−/−* mice show necrotic muscle fibers but have only mild muscle functional impairment, thus representing a model of the less severe BM [53, 54]. In spite of the mild phenotype, the *Col6a1 −/−* mice have revealed significant insights into the pathogenesis of collagen VI muscular dystrophies. *Col6a1−/−* mice show mitochondrial defects, in particular dysfunction of the mitochondrial permeability transition pore (PTP; [54]) and defective autophagy [55]. Further studies in patient biopsy samples confirmed a common pathogenesis in collagen VI myopathies of muscle cell death and defective mitochondria and autophagy [11]. The mitochondrial PTP defects led to investigations of cyclosporin A (CsA), a potent PTP inhibitor, as a potential pharmacological therapy for collagen VI myopathies [11, 54]. Treatment with CsA improves the mitochondrial defects and associated apoptosis in the mouse *Col6a1 −/−* mouse model and in muscle cell cultures from UCMD patients [11, 54]. However, because of the immunosuppressive side effects of CsA, alternative therapies continue to be pursued.

Recent studies have turned to the zebrafish collagen VI myopathy model to investigate CsA-related drugs as therapies. The zebrafish model uses an antisense morpholino (MO) to specifically remove an N-terminal domain of Col6a1, mimicking a common, severe mutation found in human UCMD patients [19]. The zebrafish *col6a1*-MO embryos exhibit motor defects and skeletal muscle structural defects, including abnormal mitochondria, consistent with a severe UCMD-like phenotype [19]. One significant advantage of the zebrafish model for investigating drug therapies is that the fish show both clear muscle structural and functional defects, whereas the mouse *Col6a1* knockout model has very subtle functional defects. Treating zebrafish *col6a1*-MO embryos with CsA improves the UCMD-like mitochondrial, apoptosis, and motor defects [19]. However, CsA treatments did not improve overall muscle fiber integrity [19]. A more recent study has tested whether additional CsA-related drugs can ameliorate the zebrafish *col6a1*-MO phenotype, in particular NIM811, which lacks immunosuppressive activity, and FK506 [56]. When *col6a1*-MO zebrafish are treated with NIM811, the treated larvae show greatly improved muscle structure, mitochondrial structure, and motor function [56]. Importantly, this study performed a side-by-side comparison of NIM811 with CsA and FK506 on *col6a1*-MO zebrafish larvae, and the results showed that NIM811 performed much better than the other drugs in ameliorating the zebrafish *col6a1*-MO phenotype [56]. This study went on to confirm the beneficial effects of NIM811 on *Col6a1*−/− mice and UCMD and BM patient cell culture models [56]. NIM811 is now a promising potential therapy for UCMD and BM patients. Further studies will likely investigate whether there are additional compounds that could provide benefits for collagen VI myopathies without significant side effects. The zebrafish *col6a1* model offers an excellent platform with which to perform initial testing of additional drugs.

10.5   Zebrafish as a Preclinical Drug Testing Model for Muscle Disease Therapies

To increase the probability of success for translation of pharmacological therapies, preclinical studies should be performed in multiple animal models [17]. Mammalian animal models, particularly mouse and dog models, have been extensively used to identify and assess pharmacological therapies for muscle diseases (Kornegay et al. [17]; please also see Chaps. 10–20 in this book). Zebrafish have already been demonstrated to be an excellent model for screening to identify new muscle disease drug therapies [24, 36, 43]. As a complement to mammalian models, the zebrafish system has great potential to also serve as a preclinical translation model for evaluating optimal drug therapies for DMD and other muscle diseases.

One advantage of using zebrafish for assessing pharmacological approaches for muscle diseases is the ease of performing side-by-side drug comparison and combination studies. Through the use of animal DMD models, many promising alternative pharmacological approaches have already been identified that benefit DMD by modulating pathological mechanisms downstream of the DMD mutation [6, 14, 17]. The current challenge is to determine how best to translate preclinical pharmacological studies to DMD patients. One critical issue is that, because many DMD patients take corticosteroids as well as other medications, drug-drug interactions are a concern. New pharmacological therapies should be compared to and tested for efficacy in the presence of these current treatments [10, 50, 57]. In mammalian animal models, drug comparison and combination studies are possible but face many challenges [57]. Zebrafish are an ideal model for testing new drug combinations and for optimizing the timing and relative doses of drugs alone and in combinations because of the ability to obtain many animals and rapidly test the effects of drug treatments.

In order to ensure rigorous, reliable, and reproducible results from drug treatment studies in zebrafish, as in any animal model, it is important to establish standard protocols for drug treatments and assessment of muscle disease amelioration. There are several variables that need to be considered when performing drug treatment studies with zebrafish. These variables include drug source, drug stock storage, embryo bath medium, vehicle used, drug concentration, density of embryos, genetic background of fish, timing of treatments, frequency of treatment/drug replacement, and whether or not the zebrafish embryos are dechorionated (egg shells removed) before drug treatment. One factor that must be incorporated into zebrafish drug treatment studies is genotyping of animals following treatments [24, 25, 36]. For any zebrafish drug treatments that show promise for testing in mammalian models, having another zebrafish laboratory independently confirm critical results would directly demonstrate reproducibility prior to moving to mammalian models. Standardizing these variables and approaches will further strengthen the use of zebrafish as a platform for assessing the therapeutic efficacy of muscle disease drug treatments.

10.6   Future Prospects

Zebrafish will continue to provide a critical model for investigating pharmacological therapies for DMD. The zebrafish system also provides outstanding potential as a screening tool for drug discovery for additional muscle diseases. Many zebrafish models of human muscle diseases (reviewed in Ref. [3, 4]), identified through forward genetic screens, are already available for use in chemical screens, similar to the screens performed with the *dmd* strain [24, 36, 43]. Now, genome editing technologies, in particular TALENs and the CRISPR-Cas9 system [38, 39], allow the potential to engineer zebrafish models of human muscle diseases that have not yet been identified through forward genetics. In particular, a very promising use of genome editing approaches will be to engineer zebrafish strains with muscle disease gene alleles that mimic specific mutations associated with muscle diseases in human patients [38, 39]. Genome-edited zebrafish strains would test whether specific patient-associated DNA sequence variants do indeed lead to muscle disease in an animal model. These engineered strains could then be used in chemical screens to directly screen for drugs that ameliorate the zebrafish muscle phenotypes caused by the human muscle disease-associated mutations. Thus, the zebrafish model is well poised to continue to provide significant insight into drug discovery, drug treatments, and the amelioration of muscle degeneration characteristic of DMD and other muscle diseases.

Acknowledgments Funding for work on zebrafish skeletal muscle disease in the Maves lab comes from the Seattle Children's Research Institute Myocardial Regeneration Initiative and NIH R03 AR065760. EP was supported by the University of Washington School of Medicine Medical Student Research Training Program.

References

1. Dalkilic I, Kunkel LM. Muscular dystrophies: genes to pathogenesis. Curr Opin Genet Dev. 2003;13:231–8.
2. Wallace GQ, McNally EM. Mechanisms of muscle degeneration, regeneration, and repair in the muscular dystrophies. Annu Rev Physiol. 2009;71:37–57.
3. Berger J, Currie PD. Zebrafish models flex their muscles to shed light on muscular dystrophies. Dis Model Mech. 2012;5:726–32.
4. Gibbs EM, Horstick EJ, Dowling JJ. Swimming into prominence: the zebrafish as a valuable tool for studying human myopathies and muscular dystrophies. FEBS J. 2013;280:4187–97.
5. Kawahara G, Kunkel LM. Zebrafish based small molecule screens for novel DMD drugs. Drug Discov Today Technol. 2013;10:e91–6.
6. Maves L. Recent advances using zebrafish animal models for muscle disease drug discovery. Expert Opin Drug Discov. 2014;14:1–13.
7. Bushby K, Finkel R, Birnkrant DJ, Case LE, Clemens PR, Cripe L, Kaul A, Kinnett K, McDonald C, Pandya S, Poysky J, Shapiro F, Tomezsko J, Constantin C, DMD Care Considerations Working Group. Diagnosis and management of Duchenne muscular dystrophy, part 1: diagnosis, and pharmacological and psychosocial management. Lancet Neurol. 2010;9: 77–93.
8. Hoffman EP, Brown Jr RH, Kunkel LM. Dystrophin: the protein product of the Duchenne muscular dystrophy locus. Cell. 1987;51:919–28.
9. Rahimov F, Kunkel LM. The cell biology of disease: cellular and molecular mechanisms underlying muscular dystrophy. J Cell Biol. 2013;201:499–510.
10. Goemans N, Buyse G. Current treatment and management of dystrophinopathies. Curr Treat Options Neurol. 2014;16:287.
11. Bernardi P, Bonaldo P. Mitochondrial dysfunction and defective autophagy in the pathogenesis of collagen VI muscular dystrophies. Cold Spring Harbor Perspect Biol. 2013;5:a011387.
12. Bushby KM, Collins J, Hicks D. Collagen type VI myopathies. Adv Exp Med Biol. 2014;802:185–99.
13. Govoni A, Magri F, Brajkovic S, Zanetta C, Faravelli I, Corti S, Bresolin N, Comi GP. Ongoing therapeutic trials and outcome measures for Duchenne muscular dystrophy. Cell Mol Life Sci. 2013;70:4585–602.
14. Ruegg UT. Pharmacological prospects in the treatment of Duchenne muscular dystrophy. Curr Opin Neurol. 2013;26:577–84.
15. Fairclough RJ, Wood MJ, Davies KE. Therapy for Duchenne muscular dystrophy: renewed optimism from genetic approaches. Nat Rev Genet. 2013;14:373–8.
16. Seto JT, Bengtsson NE, Chamberlain JS. Therapy of genetic disorders-novel therapies for Duchenne muscular dystrophy. Curr Pediatr Rep. 2014;2:102–12.
17. Kornegay JN, Spurney CF, Nghiem PP, Brinkmeyer-Langford CL, Hoffman EP, Nagaraju K. Pharmacologic management of Duchenne muscular dystrophy: target identification and preclinical trials. ILAR J. 2014;55:119–49.
18. Ljubicic V, Jasmin BJ. AMP-activated protein kinase at the nexus of therapeutic skeletal muscle plasticity in Duchenne muscular dystrophy. Trends Mol Med. 2013;19:614–24.
19. Telfer WR, Busta AS, Bonnemann CG, Feldman EL, Dowling JJ. Zebrafish models of collagen VI-related myopathies. Hum Mol Genet. 2010;19:2433–44.

20. Lin YY. Muscle diseases in the zebrafish. Neuromuscul Disord. 2012;22:673–84.
21. Santoriello C, Zon LI. Hooked! Modeling human disease in zebrafish. J Clin Invest. 2012;122:2337–43.
22. Zon LI, Peterson RT. In vivo drug discovery in the zebrafish. Nat Rev Drug Discov. 2005;4:35–44.
23. Felsenfeld AL, Walker C, Westerfield M, Kimmel C, Streisinger G. Mutations affecting skeletal muscle myofibril structure in the zebrafish. Development. 1990;108:443–59.
24. Kawahara G, Karpf JA, Myers JA, Alexander MS, Guyon JR, Kunkel LM. Drug screening in a zebrafish model of Duchenne muscular dystrophy. Proc Natl Acad Sci USA. 2011;108: 5331–6.
25. Johnson NM, Farr, GH 3rd, Maves L. The HDAC inhibitor TSA ameliorates a zebrafish model of Duchenne muscular dystrophy. PLoS Curr. 2013;5. pii: ecurrents.md.8273cf41db10e2d15d d3ab827cb4b027
26. Smith LL, Beggs AH, Gupta VA. Analysis of skeletal muscle defects in larval zebrafish by birefringence and touch-evoke escape response assays. J Vis Exp. 2013;82, e50925.
27. Bassett DI, Bryson-Richardson RJ, Daggett DF, Gautier P, Keenan DG, Currie PD. Dystrophin is required for the formation of stable muscle attachments in the zebrafish embryo. Development. 2003;130:5851–60.
28. Goody MF, Kelly MW, Reynolds CJ, Khalil A, Crawford BD, Henry CA. NAD+ biosynthesis ameliorates a zebrafish model of muscular dystrophy. PLoS Biol. 2012;10, e1001409.
29. Berger J, Sztal T, Currie PD. Quantification of birefringence readily measures the level of muscle damage in zebrafish. Biochem Biophys Res Commun. 2012;423:785–8.
30. Winder SJ, Lipscomb L, Angela Parkin C, Juusola M. The proteasomal inhibitor MG132 prevents muscular dystrophy in zebrafish. PLoS Curr. 2011;3:RRRN1286.
31. Kimmel CB, Patterson J, Kimmel RO. The development and behavioral characteristics of the startle response in the zebra fish. Dev Psychobiol. 1974;7:47–60.
32. Saint-Amant L, Drapeau P. Time course of the development of motor behaviors in the zebrafish embryo. J Neurobiol. 1998;37:622–32.
33. Dowling JJ, Arbogast S, Hur J, Nelson DD, McEvoy A, Waugh T, Marty I, Lunardi J, Brooks SV, Kuwada JY, Ferreiro A. Oxidative stress and successful antioxidant treatment in models of RYR1-related myopathy. Brain. 2012;135:1115–27.
34. Hirata H, Watanabe T, Hatakeyama J, Sprague SM, Saint-Amant L, Nagashima A, Cui WW, Zhou W, Kuwada JY. Zebrafish relatively relaxed mutants have a ryanodine receptor defect, show slow swimming and provide a model of multi-minicore disease. Development. 2007;134: 2771–81.
35. Horstick EJ, Gibbs EM, Li X, Davidson AE, Dowling JJ. Analysis of embryonic and larval zebrafish skeletal myofibers from dissociated preparations. J Vis Exp. 2013;81, e50259.
36. Kawahara G, Gasperini MJ, Myers JA, Widrick JJ, Eran A, Serafini PR, Alexander MS, Pletcher MT, Morris CA, Kunkel LM. Dystrophic muscle improvement in zebrafish via increased heme oxygenase signaling. Hum Mol Genet. 2014;23:1869–78.
37. Granato M, van Eeden FJ, Schach U, Trowe T, Brand M, Furutani-Seiki M, Haffter P, Hammerschmidt M, Heisenberg CP, Jiang YJ, Kane DA, Kelsh RN, Mullins MC, Odenthal J, Nüsslein-Volhard C. Genes controlling and mediating locomotion behavior of the zebrafish embryo and larva. Development. 1996;123:399–413.
38. Blackburn PR, Campbell JM, Clark KJ, Ekker SC. The CRISPR system--keeping zebrafish gene targeting fresh. Zebrafish. 2013;10:116–8.
39. Campbell JM, Hartjes KA, Nelson TJ, Xu X, Ekker SC. New and TALENted genome engineering toolbox. Circ Res. 2013;113:571–87.
40. Kettleborough RN, Busch-Nentwich EM, Harvey SA, Dooley CM, de Bruijn E, van Eeden F, Sealy I, White RJ, Herd C, Nijman IJ, Fényes F, Mehroke S, Scahill C, Gibbons R, Wali N, Carruthers S, Hall A, Yen J, Cuppen E, Stemple DL. A systematic genome-wide analysis of zebrafish protein-coding gene function. Nature. 2013;496:494–7.
41. Bradford Y, Conlin T, Dunn N, Fashena D, Frazer K, Howe DG, Knight J, Mani P, Martin R, Moxon SA, Paddock H, Pich C, Ramachandran S, Ruef BJ, Ruzicka L, Bauer Schaper H,

Schaper K, Shao X, Singer A, Sprague J, Sprunger B, Van Slyke C, Westerfield M. ZFIN: enhancements and updates to the Zebrafish Model Organism Database. Nucleic Acids Res. 2011;39:D822–9.
42. Berger J, Berger S, Hall TE, Lieschke GJ, Currie PD. Dystrophin-deficient zebrafish feature aspects of the Duchenne muscular dystrophy pathology. Neuromuscul Disord. 2010;20: 826–32.
43. Waugh TA, Horstick E, Hur J, Jackson SW, Davidson AE, Li X, Dowling JJ. Fluoxetine prevents dystrophic changes in a zebrafish model of Duchenne muscular dystrophy. Hum Mol Genet. 2014;23:4651–62.
44. Guyon JR, Mosley AN, Zhou Y, O'Brien KF, Sheng X, Chiang K, Davidson AJ, Volinski JM, Zon LI, Kunkel LM. The dystrophin associated protein complex in zebrafish. Hum Mol Genet. 2003;12:601–15.
45. Steffen LS, Guyon JR, Vogel ED, Beltre R, Pusack TJ, Zhou Y, Zon LI, Kunkel LM. Zebrafish orthologs of human muscular dystrophy genes. BMC Genomics. 2007;8:79.
46. Li M, Andersson-Lendahl M, Sejersen T, Arner A. Muscle dysfunction and structural defects of dystrophin-null sapje mutant zebrafish larvae are rescued by ataluren treatment. FASEB J. 2014;28:1593–9.
47. Adamo CM, Dai DF, Percival JM, Minami E, Willis MS, Patrucco E, Froehner SC, Beavo JA. Sildenafil reverses cardiac dysfunction in the mdx mouse model of Duchenne muscular dystrophy. Proc Natl Acad Sci USA. 2010;107:19079–83.
48. Asai A, Sahani N, Kaneki M, Ouchi Y, Martyn JA, Yasuhara SE. Primary role of functional ischemia, quantitative evidence for the two-hit mechanism, and phosphodiesterase-5 inhibitor therapy in mouse muscular dystrophy. PLoS ONE. 2007;2, e806.
49. Percival JM, Whitehead NP, Adams ME, Adamo CM, Beavo JA, Froehner SC. Sildenafil reduces respiratory muscle weakness and fibrosis in the mdx mouse model of Duchenne muscular dystrophy. J Pathol. 2012;228:77–87.
50. Nelson MD, Rader F, Tang X, Tavyev J, Nelson SF, Miceli MC, Elashoff RM, Sweeney HL, Victor RG. PDE5 inhibition alleviates functional muscle ischemia in boys with Duchenne muscular dystrophy. Neurology. 2014;82:2085–91.
51. Leung DG, Herzka DA, Thompson WR, He B, Bibat G, Tennekoon G, Russell SD, Schuleri KH, Lardo AC, Kass DA, Thompson RE, Judge DP, Wagner KR. Sildenafil does not improve cardiomyopathy in Duchenne/Becker muscular dystrophy. Ann Neurol. 2014;76:541–9.
52. Witting N, Kruuse C, Nyhuus B, Prahm KP, Citirak G, Lundgaard SJ, von Huth S, Vejlstrup N, Lindberg U, Krag TO, Vissing J. Effect of sildenafil on skeletal and cardiac muscle in Becker muscular dystrophy. Ann Neurol. 2014;76:550–7.
53. Bonaldo P, Braghetta P, Zanetti M, Piccolo S, Volpin D, Bressan GM. Collagen VI deficiency induces early onset myopathy in the mouse: an animal model for Bethlem myopathy. Hum Mol Genet. 1998;7:2135–40.
54. Irwin WA, Bergamin N, Sabatelli P, Reggiani C, Megighian A, Merlini L, Braghetta P, Columbaro M, Volpin D, Bressan GM, Bernardi P, Bonaldo P. Mitochondrial dysfunction and apoptosis in myopathic mice with collagen VI deficiency. Nat Genet. 2003;35:367–71.
55. Grumati P, Coletto L, Sabatelli P, Cescon M, Angelin A, Bertaggia E, Blaauw B, Urciuolo A, Tiepolo T, Merlini L, Maraldi NM, Bernardi P, Sandri M, Bonaldo P. Autophagy is defective in collagen VI muscular dystrophies, and its reactivation rescues myofiber degeneration. Nat Med. 2010;16:1313–20.
56. Zulian A, Rizzo E, Schiavone M, Palma E, Tagliavini F, Blaauw B, Merlini L, Maraldi NM, Sabatelli P, Braghetta P, Bonaldo P, Argenton F, Bernardi P. NIM811, a cyclophilin inhibitor without immunosuppressive activity, is beneficial in collagen VI congenital muscular dystrophy models. Hum Mol Genet. 2014;23:5353–63.
57. Janssen PM, Murray JD, Schill KE, Rastogi N, Schultz EJ, Tran T, Raman SV, Rafael-Fortney JA. Prednisolone attenuates improvement of cardiac and skeletal contractile function and histopathology by lisinopril and spironolactone in the mdx mouse model of Duchenne muscular dystrophy. PLoS ONE. 2014;9, e88360.

Chapter 11
miRNAs in Muscle Diseases

Diem-Hang Nguyen-Tran and Hannele Ruohola-Baker

11.1 Introduction

11.1.1 What Are miRNAs?

miRNAs are endogenously expressed small RNAs (21–23 nucleotides long) that complement with target mRNAs and mediate the repression of target gene expression [1]. miRNAs are first produced as pri-miRNAs which are then processed into pre-miRNAs by Drosha in the nucleus. Pre-miRNAs are exported into cytoplasm by Exportin 5 and processed into miRNA/miRNA* duplex by Dicer. Finally mature miRNAs and protein complexes are formed where miRNAs guide proteins to target mRNAs.

miRNAs are expressed in specific cell types and in several stages of development. There are more than 2500 mature miRNAs encoded by human genome (MirBase).

D.-H. Nguyen-Tran
Departments of Biochemistry, Biology, Bioengineering, Genome Sciences,
Institute for Stem Cell and Regenerative Medicine, University of Washington,
School of Medicine, Seattle, WA 98109, USA

Department of Neurosurgery, McKnight Brain Institute, University of Florida,
Gainesville, FL 32608, USA

H. Ruohola-Baker (✉)
Departments of Biochemistry, Biology, Bioengineering, Genome Sciences,
Institute for Stem Cell and Regenerative Medicine, University of Washington,
School of Medicine, Seattle, WA 98109, USA
e-mail: hannele@uw.edu

© Springer Science+Business Media New York 2016   295
M.K. Childers (ed.), *Regenerative Medicine for Degenerative Muscle Diseases*,
Stem Cell Biology and Regenerative Medicine,
DOI 10.1007/978-1-4939-3228-3_11

11.1.2 Muscle-Specific and Non-muscle-specific miRNAs in Skeletal Muscles

miRNAs that take crucial part in modulating muscular metabolism and cellular commitment are not necessarily expressed exclusively in muscular tissues. There are miRNAs that are expressed in muscle and other tissues and play dominant roles in regulating pathways important in muscle proliferation and differentiation [2]. Hence, miRNAs important for muscle function can be divided into two categories: muscle-specific and non-muscle-specific miRNAs.

11.1.2.1 Muscle-Specific miRNAs

Among muscle-specific miRNAs, the ones that are well studied and defined are miR-1, miR-133a/b, and miR-206 [3]. They enhance muscle differentiation or myoblast proliferation [3], while miR-208b and miR-499 have a key role in muscle fiber type identity, shifting, and determination [4].

11.1.2.2 Non-muscle-specific miRNAs

The ubiquitously expressed miRNAs mostly promote muscle differentiation (miR-24, miR-26a, miR-27b, miR-29, miR-146a, and miRNA-214) [5–13], while a few of them repress myoblast differentiation and muscle regeneration (miR-125b and miRNA-155) [14–17].

11.1.3 miRNAs That Are Upregulated and Downregulated in Muscle Diseases

miRNAs may have a significant impact on muscle diseases, as evident in their dramatic change of expression in the pathological conditions.

Having carefully analyzed the most devastating muscle disease, Duchenne muscular dystrophy (DMD), Greco et al. learned that the upregulated miRNAs in this disease are regenerative miRNAs (miR-31, miR-34c, miR-206, miR-335, miR-449, and miR-494) and inflammatory miRNAs (miR-222 and miR-223), while the downregulated ones are degenerative miRNAs (miR-29c, and miR-135a) [18].

Not only in DMD, a number of miRNAs (miR-146b, miR-221, miR-155, miR-214, and miR-222) are also consistently deregulated in other primary muscular disorders (MDs) (Becker muscular dystrophy, limb-girdle muscular dystrophy, facioscapulohumeral muscular dystrophy) [19].

The correlation between miRNA expression and MDs points out the importance of understanding and studying miRNA roles in specific muscle metabolism and processes to find a treatment for MDs.

11.2 miRNA Roles in Muscle Regeneration (Proliferation and Differentiation)

As stated in previous sections, regenerative miRNAs are upregulated in DMD. This suggests that regeneration is boosted up in DMD to make up for the wasted muscles. How do miRNAs then contribute to muscle regeneration?

In muscle disorders, muscle is injured/wasted, generating a need for replacing degenerated muscle through a process called muscle regeneration. When there is a myotrauma, satellite cells, the skeletal muscle stem cells in the adult body, are activated to proliferate and to generate differentiating myoblasts. miRNAs regulate all these stages from keeping satellite cells in their quiescence state to myoblast proliferation and differentiation.

11.2.1 miRNAs in Maintaining Satellite Cell Quiescent State

In 2012, Crist et al. showed that miR-31 maintains satellite cells in quiescent state by downregulating Myf5 [20]. Myf5 is a myogenic determination protein that activates myogenic program. In quiescent satellite cells, Myf5 is still transcribed but is not translated because Myf5 mRNAs are sequestrated along with its antagonist, miR-31, in mRNP granules. Once satellite cells are activated, mRNP granules are dissolved, releasing Myf5 mRNAs from the concentrated miR-31, thus allowing their translation and protein expression [20].

11.2.2 miRNAs in Muscle Proliferation

Once satellite cells get activated, they start to proliferate. Several miRNAs have been found to positively regulate this proliferation process. One of these miRNAs is miR-133 that represses the gene expression of SRF (serum response factor), thereby enhances myoblast proliferation and inhibits differentiation [21, 22]. Recently, McFarlane et al. have added another miRNA, miR-27a/b, to the myoblast proliferative category [23]. The authors stated that miR-27a/b enhances myoblast proliferation by targeting myostatin, an inhibitor of myogenesis. miR-27a/b overexpression leads to myostatin downregulation and then skeletal muscle hypertrophy as a consequence. On the other hand, when miR-27a/b is downregulated, myostatin is highly expressed, leading to reduced myoblast proliferation [23].

11.2.3   miRNAs in Muscle Differentiation

Muscle differentiation follows muscle proliferation to create mature muscle cells [24]. A lot of miRNAs have been shown to be involved in this process, but the most well-known ones are miR-1 and miR-206. miR-1 and miR-206 target not only one but a number of genes to enhance muscle differentiation. First, they repress HDAC4 that is supposed to inhibit MEF2, a myocyte enhancer factor. Thus, miR-1/miR-206 release MEF2 to promote muscle differentiation [21, 25]. Moreover, MiR-1/miR-206 and miR-486 also repress Pax7 [25, 26]. Pax7 is a quiescence marker and also a paired box transcription factor that downregulates the level of MyoD, a key protein in muscle differentiation. Therefore, a miR-1/miR-206- and miR-486-inhibiting action on Pax7 leads to muscle differentiation. Furthermore, expression miR-206 and -486 is induced by MyoD [26].

One more target of miR-206 was pointed out by Kim et al. [27]. Using in vitro experiment, the authors showed that miR-206 inhibits myoblast proliferation by downregulation of a subunit of DNA polymerase alpha that is required for cell proliferation [27]. These researchers suggested that miR-206 promotes muscle differentiation by indirectly suppressing myogenic repressor (MyoR) [27]. MyoR inhibits myogenic transcription factor MyoD by binding to and blocking MyoD target DNA sequences [28]. MyoR was also shown to be targeted by miR-378 [28].

Another miRNA acting in muscle differentiation is miR-29. This miRNA positively regulates muscular differentiation by targeting Yin Yang 1 (YY1), a negative regulator of muscle genes, in the C2C12 myoblast [13]. In a more recent research, miR-29 was shown to repress Akt, a protein kinase that is responsive to growth factor cell signaling. Because Akt role is to increase myoblast proliferation and reduce differentiation, decreased Akt level leads to increase myoblast differentiation as a consequent role of miRNA-29 [29].

The number of miRNAs identified being involved in muscle differentiation expands with time. Dey et al. found that miR-675-3p and miR-675-5p induce myoblast differentiation by inhibiting Smads and Cdc6 (a DNA replication initiation factor) [30].

In addition to the mRNAs that have positive roles in myoblast differentiation, there are several miRNAs that negatively regulate this process. miR-23a suppresses muscle differentiation by targeting fast myosin heavy chain 1, 2, and 4 transcripts [31]. miR-186 directly targets myogenin that has a key role in regulating myogenic differentiation, thereby inhibiting muscle cell differentiation [32].

Some other factors also use miRNAs to mediate the control of myoblast differentiation. de la Garza-Rodea AS et al. (2014) reported a case where S1P lyase, an enzyme that irreversibly degrades sphingosine-1-phosphate, upregulates the expression of miR-1, miR-206, and miR-486, thereby enhancing myoblast differentiation [33].

Also involving in S1P and miRNAs, we have recently shown that an increase in S1P level leads to an upregulation of miR-29, a positive factor in muscle differentiation [34–36].

11.3   miRNA Role in Muscle Fibrosis

In muscle regeneration, correct orientation is maintained based on scaffolds made of basement membrane of necrotic fibers. Moreover, muscle repair needs increased levels of extracellular matrix (ECM) components, including fibronectin, collagen I and II, and proteoglycan. Fibroblasts are the major cells producing ECM [37]. After the regeneration process, matrix metalloproteases will degrade ECM components [38]. In muscular dystrophies, degeneration exceeds regeneration resulting in unrepaired muscles. This leads to accumulation of ECM components that causes permanent fibrosis. Fibrotic tissues can be detected by staining collagen, the ECM protein, with picrosirius red.

miR-29 can ameliorate fibrosis by repressing the expression of collagen and elastin, the causal agents of fibrosis [34, 39–41]. In DMD mouse disease model (mdx mice) and patient biopsies, miR-29 is downregulated [39, 40] while collagen and elastin are upregulated [39]. In 2010, Cacchiarelli et al. upregulated miR-29 level in the gastrocnemius of mdx mice by electroporating miR-29a/b constructs to the muscles to increase expression and observed a decrease in mRNA levels of both collagen and elastin [39]. In 2014, Meadows et al. employed intramuscular injection to deliver miR-29 in adeno-associated viral vector into mouse muscle and saw a downregulation of muscle fibrotic, denoting by a reduced Sirius Red-stained section [40]. Also in 2014, the Ruohola-Baker group studied the ameliorating effects of S1P on DMD and saw an increase of miR-29 level along with a decrease in Col1α1, a constitutive component of collagen (the causal agent of fibrosis), and a decrease of Sirius Red-stained sections in mxd muscles, as a result of increased S1P level [34]. miR-29 is a direct target of HDAC2 [39] and is suggested as an anti-fibrotic agent for muscle diseases [39, 40].

11.4   miRNA Roles in Muscle Types

There are two main types of muscles: slow and fast twitch muscles. Fast twitch muscles have quick and short contraction time, while slow twitch muscles have long contraction time (about five times longer). We can distinguish slow and fast twitch muscles based on their (1) distinct physiological characteristics, (2) protein expression (the myosin heavy chain—MyHC isoforms), and (3) metabolic activity. We are going to delve on the differences of the two types of muscles regarding protein expression and metabolic mechanism in this chapter. To analyze slow and fast fibers, their myosin heavy chain (MHC) isoforms are usually investigated. The MYH7 isoform of MHC is expressed in slow muscles while MYH4 is exclusively expressed in fast muscles [42–45]. Regarding metabolic mechanisms, slow fibers employ mainly fatty acids for their energy, whereas fast twitch muscles rely on glucose [46]. In our body, all muscle tissues are a mixture of slow and fast twitch fibers, and the percentage contribution of each muscle fiber type characterizes the muscle tissue type.

In dystrophic context, slow twitch fibers are proven to be less susceptible to injury than fast twitch muscles [47]. And a number of researchers view the switching to slow twitch muscles as beneficial for alleviating muscular dystrophy [48, 49]. This intrigues us to understand why slow twitch muscles are more stable than fast twitch ones. We came up with a number of hypotheses that could explain the injury-resistant characteristic of slow twitch muscles.

First, the relative stability of slow twitch muscles might be conferred by their specific gene expression. The difference in gene expression of the two main muscle types has been shown in microarray data [50]. It is possible that the genes expressed in slow twitch muscle compensate some functions of dystrophin protein, increase cell survival-signaling pathway, or compensate structural proteins for muscle firmness. For example, in slow twitch muscles, an increased in utrophin gene expression was observed. Utrophin has been noted as a beneficial gene for dystrophic muscles [51–53]. Second, the characteristic of being insulin sensitive may help slow twitch muscles get better use of the body's energy fuel source [54–56]. Third, fatty acids are a higher energy source than glucose. So, slow fibers produce more energy than fast ones. To be specific, anaerobic and aerobic glycolysis yield only 2 and 36 ATPs, respectively, while oxidation of a short fatty acid (8-carbon) yields 48 ATPs and a longer fatty acid (three chains of 8-carbon) yields up to 144 ATPs.

It sounds like slow switch muscles really have advantage over fast twitch muscles in coping with the disadvantage conditions of muscle dystrophies. Can fast fibers shift to slow fibers in an adult body?

In neonatal stage, fast or slow fiber formation is determined by developmental cues [43]. After birth, the adult myofiber phenotype can be changed [44, 57]. The motor innervation pattern is in charge of regulating the switch between fiber type components. To shift preexisting fast to slow twitch fibers, muscle repetitive use, low frequency motor neuron stimulation along with high concentration of intracellular calcium can be employed. To switch slow to fast twitch fibers, we need to use phasic, high-frequency myofiber stimulation and decreased neuromuscular activity [44, 58]. In addition to motor innervation pattern, muscle fiber types can be modified by gene regulators. An increase in fatty acid oxidation and slow-fiber-specific gene expression was obtained with nuclear factor of activated T cells (NFAT), calcineurin, Ca2+/calmodulin-dependent protein kinases II (CaMK), peroxisome proliferator-activated receptor-γ coactivator (PGC-1α), and peroxisome proliferator-activated receptor delta (PPARδ) [59–63]. In addition to the abovementioned proteins, MEF2, a myocyte enhancer factor 2-transcription factor, induces the formation of slow twitch muscle by responding to a calcium-dependent pathway [63]. For fast fiber formation, a potential-inducing factor is peroxisome proliferator-activated receptor-γ coactivator 1β (PGC-1β) [64].

There have been several publications pointing out shifting to slow twitch muscles as an ameliorative effect of some potential therapeutic agents in treating muscle diseases. In 2012, von Maltzahn et al. used Wnt7a, a secreted protein that belongs to the Wnt signaling pathway [65], for an ameliorative treatment for DMD.

Besides the positive effects on muscle regeneration, myofiber hypertrophy and muscle strength, these authors also recognized a shift toward slow twitch fibers as evident in an increase in slow MHC in mdx muscles and human primary myotubes that have Wnt7a upregulated [49]. In 2014, an increase in fatty acid oxidation mechanism, suggesting a shift to slow muscles, was also observed as a beneficial effect of S1P on ameliorating DMD pathologies [34].

Along with the abovementioned small molecule, miR-208b and miR-499 have been shown to be involved in muscle type regulation [4]. Double knockout of both miR-208b and miR-499 in mouse soleus results in a substantial loss of slow myofibers. Overexpression of miR-499 leads to a shift from slow twitch to fast twitch muscles in mouse soleus, TA, and EDL. The shift to slow myofibers is accompanied with enhanced muscle endurance, as denoted by treadmill running test [4]. miR-208b and miR-499 act by targeting repressors of slow muscles genes (Sox6, Purβ, and Sp3) [66–73]. In addition, miR-208b and miR-499 also repress HP-1β, a corepressor of MEF2 [74]. MEF2 is involved in activating slow fiber gene expression [63].

11.5   miRNAs and Dystrophin Expression Regulation

The absence or dysfunction of dystrophin is believed to be the main cause of muscular disorders. In 2011, Cacchiarelli et al. discovered a miRNA that represses dystrophin expression: miR-31. miR-31 has its target site in the 3' untranslated region of the dystrophin mRNA. In the presence of miR-31, dystrophin expression is reduced. However, if the miR-31 binding site in 3'UTR of dystrophin mRNA is blocked or mutated, this gene expression is not modulated. Reducing miR-31 level also further improves the dystrophin gene expression in an exon-skipping therapy tried with DMD myoblasts of patients with exon 48–50 deletion [75].

11.6   miRNAs and the Cell Response to Oxidative Damage

The key antioxidant molecule in charge of protecting the cells against oxidative stress is the reduced form of glutathione (GSH). In the cells, the high ratio of GSH over its oxidized state GSSG is ensured by NADPH, which is maintained by an enzyme in the pentose phosphate pathway, and the glucose-6-phosphate dehydrogenase (G6PD) [39]. Interestingly, G6PD mRNA is a direct target of the miR-1 family that is downregulated in DMD patients [76]. Collectively, miR-1 indirectly downregulates GSH level, thus making the cells more susceptible to oxidative damage [39]. In DMD patients, miR-1 is downregulated so that GSH is enriched, presumably protecting the cells from oxidative stress.

11.7 The Use of miRNAs in Muscle Diseases

11.7.1 miRNA as Diagnostic Markers

In muscle diseases, the membranes of damaged muscles are weak, allowing intercellular components to leak out to blood circulation. One of such components is creatine kinase (CK). Serum CK has been used as a biomarker for muscular dystrophies [77]. However, there are several issues making it an unreliable marker. First, CK levels in serum do not correlate well with the disease severity and may even decrease when disease progresses [76, 77]. CK levels are shown to increase also in normal individuals after exercise or muscle injury [76]. It is hard to differentiate DMD from BDM (Becker's muscular dystrophy) with CK test because CK level is also significantly raised in BMD patients [76]. In addition, CK test needs special care since CK activity is very sensitive to stress conditions [77]. Thus, a more reliable biomarker for muscle dystrophies diagnostic is required.

A group of regenerative and muscle fiber-related miRNAs (miR-1/miR-206/miR-133/miR-499/miR-208a/miR-208b) has been nominated as valuable serum markers in replacing CK in regard to diagnosis and muscle disease progression [76, 77]. The serum level of all six myomiRs is elevated in young DMD patients (less than 10 years old). According to Cacchiarelli et al., among the six miRNAs, miR-499 is the best biomarker to distinguish DMD patients from healthy individuals. Especially, miR-206, miR-499, miR-208b, and miR-133 are higher in DMD than in BMD; thus they can be used to distinguish between DMD and BMD. Among these miRNAs, miR-208b is the best indicator to separate DMD and BMD. Also, in DMD patients aged 2–6 years, the level of miR-206/miR-499/miR-208b is positively correlated with age. In mdx mice, serum miRNA-1/miR-206 level is 2–40-fold higher than WT mice [77]. Remarkably, the levels of these miRNAs are rescued almost to WT level in mdx mice undergone exon-skipping treatment [77].

11.7.2 miRNAs as Therapeutic Agents/Targets

The information of functions of miRNAs in muscles and muscle diseases leads us in using mRNAs as potential therapeutic agents for muscle dystrophies. For example, we can boost up regenerative miRNAs for enhancing muscle regeneration to replace damaged muscles. We can suppress fibrosis, a prominent pathology feature of DMD, by overexpressing miR-29. We can induce the shift to the beneficial slow fiber by increasing miR-208b and miR-499. We can enhance the cells resistance to oxidative damage by suppressing miR-1. And, importantly, we can enhance dystrophin expression by inhibiting miR-31.

11.8 Perspective

Even though being an emerging field and still in its infancy [21], miRNAs have been found to be involved in many cellular processes and diseases, such as cancers [21], diabetes [78], heart disease [79], and muscular disorders. Powerfully, miRNAs are predicted to regulate at least one third of human protein-coding genes [21]. Therefore, miRNAs are a big resource for us to discover and manipulate to control many disease pathways or biological processes. This task is not simple since our genome encodes for more than 2500 miRNAs, each of which can have multiple targets, and thus being involved in many pathways or biology function. On the other hand, one pathway or biology function can also be regulated by many miRNAs [21]. Therefore, caution must be taken when using miRNAs as a tool in disease treatments as they can cause off-target effects. For example, miR-1 can enhance myogenic differentiation [21, 25], but it also can downregulate the main antioxidative molecule [39]. Thus, before clinical treatment, the pathways regulated by the potential miRNA must be investigated thoroughly. On the other hand, miRNAs are feasible for in vivo delivery because they are stable and relatively small [21]. Therefore, despite its complexity in function, miRNAs are still notable candidates for therapeutic approach.

In this chapter, we have discussed the important roles of miRNAs in normal muscle development and muscular pathology, as well as their signature expression patterns in muscle diseases. In the future these findings are expected to lead to diagnostics and therapeutic approaches for muscle diseases. miRNAs have proven to be reliable biomarkers for the diagnostic of muscular disease. Further, they give an insight on developing therapies for these diseases. We expect many new exciting findings in miRNAs in muscle processes, diagnostics, and treatments in the future.

References

1. Broderick JA, Zamore PD. MicroRNA therapeutics. Gene Ther. 2011;18:1104–10.
2. Wang XH. MicroRNA in myogenesis and muscle atrophy. Curr Opin Clin Nutr Metab Care. 2013;16:258–66.
3. Erriquez D, Perini G, Ferlini A. Non-coding RNAs in muscle dystrophies. Int J Mol Sci. 2013;14:19681–704.
4. van Rooij E, Quiat D, Johnson BA, Sutherland LB, Qi X, Richardson JA, Kelm Jr RJ, Olson EN. A family of microRNAs encoded by myosin genes governs myosin expression and muscle performance. Dev Cell. 2009;17:662–73.
5. Caretti G, Di Padova M, Micales B, Lyons GE, Sartorelli V. The Polycomb Ezh2 methyltransferase regulates muscle gene expression and skeletal muscle differentiation. Genes Dev. 2004;18:2627–38.
6. Conboy IM, Rando TA. The regulation of Notch signaling controls satellite cell activation and cell fate determination in postnatal myogenesis. Dev Cell. 2002;3:397–409.
7. Crist CG, Montarras D, Pallafacchina G, Rocancourt D, Cumano A, Conway SJ, Buckingham M. Muscle stem cell behavior is modified by microRNA-27 regulation of Pax3 expression. Proc Natl Acad Sci USA. 2009;106:13383–7.

8. Dey BK, Gagan J, Yan Z, Dutta A. miR-26a is required for skeletal muscle differentiation and regeneration in mice. Genes Dev. 2012;26:2180–91.
9. Flynt AS, Li N, Thatcher EJ, Solnica-Krezel L, Patton JG. Zebrafish miR-214 modulates Hedgehog signaling to specify muscle cell fate. Nat Genet. 2007;39:259–63.
10. Kuang W, Tan J, Duan Y, Duan J, Wang W, Jin F, Jin Z, Yuan X, Liu Y. Cyclic stretch induced miR-146a upregulation delays C2C12 myogenic differentiation through inhibition of Numb. Biochem Biophys Res Commun. 2009;378:259–63.
11. Li Z, Hassan MQ, Jafferji M, Aqeilan RI, Garzon R, Croce CM, van Wijnen AJ, Stein JL, Stein GS, Lian JB. Biological functions of miR-29b contribute to positive regulation of osteoblast differentiation. J Biol Chem. 2009;284:15676–84.
12. Sun Q, Zhang Y, Yang G, Chen X, Zhang Y, Cao G, Wang J, Sun Y, Zhang P, Fan M, et al. Transforming growth factor-beta-regulated miR-24 promotes skeletal muscle differentiation. Nucleic Acids Res. 2008;36:2690–9.
13. Wang H, Garzon R, Sun H, Ladner KJ, Singh R, Dahlman J, Cheng A, Hall BM, Qualman SJ, Chandler DS, et al. NF-kappaB-YY1-miR-29 regulatory circuitry in skeletal myogenesis and rhabdomyosarcoma. Cancer Cell. 2008;14:369–81.
14. Erbay E, Park IH, Nuzzi PD, Schoenherr CJ, Chen J. IGF-II transcription in skeletal myogenesis is controlled by mTOR and nutrients. J Cell Biol. 2003;163:931–6.
15. Ge Y, Sun Y, Chen J. IGF-II is regulated by microRNA-125b in skeletal myogenesis. J Cell Biol. 2011;192:69–81.
16. Ge Y, Wu AL, Warnes C, Liu J, Zhang C, Kawasome H, Terada N, Boppart MD, Schoenherr CJ, Chen J. mTOR regulates skeletal muscle regeneration in vivo through kinase-dependent and kinase-independent mechanisms. Am J Physiol Cell Physiol. 2009;297:C1434–44.
17. Seok HY, Tatsuguchi M, Callis TE, He A, Pu WT, Wang DZ. miR-155 inhibits expression of the MEF2A protein to repress skeletal muscle differentiation. J Biol Chem. 2011; 286:35339–46.
18. Greco S, De Simone M, Colussi C, Zaccagnini G, Fasanaro P, Pescatori M, Cardani R, Perbellini R, Isaia E, Sale P, et al. Common micro-RNA signature in skeletal muscle damage and regeneration induced by Duchenne muscular dystrophy and acute ischemia. FASEB J. 2009;23:3335–46.
19. Eisenberg I, Eran A, Nishino I, Moggio M, Lamperti C, Amato AA, Lidov HG, Kang PB, North KN, Mitrani-Rosenbaum S, et al. Distinctive patterns of microRNA expression in primary muscular disorders. Proc Natl Acad Sci USA. 2007;104:17016–21.
20. Crist CG, Montarras D, Buckingham M. Muscle satellite cells are primed for myogenesis but maintain quiescence with sequestration of Myf5 mRNA targeted by microRNA-31 in mRNP granules. Cell Stem Cell. 2012;11:118–26.
21. Chen JF, Callis TE, Wang DZ. microRNAs and muscle disorders. J Cell Sci. 2009;122: 13–20.
22. Chen JF, Mandel EM, Thomson JM, Wu Q, Callis TE, Hammond SM, Conlon FL, Wang DZ. The role of microRNA-1 and microRNA-133 in skeletal muscle proliferation and differentiation. Nat Genet. 2006;38:228–33.
23. McFarlane C, Vajjala A, Arigela H, Lokireddy S, Ge X, Bonala S, Manickam R, Kambadur R, Sharma M. Negative auto-regulation of myostatin expression is mediated by Smad3 and microRNA-27. PLoS ONE. 2014;9, e87687.
24. Alberts BA, Johnson A, Lewis J, Raff M, Robert K, Walter P. Molecular biology of the cell. New York: Garland Science; 2002.
25. Chen JF, Tao Y, Li J, Deng Z, Yan Z, Xiao X, Wang DZ. microRNA-1 and microRNA-206 regulate skeletal muscle satellite cell proliferation and differentiation by repressing Pax7. J Cell Biol. 2010;190:867–79.
26. Dey BK, Gagan J, Dutta A. miR-206 and -486 induce myoblast differentiation by downregulating Pax7. Mol Cell Biol. 2011;31:203–14.
27. Kim HK, Lee YS, Sivaprasad U, Malhotra A, Dutta A. Muscle-specific microRNA miR-206 promotes muscle differentiation. J Cell Biol. 2006;174:677–87.

28. Gagan J, Dey BK, Layer R, Yan Z, Dutta A. MicroRNA-378 targets the myogenic repressor MyoR during myoblast differentiation. J Biol Chem. 2011;286:19431–8.
29. Wei W, He HB, Zhang WY, Zhang HX, Bai JB, Liu HZ, Cao JH, Chang KC, Li XY, Zhao SH. miR-29 targets Akt3 to reduce proliferation and facilitate differentiation of myoblasts in skeletal muscle development. Cell Death Dis. 2013;4:e668.
30. Dey BK, Pfeifer K, Dutta A. The H19 long noncoding RNA gives rise to microRNAs miR-675-3p and miR-675-5p to promote skeletal muscle differentiation and regeneration. Genes Dev. 2014;28:491–501.
31. Wang L, Chen X, Zheng Y, Li F, Lu Z, Chen C, Liu J, Wang Y, Peng Y, Shen Z, et al. MiR-23a inhibits myogenic differentiation through down regulation of fast myosin heavy chain isoforms. Exp Cell Res. 2012;318:2324–34.
32. Antoniou A, Mastroyiannopoulos NP, Uney JB, Phylactou LA. miR-186 inhibits muscle cell differentiation through myogenin regulation. J Biol Chem. 2014;289:3923–35.
33. de la Garza-Rodea AS, Baldwin DM, Oskouian B, Place RF, Bandhuvula P, Kumar A, Saba JD. Sphingosine phosphate lyase regulates myogenic differentiation via S1P receptor-mediated effects on myogenic microRNA expression. FASEB J. 2014;28:506–19.
34. Nguyen-Tran DH, Hait NC, Sperber H, Qi J, Fischer K, Ieronimakis N, Pantoja M, Hays A, Allegood J, Reyes M, et al. Molecular mechanism of sphingosine-1-phosphate action in Duchenne muscular dystrophy. Dis Model Mech. 2014;7:41–54.
35. Ieronimakis N, Pantoja M, Fischer KA, Dosey TL, Qi J, Hays A, Hoofnagle AN, Sadilek M, Chamberlain JS, Ruohola-Baker H, Reyes M. Increased Sphingosine-1-Phosphate ameliorates disease pathology in mdx mice after acute injury. Skelet Muscle. 2013;3(1):20. PMID:23915702.
36. Pantoja M, Fischer KA, Ieronimakis N, Reyes M, Ruohola-Baker H. Genetic elevation of Sphingosine 1-Phosphate suppresses dystrophic muscle phenotypes in Drosophila. Development. 2013;140:136–46. PMID: 23154413.
37. Mann CJ, Perdiguero E, Kharraz Y, Aguilar S, Pessina P, Serrano AL, Munoz-Canoves P. Aberrant repair and fibrosis development in skeletal muscle. Skelet Muscle. 2011;1:21.
38. Serrano AL, Munoz-Canoves P. Regulation and dysregulation of fibrosis in skeletal muscle. Exp Cell Res. 2010;316:3050–8.
39. Cacchiarelli D, Martone J, Girardi E, Cesana M, Incitti T, Morlando M, Nicoletti C, Santini T, Sthandier O, Barberi L, et al. MicroRNAs involved in molecular circuitries relevant for the Duchenne muscular dystrophy pathogenesis are controlled by the dystrophin/nNOS pathway. Cell Metab. 2010;12:341–51.
40. Meadows E, Kota J, Malik V, Clark R, Sahenk Z, Harper S, Mendell J. MicroRNA-29 overexpression delivered by adeno-associated virus suppresses fibrosis in mdx: utrn+/− mice (S61. 003). Neurology. 2014;82:S61.003–S061. 003 %@ 0028–3878.
41. Wang L, Zhou L, Jiang P, Lu L, Chen X, Lan H, Guttridge DC, Sun H, Wang H. Loss of miR-29 in myoblasts contributes to dystrophic muscle pathogenesis. Mol Ther. 2012;20:1222–33.
42. Pette D, Staron RS. Myosin isoforms, muscle fiber types, and transitions. Microsc Res Tech. 2000;50:500–9.
43. Schiaffino S, Reggiani C. Molecular diversity of myofibrillar proteins: gene regulation and functional significance. Physiol Rev. 1996;76:371–423.
44. Simmons BJ, Cohen TJ, Bedlack R, Yao TP. HDACs in skeletal muscle remodeling and neuromuscular disease. Handb Exp Pharmacol. 2011;206:79–101.
45. Soares LE, Brugnera Junior A, Zanin FA, Pacheco MT, Martin AA. Effects of treatment for manipulation of teeth and Er:YAG laser irradiation on dentin: a Raman spectroscopy analysis. Photomed Laser Surg. 2007;25:50–7.
46. Campbell NA, Reece JB, Lawrence MG. Biology. Menlo Park: Addison Wesley Longman; 1999.
47. Webster C, Silberstein L, Hays AP, Blau HM. Fast muscle fibers are preferentially affected in Duchenne muscular dystrophy. Cell. 1988;52:503–13.
48. Selsby JT, Morine KJ, Pendrak K, Barton ER, Sweeney HL. Rescue of dystrophic skeletal muscle by PGC-1alpha involves a fast to slow fiber type shift in the mdx mouse. PLoS ONE. 2012;7, e30063.

49. von Maltzahn J, Renaud JM, Parise G, Rudnicki MA. Wnt7a treatment ameliorates muscular dystrophy. Proc Natl Acad Sci USA. 2012;109:20614–9.
50. Chemello F, Bean C, Cancellara P, Laveder P, Reggiani C, Lanfranchi G. Microgenomic analysis in skeletal muscle: expression signatures of individual fast and slow myofibers. PLoS ONE. 2011;6, e16807.
51. Fairclough RJ, Wood MJ, Davies KE. Therapy for Duchenne muscular dystrophy: renewed optimism from genetic approaches. Nat Rev Genet. 2013;14:373–8.
52. Gramolini AO, Belanger G, Thompson JM, Chakkalakal JV, Jasmin BJ. Increased expression of utrophin in a slow vs. a fast muscle involves posttranscriptional events. Am J Physiol Cell Physiol. 2001;281:C1300–9.
53. Tinsley J, Deconinck N, Fisher R, Kahn D, Phelps S, Gillis JM, Davies K. Expression of full-length utrophin prevents muscular dystrophy in mdx mice. Nat Med. 1998;4:1441–4.
54. Guan JS, Haggarty SJ, Giacometti E, Dannenberg JH, Joseph N, Gao J, Nieland TJ, Zhou Y, Wang X, Mazitschek R, et al. HDAC2 negatively regulates memory formation and synaptic plasticity. Nature. 2009;459:55–60.
55. Ryder JW, Bassel-Duby R, Olson EN, Zierath JR. Skeletal muscle reprogramming by activation of calcineurin improves insulin action on metabolic pathways. J Biol Chem. 2003;278: 44298–304.
56. Song XM, Ryder JW, Kawano Y, Chibalin AV, Krook A, Zierath JR. Muscle fiber type specificity in insulin signal transduction. Am J Physiol. 1999;277:R1690–6.
57. Bassel-Duby R, Olson EN. Signaling pathways in skeletal muscle remodeling. Annu Rev Biochem. 2006;75:19–37.
58. Olson EN, Williams RS. Calcineurin signaling and muscle remodeling. Cell. 2000;101: 689–92.
59. Chin ER, Olson EN, Richardson JA, Yang Q, Humphries C, Shelton JM, Wu H, Zhu W, Bassel-Duby R, Williams RS. A calcineurin-dependent transcriptional pathway controls skeletal muscle fiber type. Genes Dev. 1998;12:2499–509.
60. Delling U, Tureckova J, Lim HW, De Windt LJ, Rotwein P, Molkentin JD. A calcineurin-NFATc3-dependent pathway regulates skeletal muscle differentiation and slow myosin heavy-chain expression. Mol Cell Biol. 2000;20:6600–11.
61. Lin J, Wu H, Tarr PT, Zhang CY, Wu Z, Boss O, Michael LF, Puigserver P, Isotani E, Olson EN, et al. Transcriptional co-activator PGC-1 alpha drives the formation of slow-twitch muscle fibres. Nature. 2002;418:797–801.
62. Naya FJ, Mercer B, Shelton J, Richardson JA, Williams RS, Olson EN. Stimulation of slow skeletal muscle fiber gene expression by calcineurin in vivo. J Biol Chem. 2000;275:4545–8.
63. Wu H, Naya FJ, McKinsey TA, Mercer B, Shelton JM, Chin ER, Simard AR, Michel RN, Bassel-Duby R, Olson EN, et al. MEF2 responds to multiple calcium-regulated signals in the control of skeletal muscle fiber type. EMBO J. 2000;19:1963–73.
64. Arany Z, Lebrasseur N, Morris C, Smith E, Yang W, Ma Y, Chin S, Spiegelman BM. The transcriptional coactivator PGC-1beta drives the formation of oxidative type IIX fibers in skeletal muscle. Cell Metab. 2007;5:35–46.
65. Kuroda K, Kuang S, Taketo MM, Rudnicki MA. Canonical Wnt signaling induces BMP-4 to specify slow myofibrogenesis of fetal myoblasts. Skelet Muscle. 2013;3:5.
66. Adolph EA, Subramaniam A, Cserjesi P, Olson EN, Robbins J. Role of myocyte-specific enhancer-binding factor (MEF-2) in transcriptional regulation of the alpha-cardiac myosin heavy chain gene. J Biol Chem. 1993;268:5349–52.
67. Azakie A, Fineman JR, He Y. Sp3 inhibits Sp1-mediated activation of the cardiac troponin T promoter and is downregulated during pathological cardiac hypertrophy in vivo. Am J Physiol Heart Circ Physiol. 2006;291:H600–11.
68. Gupta M, Sueblinvong V, Raman J, Jeevanandam V, Gupta MP. Single-stranded DNA-binding proteins PURalpha and PURbeta bind to a purine-rich negative regulatory element of the alpha-myosin heavy chain gene and control transcriptional and translational regulation of the gene expression. Implications in the repression of alpha-myosin heavy chain during heart failure. J Biol Chem. 2003;278:44935–48.

69. Hagiwara N, Ma B, Ly A. Slow and fast fiber isoform gene expression is systematically altered in skeletal muscle of the Sox6 mutant, p100H. Dev Dyn. 2005;234:301–11.
70. Hagiwara N, Yeh M, Liu A. Sox6 is required for normal fiber type differentiation of fetal skeletal muscle in mice. Dev Dyn. 2007;236:2062–76.
71. Ji J, Tsika GL, Rindt H, Schreiber KL, McCarthy JJ, Kelm Jr RJ, Tsika R. Puralpha and Purbeta collaborate with Sp3 to negatively regulate beta-myosin heavy chain gene expression during skeletal muscle inactivity. Mol Cell Biol. 2007;27:1531–43.
72. Tsika G, Ji J, Tsika R. Sp3 proteins negatively regulate beta myosin heavy chain gene expression during skeletal muscle inactivity. Mol Cell Biol. 2004;24:10777–91.
73. von Hofsten J, Elworthy S, Gilchrist MJ, Smith JC, Wardle FC, Ingham PW. Prdm1- and Sox6-mediated transcriptional repression specifies muscle fibre type in the zebrafish embryo. EMBO Rep. 2008;9:683–9.
74. Zhang CL, McKinsey TA, Olson EN. Association of class II histone deacetylases with heterochromatin protein 1: potential role for histone methylation in control of muscle differentiation. Mol Cell Biol. 2002;22:7302–12.
75. Cacchiarelli D, Incitti T, Martone J, Cesana M, Cazzella V, Santini T, Sthandier O, Bozzoni I. miR-31 modulates dystrophin expression: new implications for Duchenne muscular dystrophy therapy. EMBO Rep. 2011;12:136–41.
76. Li X, Li Y, Zhao L, Zhang D, Yao X, Zhang H, Wang YC, Wang XY, Xia H, Yan J, et al. Circulating muscle-specific miRNAs in Duchenne muscular dystrophy patients. Mol Ther Nucleic Acids. 2014;3, e177.
77. Cacchiarelli D, Legnini I, Martone J, Cazzella V, D'Amico A, Bertini E, Bozzoni I. miRNAs as serum biomarkers for Duchenne muscular dystrophy. EMBO Mol Med. 2011;3:258–65.
78. Guay C, Regazzi R. Circulating microRNAs as novel biomarkers for diabetes mellitus. Nat Rev Endocrinol. 2013;9:513–21.
79. Ono K, Kuwabara Y, Han J. MicroRNAs and cardiovascular diseases. FEBS J. 2011; 278:1619–33.

Chapter 12
Canine-Inherited Dystrophinopathies and Centronuclear Myopathies

Joe N. Kornegay and Martin K. Childers

12.1 Introduction

Diseases that primarily affect muscle (myopathies) can be divided into acquired and inherited conditions, with each of these groups further categorized based on their underlying pathogenesis [1]. The inherited myopathies include channelopathies, myotonias, metabolic and mitochondrial conditions, congenital myopathies, and muscular dystrophies [1, 2]. For the sake of this chapter, we will focus on the congenital myopathies and muscular dystrophies, with special emphasis on Duchenne muscular dystrophy (DMD) and the centronuclear (congenital) myopathies (CNM). The muscular dystrophies are characterized by progressive myofiber necrosis and loss, with subsequent fibrosis and fatty deposition. These conditions were originally classified based primarily on their pattern of inheritance and the distribution of muscle involvement [3]. Muscular dystrophies are now associated with mutations in specific genes that code for proteins of the dystrophin-glycoprotein complex [4, 5]. Congenital myopathies occur subsequent to mutations of genes whose proteins are involved in membrane remodeling [6]. They have been distinguished from the dystrophies based on their relatively nonprogressive histologic nature and the presence

J.N. Kornegay, DVM, PhD (✉)
Department of Veterinary Integrative Biosciences (Mail Stop 4458), College
of Veterinary Medicine, Texas A&M University, College Station, TX 77843-4458, USA
e-mail: jkornegay@cvm.tamu.edu

M.K. Childers, DO, PhD
Department of Rehabilitation Medicine, University of Washington,
Seattle, WA, USA

Institute for Stem Cell and Regenerative Medicine, School of Medicine,
University of Washington, Seattle, WA, USA
e-mail: mkc8@uw.edu

© Springer Science+Business Media New York 2016   309
M.K. Childers (ed.), *Regenerative Medicine for Degenerative Muscle Diseases*,
Stem Cell Biology and Regenerative Medicine,
DOI 10.1007/978-1-4939-3228-3_12

of specific structures such as cores and nemaline rods within muscle fibers [7, 8]. This distinction has been blurred by recognition of similar structures in the dystrophies [4]. As with the muscular dystrophies, genetic mutations have been linked to many of the congenital myopathies. However, for the most part, the associations are not as straightforward as those for the dystrophies [9].

Our understanding of inherited human muscle diseases has been greatly facilitated by studies in animal models. By definition, a biomedical model is "a surrogate for a human being, or a human biologic system, that can be used to understand normal and abnormal function from gene to phenotype and to provide a basis for preventive or therapeutic intervention in human diseases" [10]. Biological models can be divided into two broad classes, depending on whether the modeling is based in *analogy* or *homology* [11]. Modeling by analogy implies a point-by-point relationship between one structure and process to another and requires that there are similarities between the structures and processes being compared. On the other hand, homologous models have a shared evolutionary history and matching DNA makeup. To be functionally useful, models by homology must also be good models by analogy.

Genetic models may either occur naturally (spontaneously) or be produced through genetic manipulation (genetically modified [transgenic or gene knockdown/out] animals). While genetically engineered mice have provided a powerful tool to study the molecular pathogenesis of disease [12, 13], results do not necessarily extrapolate to humans, presumably due to differences between murine and human size and physiology [14]. Additional questions have been raised because of the mouse's small size and whether variables influenced by scale, such as cell migration or drug diffusion, can be appropriately modeled [15]. These shortcomings are partially obviated with canine models, which have been used extensively to study disease pathogenesis and treatment efficacy [16, 17]. This trend toward the use of dogs as models will likely accelerate with the completed sequencing of the canine genome [18].

A variety of spontaneous acquired and inherited myopathies have been described in dogs, providing the opportunity to study the underlying pathogenesis of analogous human conditions and/or develop preclinical models. Acquired myopathies include both infectious [19] and immune-mediated inflammatory conditions [20], as well as metabolic and endocrine diseases [21]. Inherited muscular dystrophies and congenital myopathies, arising from mutations in genes coding for several components of the dystrophin-glycoprotein complex and proteins involved in membrane remodeling, respectively, have been described in dogs (Table 12.1) [35, 36]. This chapter focuses on diseases that are homologous to Duchenne muscular dystrophy (DMD) and the centronuclear myopathies (CNM) of human beings.

12.2   Canine *DMD* Gene Mutations (Dystrophinopathies)

Duchenne muscular dystrophy is an X-linked recessive disorder that affects ~1 in 5000 newborn human males [37] in whom absence of the protein dystrophin causes progressive degeneration of skeletal and cardiac muscle [38]. Both spontaneous and genetically modified animal models of DMD have been used to define pathogenetic mechanisms and establish efficacy of experimental therapies. Spontaneous mammalian models have been identified in mice [39, 40], cats [41, 42], pigs [43], and multiple dog breeds [22, 44]. Most preclinical studies have been conducted in the *mdx* mouse model, which results from a spontaneous nonsense point mutation in exon 23 [45]. While genetically homologous to DMD, the mdx mouse is not analogous, as affected mice having a relatively mild phenotype. This phenotype can be

Table 12.1 Inherited canine dystrophinopathies and centronuclear myopathies with confirmed mutations

Disease	Breed	Gene/mutation	Clinical signs	Reference(s)
X-linked dystrophinopathy	Golden retriever	*DMD*—base change (A–G) in the 3′ splice site of intron 6, with deletion of exon 7 in the mRNA transcript	Weakness and respiratory difficulty possible at birth; stunted growth; progressive weakness and contractures, especially over the ages of 3–6 months; respiratory and cardiac involvement	[22, 23]
	German shorthaired pointer	*DMD*—large deletion encompassing entire gene		[24]
	Pembroke Welsh corgi	*DMD*—repetitive element-1 (LINE-1) insertion in intron 13		[25]
	Cavalier King Charles spaniel	*DMD*—base change (G–T) in the 5′ splice site of intron 50, with deletion of exon 50 in the mRNA transcript		[26]
	Rottweiler	*DMD*—base change (G–T) in exon 58 resulting in stop codon		[27]
	Cocker spaniel	*DMD*—deletion of four nucleotides in exon 65, resulting in stop codon		[22]
	Tibetan terrier	*DMD*—deletion of exons 8–29		[22]
	Labrador retriever	*DMD*—184 nucleotide [pseudoexon] insertion between exons 19 and 20, resulting in stop codon		[22]

(continued)

Table 12.1 (continued)

Disease	Breed	Gene/mutation	Clinical signs	Reference(s)
X-linked myotubular myopathy	Labrador retriever	*MTM1*—base change (C–A) in exon 7	Stunted growth, muscle atrophy, and pelvic limb weakness at 7 weeks. Patellar hyporeflexia, dysphagia, and hoarse bark; paradoxical respiration at 10 weeks. Progressive weakness and muscle atrophy, with loss of ambulation, by 4–6 months	[28, 29]
Hereditary (centronuclear) myopathy of Labradors	Labrador retriever	*PTPLA*—short interspersed repeat element (SINE) insertion in exon 2	Abnormal head and neck posture; stiff, hopping gait; muscle atrophy; signs stabilize somewhat at 1 year	[30–32]
Inherited (centronuclear) myopathy of Great Danes	Great Dane	*BIN1*—base change (A–G) in the 3′ splice site of intron 10, with deletion of exon 11 in the mRNA transcript	Weakness, muscle atrophy, exercise intolerance, trembling, and characteristic posture with the pelvic limbs held under the body. Onset of signs at 6–19 months (median of 7), with variable progression	[33, 34]

exaggerated through further genetic mutation, most notably by knocking out the utrophin gene to produce so-called double-knockout (dko) mice [46]. However, the clinical relevancy of knockout models is complicated by the potential introduction of additional biochemical or biological differences independent of the absence of dystrophin. Dystrophic dogs more closely mimic the DMD phenotype, with signs occurring early and progressing [22]. Although dystrophin-deficient muscular dystrophy has been characterized clinically in a number of dog breeds, few have been studied at the molecular level [44]. The most extensively studied canine condition was originally recognized in golden retrievers and has since been outbred. Affected dogs have a single base change in the 3′ consensus splice site of intron 6 of the *DMD* gene [23]. As a result, exon 7 is skipped during mRNA processing, and the dystrophin reading frame is terminated within its N-terminal domain in exon 8. We use the term golden retriever muscular dystrophy (GRMD) to refer to dogs with the original splice site mutation regardless of crossbreeding.

12.3 Centronuclear Myopathies

The term centronuclear myopathy refers to a subset of congenital myopathies characterized by skeletal muscle weakness and a preponderance of muscle fibers with central nuclei and/or cores on biopsy. Among the congenital myopathies, CNMs have the most diverse mammalian models available for study (Table 12.2). Affected CNM patients exhibit a range of mutations in genes coding for proteins involved in membrane remodeling and a diverse clinical phenotype [6, 54]. The most common CNM, X-linked myotubular myopathy (XLMTM), occurs due to mutations in the myotubularin 1 gene (*MTM1*) and affects about 1 in 50,000 male births [55]. Affected boys have marked hypotonia, generalized muscle weakness, and respiratory failure in the immediate postnatal period [56]. Various knockout murine models have been valuable in studying disease pathogenesis but have inherent limitations for preclinical testing [6, 57].

Homologous canine forms of CNM occur due to mutations in three distinct genes, protein tyrosine phosphatase-like A (*PTPLA*) [32, 50], *MTM1* [28, 29], and bridging integrator 1 (*BIN1*) [33, 34], coding for proteins PTPLA, myotubularin, and amphiphysin, respectively. Conditions associated with mutations in the *PTPLA* and *MTM1* genes occur in Labrador retrievers but differ genetically and phenotypically. Mutations in the *PTPLA* gene are inherited in an autosomal recessive manner and cause a relatively mild myopathy. In sharp contrast, *MTM1* mutations are X-linked and associated with severe, rapidly progressive disease. The myopathy caused by *BIN1* gene mutations occurs in Great Danes, is inherited in an autosomal recessive manner, and has a variable phenotype.

12.4 Experimental Design of Canine Preclinical Studies

Our own studies in canine models of DMD and XLMTM demonstrate the need to develop treatment strategies that parallel the progression of the human disease counterparts. While DMD has a relatively protracted clinical course that extends over years, XLMTM typically presents at birth with hypotonia and respiratory insufficiency requiring ventilator support. Symptoms of both DMD and XLMTM

Table 12.2 Centronuclear myopathies

Structural features	Disease	Gene	Protein	Mammalian models	Reference(s)
Congenital myopathies with central nuclei	Myotubular myopathy	*MTM1*	Myotubularin	*MTM1* KO mouse R69C mouse XLMTM dog	[47] [48] [28]
	Centronuclear myopathy	*DNM2*	Dynamin 2	R465W mouse	[49]
		PTPLA	Protein tyrosine phosphatase-like member A	PTPLA dog	[50]
		BIN1	Amphiphysin	shRNA-Bin1 knockdown in mice IMGD dog	[51] [33]
		RYR1	Ryanodine receptor	*Ryr1I4895T/wt* (IT/+) mouse	[52]
		TTN	Titin	Ttn(mdm) mouse	[53]

relate principally to skeletal muscle disease, but the nature of clinical involvement differs dramatically. The appendicular musculature is affected early in DMD, with respiratory function becoming impaired in the later states. In contrast, many XLMTM patients are never able to walk and must receive ventilator support soon after birth.

12.4.1 GRMD Phenotype and Preclinical Trials

Based on comparative longevity studies for dogs and humans, the first year of a golden retriever's life roughly equates to 20 years of a human [58]. By extrapolation, the clinical courses of GRMD and DMD over these respective periods can be divided into quartiles, with 0–3, 3–6, 6–9, and 9–12 months of a golden retriever's life corresponding to 0–5, 5–10, 10–15, and 15–20 years of a human's. In drawing comparisons between the two conditions, there is a relatively strong parallel in the disease progression up to 6 months/10 years of age (Fig. 12.1). GRMD dogs are weak at birth and can develop a so-called fulminant form [59] that leads to early death. This peracute neonatal condition is often complicated by severe respiratory compromise [60]. Infants with DMD do not appear to experience such a devastating syndrome. Over the first 3 months of life, surviving GRMD pups have delayed weight gain [61] and are weaker, characterized by clumsiness, falling occasionally, and being pushed aside by stronger, normal littermates. These features are akin to

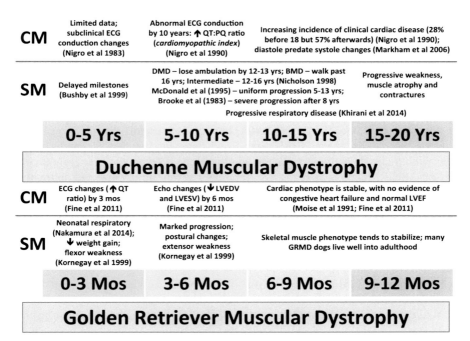

Fig. 12.1 Comparative disease course of GRMD based on the relative equivalency of the first year of a golden retriever's life and initial 20 years of a human's. The two periods are divided into quartiles, e.g., 0–3 months of GRMD paralleling 0–5 years of DMD, with signs of skeletal myopathy (SM) and cardiomyopathy (CM) listed for each period. Note: the GRMD clinical course from 0 to 6 months largely parallels that of DMD over the 0–10-year period. However, the GRMD and DMD phenotypes then dramatically diverge, with GRMD dogs often stabilizing and DMD continuing to progress. *SM* skeletal myopathy, *CM* cardiomyopathy, *LVEDV and LVSDV* left ventricular end-diastolic and systolic volumes, *LVEF* left ventricular ejection fraction

delayed motor milestones seen in DMD boys over the first several years of life [62]. Cardiac disease is subclinical, limited to abnormalities on ECG, at this early stage in GRMD [63] and DMD [64].

The GRMD skeletal muscle phenotype evolves rapidly from 3 to 6 months [61, 65], corresponding to an analogous period of progressive appendicular weakness from 5 to 10 years in DMD [66–68] (Fig. 12.1). Some GRMD dogs lose the ability to walk before 6 months [69], although loss of ambulation occurs rarely in our experience and that of others [70]. The rapid progression of the GRMD phenotype over a short 3-month period provides a convenient window for preclinical trials. Indeed, we have used this period for most of our proof-of-concept efficacy studies in GRMD [22, 71]. In contrast with the rapid course of skeletal muscle involvement, cardiac disease is largely subclinical at 6 months in GRMD [63, 72] and 10 years in DMD [73]. Changes on ECG occur early in both conditions and could, in principle, be used to monitor efficacy of treatments that have cardiac effects. By 6 months in

GRMD, left ventricular end-diastolic and systolic internal diameters and volumes are decreased on echocardiography, perhaps associated with an overall reduction in cardiac size or an early phase of hypertrophic cardiomyopathy [63]. By 10 years of age in DMD, diastolic changes may be seen, while those for systole remain normal [74]. Otherwise, variables of cardiac function, such as left ventricular ejection fraction and fractional shortening, are normal over the first 6 months in GRMD and analogous 10 years in most DMD patients.

After 6 months, the GRMD and DMD phenotypes markedly diverge (Fig. 12.1). Skeletal muscle disease in GRMD stabilizes somewhat between 6 months and a year, with many dogs living well into adulthood. As an example, our current colony of ~30 GRMD dogs at Texas A&M includes seven, mildly affected, untreated 5-year-old dystrophic dogs, an age that corresponds to ~42 years in humans [58]. In addition, we have a 9-year-old and two 7-year-old GRMD dogs with mild phenotypes. Most of these dogs are heterozygous males that were used in breeding at earlier ages. Our oldest GRMD dog, *Scrappy*, lived to almost 13 years of age, when he was euthanized because of slowly progressive difficulty in walking. Others have reported long-term survival in GRMD, although this seems to have been exceptional [75] compared to our own experience with a number of long-lived dystrophic dogs.

While symptoms related to the appendicular musculature predominate in the early stages of DMD, respiratory failure had long been considered the main cause of mortality, ultimately accounting for up to 80 % of deaths [76, 77]. This has changed with the advent of aggressive respiratory management, as DMD patients now live into their 30s, with some studies showing that cardiac disease is now the principal cause of death [78]. As in people, GRMD dogs often die due to respiratory failure [79]. Per the discussion above, some puppies develop a so-called fulminant form of the disease in which severe respiratory compromise can lead to death in the neonatal period [59]. With maturation, GRMD dogs continue to have respiratory difficulty, evidenced most notably by increased respiratory rate and abdominal breathing. However, as with other features of skeletal muscle disease, respiratory function, as judged clinically, is typically stable over the 6–12-month period.

Consistent with skeletal muscle involvement, cardiac indices in GRMD dogs are proportionally stable between 6 and 12 months. Clinical cardiac disease, with associated functional changes on echocardiography, typically is not seen until at least 3 years, if at all [22, 72]. We have reported on GRMD dogs treated for dilated cardiomyopathy at 9 years of age [22]. In contrast to GRMD, the DMD cardiac phenotype progresses markedly from 10 to 20 years, with most DMD patients having clinical involvement by 18 years of age [73].

The relatively benign course of GRMD over the 6–12-month period, and extending into later life, differs radically from the progressive disease seen over the ages of 10–20 years in DMD [66, 73, 80]. A discussion of the mechanisms responsible for the divergent DMD and GRMD disease course over analogous ages is beyond the scope of this chapter. Genetic mechanisms, including modifier genes, which drive processes such as fibrosis and fatty change, likely, play an important role [44].

12.4.2   Biomarkers in GRMD

A variety of outcome parameters, in many cases modeled after analogous procedures used to assess DMD patients, have been developed (Table 12.3) and utilized in preclinical trials [22]. To counter the effects of well-established phenotypic variation in the GRMD model, baseline testing can first be done at 3 months and then repeated at the end of treatment. Thus, the relative longitudinal effect of treatments can be established in variably affected dogs.

12.4.2.1   In Vivo Pelvic Limb Muscle Physiology in GRMD

For the sake of preclinical studies, we have focused particularly on assessment of contractile properties of distal pelvic limb muscles [61]. Briefly, the pelvic limb torque of anesthetized dogs is measured by securing the paw to a pedal mounted on the shaft of a servomotor that also functions as a force transducer. Because there is marked differential involvement of muscles in GRMD, we have studied torque generated by both tibiotarsal joint flexor and extensor muscles through selective percutaneous stimulation of the peroneal and tibial nerves, respectively. Power analysis of data collected from natural history studies suggested that flexion torque would be more useful than extension torque to document therapeutic benefit in GRMD dogs. Extension values in this initial study varied more markedly, necessitating larger group sizes to establish significance. Since this original study, the inbreeding coefficient of sire-dam pairs has been monitored more closely, with an associated reduction of disease severity. A subsequent trial of prednisone in GRMD dogs showed a significant improvement in extension torque using a treatment group size of six GRMD dogs [82].

12.4.2.2   Eccentric Contractions in GRMD

Muscle injury is more pronounced after lengthening (eccentric) contractions versus shortening (concentric) contractions [83]. Since dystrophin plays a key role in protecting the sarcolemmal membrane from mechanical damage, eccentric contractions induce abnormally high levels of muscle injury in dystrophin-deficient muscles. We developed a method to measure the degree of muscle injury by measuring the loss of force (force deficit or decrement) following eccentric contractions in anesthetized GRMD dogs and use this as a functional readout [84, 85]. In the absence of fatigue, the force deficit is a reliable and quantitative physiologic measure of muscle damage [86]. The force deficit or drop (Fig. 12.2), defined as the difference between maximum isometric force (P_0) before and after repeated contractions (exercise), is expressed as a percent change of the non-exercised value for P_0: force deficit $= ([P_0 - P_{0\ \text{after exercise}}]/P_0) \times 100$. Because skeletal muscle damage caused by exercise peaks after 3 days in normal humans, we compared the force

Table 12.3 Outcome parameters in GRMD dogs[a]

Test	Age (mos)	Normal dogs	GRMD dogs	Significance (p-value)	Reference
Body weight (kg)	3	10.65 ± 1.75	7.47 ± 1.21	<0.01	[61]
	6	20.24 ± 2.30	12.86 ± 3.08	<0.01	
	12	23.17 ± 1.70	18.23 ± 3.22	<0.01	
TTJ tetanic flexion (normalized N/kg)	3	0.486 ± 0.142	0.200 ± 0.094	<0.01	[61]
	6	0.825 ± 0.256	0.469 ± 0.183	<0.01	
	12	1.10 ± 0.27	0.550 ± 0.200	<0.01	
TTJ tetanic extension (normalized N/kg)	3	2.55 ± 0.28	1.32 ± 0.43	<0.01	[61]
	6	2.95 ± 0.53	0.965 ± 0.506	<0.01	
	12	2.98 ± 0.28	1.34 ± 0.58	<0.01	
Speed (m/sec)	2	1.77 ± 0.29	1.02 ± 0.24	0.001	[81]
	9	2.61 ± 0.18	0.88 ± 0.46	0.0001	
Stride Length/height at withers	2	1.97 ± 0.26	1.35 ± 0.27	0.0001	[81]
	9	1.99 ± 0.06	0.92 ± 0.31	0.0001	
LVID (d) (cm)[b]	3	3.0 ± 0.0	2.5 ± 0.2	<0.01	[63]
	6	3.6 ± 0.1	2.9 ± 0.1	<0.01	
	12	4.1 ± 0.2	3.2 ± 0.1	<0.01	
Fraction shortening (%)[b]	3	36.9 ± 2.5	35.0 ± 1.2	NS	[63]
	6	33.7 ± 0.5	39.3 ± 2.6	NS	
	12	32.3 ± 1.7	38.8 ± 6.5	NS	
EKG lead 2 Q/R ratio[b]	3	0.2 ± 0.0	0.4 ± 0.1	<0.01	[63]
	6	0.3 ± 0.0	0.7 ± 0.0	<0.01	
	12	0.3 ± 0.0	0.6 ± 0.1	<0.01	
Serum CK[c]	2 days	700 (200–800)[d]	18,900 (2300–39,500)	ND	[59]
	6 weeks	300 (200–500)	8200 (6500–162,100)	ND	
	6	400 (400–400)	32,400 (30,300–42,100)	ND	

ND not determined

From Kornegay JN, et al. Pharmacologic management of Duchenne muscular dystrophy: target identification and preclinical trials. ILAR J. 2014;55:119–49

[a]Listed tests were assessed longitudinally; *TTJ* tibiotarsal joint, *N/kg* Newtons/kg, *LVID (d)* left ventricular diameter at diastole, *CK* creatine kinase

[b]Cardiac data were from GRMD dogs crossbred with Labrador retrievers carrying a different DMD gene mutation

[c]CK results for CXMD (GRMD) dogs were from male "small-breed" dogs

[d]Median and range for CK values

deficit between GRMD and normal dogs 3 days after experimental contractions. Our data indicated that eccentric contractions induce significantly greater force deficit in muscles of the cranial (anterior) tibial compartment in GRMD dogs compared with controls [84]. Subsequent histological staining of esterase (a nonspecific marker of myofiber inflammation) measured 3 days after contractions in the

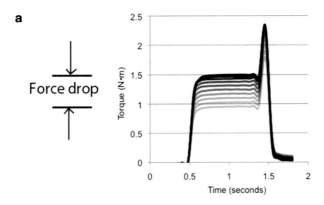

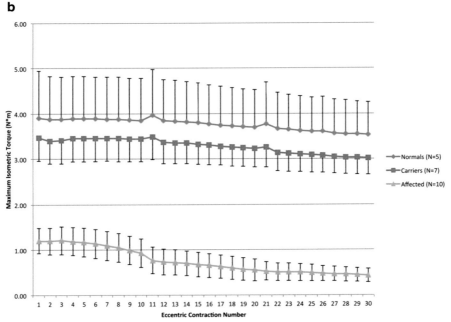

Fig. 12.2 Dystrophin-deficient muscles exhibit abnormally large force drop after repeated eccentric contractions. (**a**) A representative torque tracing from an affected GRMD dog undergoing repeated eccentric contractions. The *darkest line* displays the first contraction while the *lightest line* shows the tenth contraction. Following eccentric (lengthening) contractions, GRMD dogs rapidly lose force (*force drop*). Normal dogs (*not shown*) do not display such a rapid decline. (**b**) Isometric torque was measured immediately before and after a forced stretch was applied in contracting muscles of affected GRMD dogs. The graph displays a gradual decline in maximum isometric torque (Y-axis) most notably in affected dogs (~60 % force drop) compared to normal or carriers (~10 % force drop) (Modified from Tegler et al. [85])

muscles of the GRMD dogs positively correlated with force data. Together, our findings indicated that force drop following eccentric contractions can be used to measure the amount of muscle damage incurred in the limb muscles of GRMD dogs. Future preclinical trials might exploit this method to test compounds thought to protect the muscle membrane from mechanical damage.

12.4.2.3 Gait Assessment

Accelerometry has been shown to objectively distinguish gait abnormalities in GRMD dogs, with progressive changes seen as early as 2 months of age [81, 87]. Our group [88] and others [89] have utilized video gait analysis in GRMD dogs. We used a two-dimensional kinematic analysis, focused on the stifle (knee) and hock (ankle) angles of the pelvic limb. Retroreflective markers (Fig. 12.3, lower panel) were placed on anatomical landmarks of the pelvic limb. Digital video of sagittal plane motion was collected during overground walking at the dog's self-selected pace with a single camera. During filming, the dogs were led on a leash by the same experienced handler. Data were collected on both sides of each dog. Trials where dogs increased or decreased speed, stopped, or walked out of the calibrated sagittal plane were excluded from analysis. Digitizing software and analysis programs were used to calculate the joint angles of the stifle and hock. The main findings indicated that (1) GRMD dogs walked slower than controls; (2) at the stifle joint, both groups displayed similar range of motion (ROM), but compared to controls, GRMD dogs walked with the stifle joint more extended; and (3) at the hock joint, GRMD dogs displayed less ROM and walked with the joint less flexed.

Given the prominent role that the 6-minute walk test (6MWT) now plays in DMD clinical trials [90], we have begun assessing this metric in GRMD dogs and correlating results with other functional outcome parameters. Dogs are first conditioned/trained for several sessions beginning at ~8 weeks of age and then evaluated at monthly intervals. Our preliminary data indicate that affected dogs walk shorter distances than normal/carrier dogs (Van Wie and Kornegay, unpublished observations, 2014).

12.4.2.4 Respiratory Assessment in GRMD

A number of GRMD preclinical trials have been done to assess genetic, cellular, and pharmacologic treatments [22, 71, 91]. These studies have utilized a variety of outcome parameters, including measures of skeletal muscle and cardiac function. None of these trials have assessed respiratory function in treated dogs. In anticipation of assessing respiratory function in GRMD preclinical trials, we assessed arterial blood gas values, tidal breathing spirometry, and respiratory inductance plethysmography (RIP) in a group of age-matched adult affected and carrier dogs [79]. Partial pressure of carbon dioxide and bicarbonate concentration were higher in GRMD dogs but within reference ranges. Notably, GRMD dogs had markedly

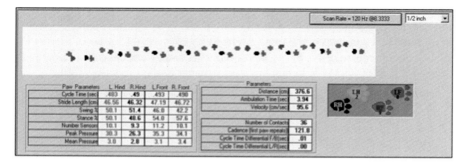

Fig. 12.3 Canine gait analysis. *Top panel*: screenshot of spatiotemporal measures such as velocity, stride length, and stance percentage. Step length, stance time, and stance percentage can be analyzed. Stride length is defined as the distance between heel strike to heel strike of the same paw. Step length is the distance between the heel strikes of opposing paws. Stance time is the amount of time a paw spends on the ground during a walk. Stance percentage indicates what fraction of a gait cycle is spent in stance. *Bottom panel*: screenshot of retroreflective markers used to identify range-of-motion excursion of the hock and stifle joints during walking gait

higher tidal breathing peak expiratory flows, and most had abnormal abdominal motion. It will now be important to extend these studies to younger GRMD dogs, in keeping with the treatment paradigm used to assess ambulation. A separate ex vivo study of GRMD respiratory function, done in collaboration with the Stedman laboratory at the University of Pennsylvania, suggested that the increased expiratory flows likely occur through diaphragmatic remodeling to increase tidal volume (a phenomenon called post-expiratory recoil) [92].

12.4.3   XLMTM Phenotype

Labrador retriever puppies with XLMTM may be somewhat stunted and display subtle muscle atrophy and pelvic limb weakness as early as 7 weeks of age [28, 29]. Patellar hyporeflexia, dysphagia, dropped jaw, and a hoarse bark appear over time. Unlike the GRMD model, little variation in the phenotype of XLMTM dogs has been noted in the research setting, despite detailed physiological and clinical observations in ~20 affected dogs since 2009 (Childers et al., unpublished observations, [93]). Moreover, the disease onset is more rapid than in GRMD and universally results in markedly progressive weakness and muscle atrophy leading to loss of ambulation, necessitating euthanasia by 4–6 months of age. Returning to the age-equivalency paradigm used for GRMD and DMD [58] and assuming a parallel between the first 3 months of a Labrador retriever's life and first 5 years of a human's, the 4–6-month course of XLMTM in Labradors is somewhat delayed compared to the progression of XLMTM in humans over a matter of days to weeks [56].

12.4.4 Biomarkers in XLMTM

12.4.4.1 In Vivo Pelvic Limb Muscle Physiology

Using the same system discussed above to measure tibiotarsal joint torque in GRMD, torque generated by XLMTM dogs at 10 weeks is ~40 % lower than normal. By 18 weeks of age, torque generated by affected dogs is only ~15 % of that of wild-type dogs [94].

12.4.4.2 Eccentric Contractions in XLMTM

In contrast to the progressive force drop observed in GRMD limb muscles subjected to repeated eccentric contractions (Fig. 12.2), XLMTM dogs display a progressive *increase* (summation) in the isometric phase (Fig. 12.4). Recent findings [94] indicate that despite a progressive increase during repeated contractions, maximum isometric torque never exceeds values observed in normal controls or carriers. These data also provide insight into the underlying pathophysiology of XLMTM. Mutant dogs generate a gradual increase in isometric torque during a series of eccentric contractions until activations cease. Following a brief rest, the next activation of eccentric contractions begins at a lower initial torque (Fig. 12.4). The pronounced torque *increase* following repeated stretch activations in XLMTM dogs is in keeping with findings in ryanodine receptor (RyR1) mutant mice [95]. Both *MTM1* mutant dogs and *MTM1* mutant mice display similar increased force summation with repetitive nerve stimulation in vivo. While the precise molecular mechanism responsible for these observations is unknown, a defect in the physiological process of converting an electrical stimulus to a mechanical response [known as excitation-contraction (E-C) coupling] likely accounts for these findings in the XLMTM dog. Affected dogs also exhibit right-shifted torque-frequency responses compared to those of the normal and carrier dogs, along with a gradual increase in torque during a series of eccentric contractions, indicating an E-C coupling impairment as a key feature in this mutant animal model.

12.4.4.3 Respiratory Assessment in XLMTM

Because of the dominant role that pulmonary dysfunction plays in human XLMTM patients, particular attention has been paid to assessment of respiratory indices [57, 96]. Affected dogs demonstrate paradoxical breathing at ~10 weeks and, when

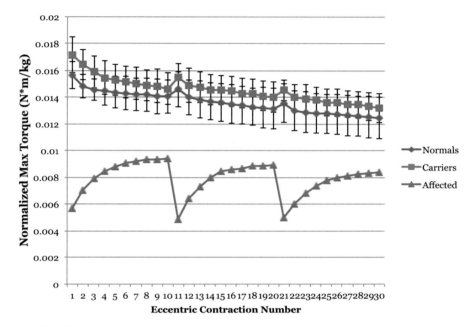

Fig. 12.4 Myotubularin-deficient muscles display abnormally low isometric torque that increases with repeated eccentric contractions. Isometric torque was measured immediately before and after a forced stretch was applied in contracting muscles in XLMTM dogs. The graph displays a gradual increase in maximum isometric torque (Y-axis) in affected XLMTM ($n=3$), but not normal ($n=6$) or carrier ($n=5$) dogs. *Note*: torque responses are normalized to body mass (n-m/kg) (Modified from Grange et al. [94])

compared to wild-type dogs on serial measurements, have reduced inspiratory and expiratory flow rates before and after dosing with doxapram, a centrally acting respiratory stimulant [96]. Reduced inspiratory flow rates suggest involvement of the diaphragm. Using ultrasonography, images of the diaphragm were acquired from normal and affected XLMTM dogs and a single affected XLMTM dog that received systemic AAV8-*MTM1* gene replacement [97]. Quantitative parameters of diaphragm structure were different among the animals. The normal diaphragm appeared thicker and less echogenic than the XLMTM one, whereas the diaphragm measurements of the *MTM1*-treated XLMTM diaphragm were comparable to those of the normal diaphragm. These findings suggest that thickness and echodensity reflect a functional response to *MTM1* gene replacement in *MTM1*-mutant diaphragm muscle and align with findings of increased peak inspiratory flow after gene replacement therapy.

12.4.4.4   Gait Assessment in XLMTM

Using the same video-based motion-capture analysis described above, gait was assessed in a small cohort of XLMTM, carrier, and normal dogs [98]. In addition to motion capture, an instrumented carpet [99] was used to assess walking gait.

Instrumented carpets with imbedded sensors allow investigators to measure spatio-temporal gait characteristics. Customized software is available for quick readouts of parameters such as overall speed of gait, step length, step width, and stride length (Fig. 12.3, upper panel). However, spatiotemporal data alone do not capture postural changes or range-of-motion (ROM) excursion; kinematic (motion-capture) analysis is required to assess these parameters.

Unlike GRMD, XLMTM dogs appear to maintain relatively normal joint range of motion, but walk slower than normal with shorter stilted strides. These differences become greater over time, suggesting that slowing of walking gait generally correlates with progressive limb weakness. Similar to the GRMD model, data indicate that weakness in XLMTM dogs causes a measureable impact on walking speed. The reduced speed seen in affected dogs appears to be characterized by shorter steps and a longer stance phase. Video analysis of step length and speed and stride length can readily distinguish between groups and are sensitive to change over time. As opposed to the GRMD model, angular kinematic data (changes in joint angles measured by video capture) do not easily distinguish differences between XLMTM and normal dogs, suggesting that joint angle measurements are not practical as outcome measures. Alternatively, spatiotemporal data captured by instrumented carpets are useful in assessing variables such as speed and stride length.

12.4.5   *XLMTM Preclinical Trials*

Systemic treatment of XLMTM dogs with an AAV8-MTM1 construct at 9 weeks of age normalized tibiotarsal joint torque and peak inspiratory flow rates 6–7 weeks after infusion [93]. Moreover, all treated dogs remained ambulatory until the study was terminated 1 year or longer after treatment, well beyond the age of 18 weeks when untreated XLMTM dogs can no longer walk. Two systemically treated male dogs remain in the XLMTM colony and have successfully bred to carrier females producing several litters of dogs, indicating the sustainability of gene replacement therapy.

12.5   Conclusions

Inherited myopathies cause debilitating limb weakness and respiratory dysfunction in affected humans. Animal models provide platforms to study underlying disease mechanisms and test potential therapies. The clinical courses of spontaneous canine conditions, such as GRMD and XLMTM, largely parallel those of homologous human diseases. This homology makes these canine myopathies particularly attractive models for studying their human counterparts. In the context of preclinical trials, particular attention must be paid to onset and progression of the canine diseases. While GRMD progresses over a relatively protracted period, the XLMTM syndrome has a much more rapid clinical course. Studies in both GRMD and XLMTM have already provided insight on the efficacy of treatments intended for human patients and hold great promise for future investigations.

References

1. Barohn RJ, Dimachkie MM, Jackson CE. A Pattern recognition approach to patients with a suspected myopathy. Neurol Clin. 2014;32(3):569–93. doi:10.1016/j.ncl.2014.04.008.
2. Rakowicz WP, Lane RJM. Myopathies. Medicine. 2004;32:119–23.
3. Wicklund MP. The muscular dystrophies. Continuum (Minneap Minn). 2013;19:1535–70. doi:10.1212/01.CON.0000440659.41675.8b.
4. Amato AA, Griggs RC. Overview of the muscular dystrophies. Handb Clin Neurol. 2011;101:1–9. doi:10.1016/B978-0-08-045031-5.00001-3.
5. Cohn RD, Campbell KP. Molecular basis of muscular dystrophies. Muscle Nerve. 2000;23:1456–71.
6. Cowling BS, Toussaint A, Muller J, Laporte J. Defective membrane remodeling in neuromuscular diseases: insights from animal models. PLoS Genet. 2012;8, e1002595. doi:10.1371/journal.pgen.1002595. Epub 2012 Apr 5.
7. Gilbreath HR, Castro D, Iannaccone ST. Congenital myopathies and muscular dystrophies. Neurol Clin. 2014;32:689–703. doi:10.1016/j.ncl.2014.04.006.
8. Iannaccone ST, Castro D. Congenital muscular dystrophies and congenital myopathies. Continuum (Minneap Minn). 2013;19:1509–34. doi:10.1212/01.
9. Laing NG, Sewry CA, Lamont P. Congenital myopathies. In: Mastaglia FL, Hilton-Jones D, editors. Handbook of clinical neurology, 3rd series, vol. 86. Edinburgh: Elsevier; 2007. p. 1–33.
10. National Research Council. Biomedical models and resources: current needs and future opportunities. Washington, DC: The National Academies Press; 1998.
11. National Research Council. Models for biomedical research. A new perspective. Washington, DC: The National Academy Press; 1985.
12. Capecchi MR. Gene targeting in mice: functional analysis of the mammalian genome for the twenty-first century. Nat Rev Genet. 2005;6:507–12.
13. Sacca R, Engle SJ, Qin W, Stock JL, McNeish JD. Genetically engineered mouse models in drug discovery research. Methods Mol Biol. 2010;602:37–54.
14. Lin JH. Applications and limitations of genetically modified mouse models in drug discovery and development. Curr Drug Metabol. 2008;9:419–38.
15. Partridge TA. The mdx mouse model as a surrogate for Duchenne muscular dystrophy. FEBS J. 2013;280:4177–86. doi:10.1111/febs.12267.
16. Schneider MR, Wolf E, Braun J, Kolb HJ, Adler H. Canine embryo-derived stem cells and models for human diseases. Hum Mol Genet. 2007;17(R1):R42–7.
17. Tsai KL, Clark LA, Murphy KE. Understanding hereditary diseases using the dog and human as companion model systems. Mamm Genome. 2007;18:444–51.
18. Lindblad-Toh K, Wade CM, Mikkelsen TS, Karlsson EK, Jaffe DB, Kamal M, et al. Genome sequence, comparative analysis and haplotype structure of the domestic dog. Nature. 2005;438:803–19.
19. Reichel MP, Ellis JT, Dubey JP. Neosporosis and hammondiosis in dogs. J Small Anim Pract. 2007;48:308–12.
20. Shelton GD. From dog to man: the broad spectrum of inflammatory myopathies. Neuromuscul Disord. 2007;17:663–70.
21. Platt SR. Neuromuscular complications in endocrine and metabolic disorders. Vet Clin North Am Small Anim Pract. 2002;32:125–46.

22. Kornegay JN, Bogan JR, Bogan DJ, Childers MK, Li J, Nghiem P, et al. Canine models of Duchenne muscular dystrophy and their use in therapeutic strategies. Mamm Genome. 2012;23:85–108. doi:10.1007/s00335-011-9382-y.

23. Sharp NJH, Kornegay JN, Van Camp SD, Herbstreith MH, Secore SL, Kettle S, et al. An error in dystrophin mRNA processing in golden retriever muscular dystrophy, an animal homologue of Duchenne muscular dystrophy. Genomics. 1992;13:115–21.

24. Schatzberg SJ, Olby NJ, Breen M, Anderson LV, Langford CF, Dickens HF, et al. Molecular analysis of a spontaneous dystrophin 'knockout' dog. Neuromuscul Disord. 1999;9:289–95.

25. Smith BF, Yue Y, Woods PR, Kornegay JN, Shin JH, Williams RR, et al. An intronic LINE-1 element insertion in the dystrophin gene aborts dystrophin expression and results in Duchenne-like muscular dystrophy in the corgi breed. Lab Invest. 2011;91:216–31. doi:10.1038/labinvest.2010.146.

26. Walmsley GL, Arechavala-Gomeza V, Fernandez-Fuente M, Burke MM, Nagel N, Holder A, et al. A duchenne muscular dystrophy gene hot spot mutation in dystrophin-deficient cavalier king charles spaniels is amenable to exon 51 skipping. PLoS ONE. 2010;5(1), e8647. doi:10.1371/journal.pone.0008647.

27. Winand N, Pradham D, Cooper B. Molecular characterization of severe Duchenne-type muscular dystrophy in a family of Rottweiler dogs. In: Molecular mechanisms of neuromuscular disease. Tucson, AZ: Muscular Dystrophy Association; 1994.

28. Beggs AH, Böhm J, Snead E, Kozlowski M, Maurer M, Minor K, et al. MTM1 mutation associated with X-linked myotubular myopathy in Labrador Retrievers. Proc Natl Acad Sci USA. 2010;107:14697–702. doi:10.1073/pnas.1003677107.

29. Cosford KL, Taylor SM, Thompson L, Shelton GD. A possible new inherited myopathy in a young Labrador Retriever. Can Vet J. 2008;49:393–7.

30. Blot S, Tiret L, Devillaire AC, Fardeau M, Dreyfus PA. Phenotypic description of a canine centronuclear myopathy. J Neurol Sci. 2002;199:S9.

31. Kramer JW, Hegreberg GA, Bryan GM, Meyers K, Ott RL. A muscle disorder of Labrador Retrievers characterized by deficiency of type II muscle fibers. J Am Vet Med Assoc. 1976;169:817–20.

32. Maurer M, Mary J, Guillaud L, Fender M, Pelé M, Bilzer T, et al. Centronuclear myopathy in Labrador retrievers: a recent founder mutation in the PTPLA gene has rapidly disseminated worldwide. PLoS ONE. 2012;7(10), e46408. doi:10.1371/journal.pone.0046408. Epub 2012 Oct 5.

33. Böhm J, Vasli N, Maurer M, Cowling BS, Shelton GD, Kress W, et al. Altered splicing of the BIN1 muscle-specific exon in humans and dogs with highly progressive centronuclear myopathy. PLoS Genet. 2013;9(6), e1003430. doi:10.1371/annotation/22ca13f1-1ce9-4bb5-9c9e-98670f7c4240.

34. Luján-Feliu-Pascual A, Shelton GD, Targett MP, Long SN, Comerford EJ, McMillan C, et al. Inherited myopathy of great Danes. J Small Anim Pract. 2006;47:249–54.

35. Lorenz MD, Coates JR, Kent M. Handbook of veterinary neurology. 5th ed. St Louis, MO: Elsevier Saunders; 2011.

36. Shelton GD, Engvall E. Canine and feline models of human inherited muscle diseases. Neuromuscul Disord. 2005;15(2):127–38.

37. Mendell JR, Shilling C, Leslie ND, Flanigan KM, al-Dahhak R, Gastier-Foster J, et al. Evidence-based path to newborn screening for Duchenne muscular dystrophy. Ann Neurol. 2012;71:304–13. doi:10.1002/ana.23528.

38. Hoffman EP, Brown Jr RH, Kunkel LM. Dystrophin: the protein product of the Duchenne muscular dystrophy locus. Cell. 1987;51:919–28.

39. Bulfield G, Siller WG, Wight PAL, Moore KJ. X chromosome-linked muscular dystrophy (mdx) in the mouse. Proc Natl Acad Sci USA. 1984;81:1189–92.

40. Gillis JM. Understanding dystrophinopathies: an inventory of the structural and functional consequences of the absence of dystrophin in muscles of the mdx mouse. J Muscle Res Cell Motil. 1999;20:605–25.

41. Carpenter JL, Hoffman EP, Romanul FC, Kunkel LM, Rosales RK, Ma NS, et al. Feline muscular dystrophy with dystrophin deficiency. Am J Pathol. 1989;135:909–19.
42. Gaschen FP, Hoffman EP, Gorospe JR, Uhl EW, Senior DF, Cardinet 3rd GH, et al. Dystrophin deficiency causes lethal muscle hypertrophy in cats. J Neurol Sci. 1992;110:149–59.
43. Hollinger K, Yang CX, Montz RE, Nonneman D, Ross JW, Selsby JT. Dystrophin insufficiency causes selective muscle histopathology and loss of dystrophin-glycoprotein complex assembly in pig skeletal muscle. FASEB J. 2014;28:1600–9. doi:10.1096/fj.13-241141.
44. Brinkmeyer-Langford C, Kornegay JN. Comparative genomics of X-linked muscular dystrophies: the golden retriever model. Curr Genomics. 2013;14:330–42. doi:10.2174/138920291 13149990004.
45. Sicinski P, Geng Y, Ryder-Cook AS, Barnard EA, Darlison MG, Barnard PJ. The molecular basis of muscular dystrophy in the mdx mouse: a point mutation. Science. 1989;244:1578–80.
46. Deconinck AE, Rafael JA, Skinner JA, Brown SC, Potter AC, Metzinger L, et al. Utrophin-dystrophin-deficient mice as a model for Duchenne muscular dystrophy. Cell. 1997;90: 717–27.
47. Buj-Bello A, Laugel V, Messaddeq N, Zahreddine H, Laporte J, Pellissier JF, Mandel JL. The lipid phosphatase myotubularin is essential for skeletal muscle maintenance but not for myogenesis in mice. Proc Natl Acad Sci USA. 2002;99:15060–5.
48. Pierson CR, Dulin-Smith AN, Durban AN, Marshall ML, Marshall JT, Snyder AD, et al. Modeling the human MTM1 p.R69C mutation in murine Mtm1 results in exon 4 skipping and a less severe myotubular myopathy phenotype. Hum Mol Genet. 2012;21:811–25.
49. Durieux AC, Vignaud A, Prudhon B, Viou MT, Beuvin M, Vassilopoulos S, Fraysse B, Ferry A, Laine J, Romero NB, Guicheney P, Bitoun M. A centronuclear myopathy-dynamin 2 mutation impairs skeletal muscle structure and function in mice. Hum Mol Genet. 2010;19:4820–36.
50. Gentilini F, Zambon E, Gandini G, Rosati M, Spadari A, Romagnoli N, Turba ME, Gernone F. Frequency of the allelic variant of the PTPLA gene responsible for centronuclear myopathy in Labrador Retriever dogs as assessed in Italy. J Vet Diagn Invest. 2011;23:124–6.
51. Tjondrokoesoemo A, Park KH, Ferrante C, Komazaki S, Lesniak S, Brotto M, et al. Disrupted membrane structure and intracellular Ca(2)(+) signaling in adult skeletal muscle with acute knockdown of Bin1. PLoS ONE. 2011;6, e25740.
52. Zvaritch E, Kraeva N, Bombardier E, McCloy RA, Depreux F, Holmyard D, et al. Ca2+ dysregulation in Ryr1(I4895T/wt) mice causes congenital myopathy with progressive formation of minicores, cores, and nemaline rods. Proc Natl Acad Sci USA. 2009;106:21813–8.
53. Garvey SM, Rajan C, Lerner AP, Frankel WN, Cox GA. The muscular dystrophy with myositis (mdm) mouse mutation disrupts a skeletal muscle-specific domain of titin. Genomics. 2002;79:146–9.
54. Romero NB, Bitoun M. Centronuclear myopathies. Semin Pediatr Neurol. 2011;18:250–6. doi:10.1016/j.spen.2011.10.006.
55. Laporte J, Blondeau F, Buj-Bello A, Mandel JL. The myotubularin family: from genetic disease to phosphoinositide metabolism. Trends Genet. 2001;17:221–8.
56. Jungbluth H, Wallgren-Pettersson C, Laporte J. Centronuclear (myotubular) myopathy. Orphanet J Rare Dis. 2008;3:26. doi:10.1186/1750-1172-3-26.
57. Smith BK, Goddard M, Childers MK. Respiratory assessment in centronuclear myopathies. Muscle Nerve. 2014. doi:10.1002/mus.24249.
58. Patronek GJ, Waters DJ, Glickman LT. Comparative longevity of pet dogs and humans: implications for gerontology research. J Gerontol A Biol Sci Med Sci. 1997;52:B171–8.
59. Valentine BA, Cooper BJ, de Lahunta LA, O'Quinn R, Blue JT. Canine X-linked muscular dystrophy. An animal model of Duchenne muscular dystrophy: clinical studies. J Neurol Sci. 1988;88:69–81.
60. Nakamura A, Kobayashi M, Kuraoka M, Yuasa K, Yugeta N, Okada T, et al. Initial pulmonary respiration causes massive diaphragm damage and hyper-CKemia in Duchenne muscular dystrophy dog. Sci Rep. 2013;3:2183.

61. Kornegay JN, Bogan DJ, Bogan JR, Childers MK, Cundiff DD, Petroski GF, et al. Contraction force generated by tibiotarsal joint flexion and extension in dogs with golden retriever muscular dystrophy. J Neurol Sci. 1999;166:115–21.
62. Bushby KMD, Hill A, Steele JG. Failure of early diagnosis in symptomatic Duchenne muscular dystrophy. Lancet. 1999;353:557–8.
63. Fine DM, Shin JH, Yue Y, Volkmann D, Leach SB, Smith BF, et al. Age-matched comparison reveals early electrocardiography and echocardiography changes in dystrophin-deficient dogs. Neuromuscul Disord. 2011;21:453–61.
64. Nigro G. Importance of cardiological studies in the muscular dystrophies. Cardiomyology. 1983;2:209–19.
65. Kornegay JN, Sharp NJ, Schueler RO, Betts CW. Tibiotarsal joint contracture in dogs with golden retriever muscular dystrophy. Lab Anim Sci. 1994;44:331–3.
66. Brooke MH, Fenichel GM, Griggs RC, Mendell JR, Moxley R, Miller JP, et al. Clinical investigation in Duchenne dystrophy: 2. Determination of the "power" of therapeutic trials based on the natural history. Muscle Nerve. 1983;6:91–103.
67. McDonald CM, Abresch RT, Carter GT, Fowler Jr WM, Johnson ER, Kilmer DD, et al. Profiles of neuromuscular diseases. Duchenne muscular dystrophy. Am J Phys Med Rehabil. 1995;74(5 Suppl):S70–92.
68. Nicholson LVB, Johnson MA, Bushby KMD, Gardner-Medwin D, Curtis A, Ginjaar IB, et al. Integrated study of 100 patients with Xp21 linked muscular dystrophy using clinical, genetic, immunochemical, and histopathological data. Part 1. Trends across the clinical groups. J Med Genet. 1993;30:728–37.
69. Barthélémy I, Pinto-Mariz F, Yada E, Desquilbet L, Savino W, Silva-Barbosa SD, et al. Predictive markers of clinical outcome in the GRMD dog model of Duchenne muscular dystrophy. Dis Model Mech. 2014;7:1253–61. doi:10.1242/dmm.016014.
70. Duan D, Hakim CH, Ambrosio CE, Smith BF, Sweeney HL. Early loss of ambulation is not a representative clinical feature in Duchenne muscular dystrophy dogs: remarks on the article of Barthélémy et al. Dis Model Mech. 2015;8:193–4. doi:10.1242/dmm.019216.
71. Kornegay JN, Peterson JM, Bogan DJ, Kline W, Bogan JR, Dow JL, et al. NBD delivery improves the disease phenotype of the golden retriever model of Duchenne muscular dystrophy. Skelet Muscle. 2014;4:18. doi:10.1186/2044-5040-4-18.
72. Moise NS, Valentine BA, Brown CA, Erb HN, Beck KA, Cooper BJ, et al. Duchenne's cardiomyopathy in a canine model: electrocardiographic and echocardiographic studies. J Am Coll Cardiol. 1991;17:812–20.
73. Nigro G, Comi LI, Politano L, Bain RJI. The incidence and evolution of cardiomyopathy in Duchenne muscular dystrophy. Int J Cardiol. 1990;26:271–7.
74. Markham LW, Michelfelder EC, Border WL, Khoury PR, Spicer RL, Wong BL, et al. Abnormalities of diastolic function precede dilated cardiomyopathy associated with Duchenne muscular dystrophy. J Am Soc Echocardiogr. 2006;19:865–71.
75. Zatz M, Vieira NM, Zucconi E, Pelatti M, Gomes J, Vainzof M, et al. A normal life without dystrophin. Neuromuscul Disord. 2015;25:371–4.
76. Inkley SR, Oldenburg FC, Vignos Jr PJ. Pulmonary function in Duchenne muscular dystrophy related to stage of disease. Am J Med. 1974;56:297–306.
77. Laghi F, Tobin MJ. Disorders of the respiratory muscles. Am J Respir Crit Care Med. 2003;168:10–48.
78. Toussaint M, Steens M, Wasteels G, Soudon P. Diurnal ventilation via mouthpiece: survival in end-stage Duchenne patients. Eur Respir J. 2006;28:549–55.
79. DeVanna JC, Kornegay JN, Bogan DJ, Bogan JR, Dow JL, Hawkins EC. Respiratory dysfunction in unsedated dogs with golden retriever muscular dystrophy. Neuromuscul Disord. 2014;24:63–73. doi:10.1016/j.nmd.2013.10.001.
80. Khirani S, Ramirez A, Aubertin G, Boulé M, Chemouny C, Forin V, et al. Respiratory muscle decline in Duchenne muscular dystrophy. Pediatr Pulmonol. 2014;49:473–81.
81. Barthélémy I, Barrey E, Aguilar P, Uriarte A, Le Chevoir M, Thibaud JL, et al. Longitudinal ambulatory measurements of gait abnormality in dystrophin-deficient dogs. BMC Musculoskelet Disord. 2011;12:75.

82. Liu JMK, Okamura CS, Bogan DJ, Bogan JR, Childers MK, Kornegay JN. Effects of prednisone in canine muscular dystrophy. Muscle Nerve. 2004;30:767–73.

83. Armstrong RB, Warren GL, Warren JA. Mechanisms of exercise- induced muscle fibre injury. Sports Med. 1991;12:184–207.

84. Childers MK, Okamura CS, Bogan DJ, Bogan JR, Petroski GF, McDonald K, et al. Eccentric contraction injury in dystrophic canine muscle. Arch Phys Med Rehabil. 2002;83:1572–8.

85. Tegler CJ, Grange RW, Bogan DJ, Markert CD, Case D, Kornegay JN, et al. Eccentric contractions induce rapid isometric torque drop in dystrophic-deficient dogs. Muscle Nerve. 2010;42:130–2.

86. Faulkner JA, Jones DA, Round JM. Injury to skeletal muscles of mice by forced lengthening during contractions. Q J Exp Physiol. 1989;74:661–70.

87. Barthélémy I, Barrey E, Thibaud J-L, Uriarte A, Voit T, Blot S, Hogrel J-V. Gait analysis using accelerometry in dystrophin-deficient dogs. Neuromuscul Disord. 2009;19:788–96.

88. Marsh AP, Eggebeen JD, Kornegay JN, Markert CD, Childers MK. Kinematics of gait in golden retriever muscular dystrophy. Neuromuscul Disord. 2010;20:16–20.

89. Shin JH, Greer B, Hakim CH, Zhou Z, Chung YC, Duan Y, et al. Quantitative phenotyping of Duchenne muscular dystrophy dogs by comprehensive gait analysis and overnight activity monitoring. PLoS ONE. 2013;8(3), e59875.

90. McDonald CM, Henricson EK, Abresch RT, Florence JM, Eagle M, Gappmaier E, et al. The 6-minute walk test and other endpoints in Duchenne muscular dystrophy: longitudinal natural history observations over 48 weeks from a multicenter study. Muscle Nerve. 2013;48:343–56.

91. Kornegay JN, Spurney CF, Nghiem PP, Brinkmeyer-Langford CL, Hoffman EP, Nagaraju K. Pharmacologic management of Duchenne muscular dystrophy: target identification and preclinical trials. ILAR J. 2014;55:119–49. doi:10.1093/ilar/ilu011.

92. Mead AF, Petrov M, Malik AS, Mitchell MA, Childers MK, Bogan JR, et al. Diaphragm remodeling and compensatory respiratory mechanics in a canine model of Duchenne muscular dystrophy. J Appl Physiol (1985). 2014;116:807–15. doi:10.1152/japplphysiol. 00833.2013.

93. Childers MK, Joubert R, Poulard K, Moal C, Grange RW, Doering JA, et al. Gene therapy prolongs survival and restores function in murine and canine models of myotubular myopathy. Sci Transl Med. 2014;6(220), 220ra10. doi:10.1126/scitranslmed.3007523.

94. Grange RW, Doering J, Mitchell E, Holder MN, Guan X, Goddard M, et al. Muscle function in a canine model of X-linked myotubular myopathy. Muscle Nerve. 2012;46:588–91. doi:10.1002/mus.23463.

95. Yamaguchi N, Prosser BL, Ghassemi F, Xu L, Pasek DA, Eu JP, et al. Modulation of sarcoplasmic reticulum Ca2+ release in skeletal muscle expressing ryanodine receptor impaired in regulation by calmodulin and S100A1. Am J Physiol Cell Physiol. 2011;300:C998–1012.

96. Goddard MA, Mitchell EL, Smith BK, Childers MK. Establishing clinical end points of respiratory function in large animals for clinical translation. Phys Med Rehabil Clin N Am. 2012;23:75–94. doi:10.1016/j.pmr.2011.11.017.

97. Sarwal A, Cartwright MS, Walker FO, Mitchell E, Buj-Bello A, Beggs AH, et al. Ultrasound assessment of the diaphragm: preliminary study of a canine model of X-linked myotubular myopathy. Muscle Nerve. 2014;50:607–9.

98. Goddard MA, Burlingame E, Beggs AH, Buj-Bello A, Childers MK, Marsh AP, et al. Gait characteristics in a canine model of X-linked myotubular myopathy. J Neurol Sci. 2014; pii: S0022-510X(14)00564-4. doi:10.1016/j.jns.2014.08.032.

99. Light VA, Steiss JE, Montgomery RD, Rumph PF, Wright JC. Temporal-spatial gait analysis by use of a portable walkway system in healthy Labrador Retrievers at a walk. Am J Vet Res. 2010;71:997–1002.

Index

© Springer Science+Business Media New York 2016
M.K. Childers (ed.), *Regenerative Medicine for Degenerative Muscle Diseases*,
Stem Cell Biology and Regenerative Medicine,
DOI 10.1007/978-1-4939-3228-3